# Essentials of Skilled Helping

## Managing Problems, Developing Opportunities

### Gerard Egan
Professor Emeritus
Loyola University of Chicago

WADSWORTH
CENGAGE Learning

Australia • Brazil • Japan • Korea • Mexico • Singapore • Spain • United Kingdom • United States

**Essentials of Skilled Helping: Managing Problems, Developing Opportunites**
Gerard Egan

Executive Editor: Lisa Gebo

Technology Project Manager: Barry Connolly

Assistant Editor: Alma Dea Michelena

Editorial Assistant: Sheila Walsh

Marketing Manager: Caroline Concilla

Marketing Assistant: Rebecca Weisman

Marketing Communications Manager:
Tami Strang

Senior Art Director: Vernon Boes

Print Buyer: Rebecca Cross

Permissions Editor: Sarah Harkrader

Production Service and Compositor:
Pre-Press Co., Inc.

Text Designer: Lisa Henry

Illustrator: Pre-Press Co., Inc.

Cover Designer: Irene Morris/Morris Design

Cover Image: Getty Images/Glen Allison

ExamView® and ExamView Pro® are registered trademarks of FSCreations, Inc. Windows is a registered trademark of the Microsoft Corporation used herein under license. Macintosh and Power Macintosh are registered trademarks of Apple Computer, Inc. Used herein under license.

For product information and technology assistance, contact us at
**Cengage Learning Customer & Sales Support,
1-800-354-9706**.
For permission to use material from this text or product, submit all requests online at **www.cengage.com/permissions**.
Further permissions questions can be emailed to
**permissionrequest@cengage.com**.

Library of Congress Control Number: 2005929701

Student Edition
ISBN-13: 978-0-495-00487-5
ISBN-10: 0-495-00487-1

International Student Edition
(not for sale in the United States)
ISBN-13: 978-0-495-00815-6
ISBN-10: 0-495-00815-X

**Wadsworth**
20 Davis Drive
Belmont, CA 94002-3098
USA

Cengage Learning is a leading provider of customized learning solutions with office locations around the globe, including Singapore, the United Kingdom, Australia, Mexico, Brazil, and Japan. Locate your local office at: **www.cengage.com/global**.

Cengage Learning products are represented in Canada by Nelson Education, Ltd.

To learn more about Wadsworth, visit
**www.cengage.com/wadsworth**.

Purchase any of our products at your local college store or at our preferred online store **www.ichapters.com**.

Printed in the U.S.A.
6 7 8 9 10    15 14

# CONTENTS

P A R T   O N E

## Laying the Groundwork 1

C H A P T E R  1

## Introduction to Helping 2

Introduction: The Nature and Purpose of Helping 3

An Overview of the Skilled Helper Model 11

Moving from Smart to Wise: Managing the Shadow Side
of Helping 22

Exercises on Personalizing 24

C H A P T E R  2

## The Helping Relationship and the Values That Drive It 25

The Helping Relationship 26

Values As Guides and Drivers 27

Key Values in Helping: Respect, Empathy, Genuineness,
Self-Responsibility, and a Bias toward Action 29

Shadow-Side Realities in the Helping Relationship 40

Exercises On Personalizing 42

P A R T  T W O

## The Therapeutic Dialogue                                                45

C H A P T E R 3

## The Basics of Communication: Tuning In to Clients and Active Listening        46

The Importance of Dialogue in Helping    47

Tuning In to Clients: The Importance of Empathic Presence    49

Active Listening: The Foundation of Understanding    53

Processing What You Hear: The Thoughtful Search for Meaning    65

Listening to Yourself: The Helper's Internal Conversation    69

The Shadow Side of Listening to Clients    70

Exercises on Personalizing    73

C H A P T E R 4

## Communicating Empathy and Checking Understanding        74

The Three Dimensions of Responding Skills:
    Perceptiveness, Know-How, and Assertiveness    75

Sharing Empathic Highlights:
    Communicating Understanding to Clients    77

Principles for Sharing Highlights    85

Tactics for Communicating Highlights    92

The Shadow Side of Sharing Empathic Highlights    93

The Importance of Empathic Relationships    96

Exercises on Personalizing    97

C H A P T E R 5

## The Art of Probing and Summarizing        98

Using Prompts and Probes Effectively    99

The Art of Summarizing: Providing Focus and Direction    111

The Shadow Side of Communication Skills    114

Exercises on Personalizing    117

C H A P T E R 6

## Helping Clients Challenge Themselves    **118**

Challenge: The Basic Concept    119

Blind Spots    130

Linking Challenge to Action    134

The Shadow Side of Challenging    137

Exercises on Personalizing    140

C H A P T E R 7

## Challenging Skills and the Wisdom Needed to Use Them Well    **141**

Specific Challenging Skills    142

From Smart to Wise: Guidelines for Effective Challenging    157

Linking Challenge to Action    162

Challenge and the Shadow Side of Helpers    163

Exercises on Personalizing    166

P A R T    T H R E E

## The Stages and Tasks of the Helping Model    **167**

C H A P T E R 8

## Stage I: Helping Clients Tell Their Stories: An Introduction to Stage I    **168**

An Introduction to Stage I    169

Task 1: Help Clients Tell Their Stories:
   "What Are My Concerns?"    170

Task 2: Help Clients Develop New Perspectives
   and Reframe Their Stories: "What Am I Overlooking
   or Avoiding?"    188

Task 3: Help Clients Achieve Leverage by Working on
   Issues That Make a Difference:
   "What Will Help Me Most?"    192

Exercises on Personalizing    203

C H A P T E R 9

**Stage II: Helping Clients Identify, Choose, and Shape Problem-Managing Goals**    **204**

The Preferred Picture: "What Do I Need and Want?"    205

Task 1: Help Clients Discover Possibilities for a Better Future: "What Would My Life Look Like If It Were Better?"    211

Task 2: Help Clients Craft Problem-Managing Goals: "What Solutions Are Best for Me?"    222

Task 3: Help Clients Find the Incentives to Commit Themselves to a Better Future: "What Am I Willing to Pay for What I Want?"    234

Stage II and Action    240

The Shadow Side of Goal Setting    241

Exercises on Personalizing    243


C H A P T E R 1 0

**Stage III: Helping Clients Develop Strategies and Plans for Accomplishing Goals**    **244**

Introduction to Stage III    245

Task 1: Help Clients Review Possible Strategies to Achieve Goals: "What Strategies Will Help Me Get What I Want?"    246

Task 2: Help Clients Choose Strategies That Best Fit Their Resources: "What Courses of Action Make Most Sense for Me?"    252

Task 3: Help Clients Pull Strategies Together into a Manageable Plan: "How Do I Put Everything Together and Move Forward?"    261

Exercises on Personalizing    273


C H A P T E R 1 1

**The Action Arrow: Helping Clients Overcome Obstacles, Execute Plans, and Get Results**    **274**

Obstacles to Change    275

Helping Clients Become Change Agents in Their Own Lives    280

Getting Along Without a Helper    288

An Amazing Case of Resilience and Perseverance    290

The Shadow Side of Self Change    291

Exercises on Personalizing    293


REFERENCES    295

NAME INDEX    303

SUBJECT INDEX    305

**GERARD EGAN**, Ph.D. is Professor Emeritus of Psychology and Organization Development at Loyola University of Chicago. He has written over fifteen books. His books in the field of counseling and communication include *The Skilled Helper, Interpersonal Living, People in Systems* (with Michael Cowan), *TalkWorks: How to Get More Out of Life Through Better Conversations* (with Andrew Bailey), and *TalkWorks at Work: How to Become a Better Communicator and Make the Most of Your Career* (with Andrew Bailey). *The Skilled Helper*, now in its 7th edition, has been translated into both European and Asian languages. It is currently the most widely used counseling text in the world.

His other books, dealing with business and management, include *Change Agent Skills in Helping and Human Service Settings; Change Agent Skills A: Designing and Assessing Excellence; Change Agent Skills B: Managing Innovation and Change; Adding Value: A Systematic Guide to Business-Based Management and Leadership;* and *Working the Shadow Side: A Guide to Positive Behind-the-Scenes Management.* Through these writings, complemented by extensive consulting, he has created a comprehensive and integrated business-based management system focusing on strategy, operations, structure, human resource management, the managerial role itself, and leadership. The management system also includes a framework for initiating and managing change. *Working the Shadow Side* provides a framework for managing such shadow-side complexities as organizational messiness, personalities, social systems, politics, and culture. It also focuses on the messiness and politics of change management. He has consulted, lectured, and given workshops on these topics in Africa, Asia, Australia, Europe, and North America.

Dr. Egan has consulted with a variety of companies and institutions worldwide, including the Center for Educational Leadership (University of Manchester) and universities in the People's Republic of China. He works with managers and their teams as consultant, coach, and counselor, often on a longer-term basis. He has been involved in Executive Leadership Programs with a range of companies around the globe.

*E*ssentials of *Skilled Helping* is an updated, more streamlined, and more compact version of the author's book *The Skilled Helper*. It is meant for those who are thinking of becoming professional helpers, for those in human services professions who inevitably find themselves in the role of counselor, and for anyone interested in helping others. It is a book about three things—the values needed to establish authentic helping relationships with others, the communication skills needed to engage others in a collaborative helping dialogue, and a universal problem-management and opportunity-development framework that outlines the fundamental stages, tasks, and processes of helping. The model is client-centered in the sense that the starting point is the troubled person's humanity and concerns. The model is not about schools of psychology, interesting theories, or the latest fads. It is about people with problem situations and unused opportunities. But the model also recognizes that people often need to be invited to challenge themselves in a host of ways if they are to improve their lot in life. Caring tough love is also client-centered.

The helping model here is cognitive, affective, and behavioral in character, that is, it deals with the way people think, the way they come to feel and express emotions, and the way they act. While past and present thinking, emotional expression, and acting are often part of the problem situation or unused opportunities, different ways of thinking, managing emotions, and acting are pathways to a better future. Those who use the model best recognize that people in many ways create their own reality, but they also understand that reality itself has a way of biting back. While the model recognizes the pervasiveness and importance of feelings and emotions, it also stands by the importance of emotional intelligence. Competent helpers espouse individual freedom, but they do not confuse liberty with license. They also appreciate the wisdom of the French saying, "*Nos actes nous suivent,*" our actions have a way of pursuing us. While they appreciate individuality, they also recognize everyone's need for some form of community. The model as spelled out in these pages recognizes and delights in human diversity, but it remains rooted in the commonalities of our humanity.

While the model does not discount the importance of the past, it is not deterministic and refuses to subscribe to the tyranny of the past. Instead it unapologetically focuses more on possibilities for a better future than on problems and speculation about their causes. That said, the past can be both a source of learnings for a better future and a warehouse of possible solutions. "When your marriage was at its best, what was it like?"

The model appreciates the power of both goal setting and planning even when the troubled person—or the world at large—does not. Talking about goal setting and planning may evoke a yawn in some quarters, but the challenge of the helper is to make them living, vibrant, useful realities in the eyes of those seeking or needing help. Helping is about decision making. Effective helpers understand both the bright and dark side of decision making and become guides as troubled people muse about, make, glide towards, flirt with, or fall into decisions—or attempt to avoid them.

The model is frank about the obstacles to change, even life-enhancing change. Discretionary change—that is, change that would be helpful but is not demanded or imposed—has a poor track record in human affairs. Consider New Year's resolutions even among the resolute. Helping is about real change—internal, external, or both. Therefore, the model encourages small actions—small problem-managing changes—right from the beginning.

Effective helpers know that grappling with problems in living is hard work and don't hesitate to caringly invite those seeking solutions to buckle down and engage in that work. But not work for the sake of work. In the end helping is about work that produces outcomes that favorably impact the lives of help seekers.

## Acknowledgments

I would like to thank the following reviewers for their helpful insights and suggestions: Hal Cain, Private Practice; Susan Halverson, Portland State University; Jeanmarie Keim, University of Arizona South; Lynne Kellner, Fitchburg State College; Rich McGourty, Private Practice; Martin K. Munoz, University of Houston-Victoria; Catherine Pittman, Saint Mary's College; Nicholas Poulos, Onondaga Community College; and Fernelle L. Warren, Troy State University.

GERARD EGAN

# PART ONE

# Laying the Groundwork

*A*lthough the centerpiece of this book is a problem-management and opportunity-development helping model and the methods and communication skills that make it work, there is some groundwork to be laid. This includes outlining the nature and goal of helping, together with an overview of the helping process itself (Chapter 1), and describing the helping relationship and the values that should drive it (Chapter 2).

# Introduction to Helping

**Introduction: The Nature and Purpose of Helping**
Clients with Problem Situations and Unused Opportunities
The Primary Goal of Helping
A Secondary Goal
Does Helping Help?
Is Helping for Everyone?

**An Overview of the Skilled Helper Model**
Stage I: The Current Picture—Help Clients Tell Their Stories
Stage II: The Preferred Picture—Help Clients Identify, Choose, and Shape Problem-Managing Goals
Stage III: The Way Forward—Help Clients Develop Strategies and Plans for Accomplishing Goals
The Action Arrow: Help Clients Implement Their Plans and Get Results
Help Clients Evaluate Their Progress Along the Way: "How Are We Doing?"
Be Flexible in the Use of the Model

**Moving from Smart to Wise: Managing the Shadow Side of Helping**

**Exercises on Personalizing**

## Introduction: The Nature and Purpose of Helping

This chapter has two main sections: The first is an introduction to helping; the second is an overview of the problem-management helping model outlined and illustrated in the pages of this book.

Throughout history there has been a deeply embedded conviction that, under the proper conditions, some people are capable of helping others come to grips with problems in living. This conviction, of course, plays itself out differently in different cultures, but it is still a cross-cultural phenomenon. Today, this conviction is often institutionalized in a variety of formal helping professions. In Western cultures, counselors, psychiatrists, psychologists, social workers, and ministers of religion are counted among those whose formal role is to help people manage the distressing problems of life.

There is also a second set of professionals who, although they are not helpers in the formal sense, often deal with people in times of crisis and distress. Included here are organizational consultants, dentists, doctors, lawyers, nurses, probation officers, teachers, managers, supervisors, police officers, and practitioners in other service industries. Although these people are specialists in their own professions, there is still some expectation that they will help those they serve manage a variety of problem situations. For instance, teachers teach English, history, and science to students who are growing physically, intellectually, socially, and emotionally and are struggling with developmental tasks and crises. Teachers are, therefore, in a position to help their students, in direct and indirect ways, explore, understand, and deal with the problems of growing up. Managers and supervisors help workers cope with problems related to work performance, career development, interpersonal relationships in the workplace, and a variety of personal problems that affect their ability to do their jobs. This book is addressed directly to the first set of professionals and indirectly to the second.

A third set of helpers includes any and all who try to help themselves and others, including relatives, friends, acquaintances, and even strangers (on buses and planes), come to grips with problems in living. Indeed, only a small fraction of the help provided on any given day comes from helping professionals. Informal helpers abound in the social settings of life. Friends help one another through troubled times. Parents need to manage their own marital problems while helping their children grow and develop. Indeed, most people grappling with problems in living seek help, if they seek it at all, from informal sources (Swindle, Heller, Pescosolido, & Kikuzawa, 2000). In the end, of course, all of us must learn how to help ourselves cope with the problems and crises of life. Self-help in the context of some kind of caring community is the ideal. This book is about the fundamental models, methods, processes, and skills of helping. It is designed to help you become a better helper no matter which category you fall into.

Let's call the person seeking or needing help with problems in living a *client* and consider two questions. Why do people seek help in the first place? And what is the principal goal of helping?

## Clients with Problem Situations and Unused Opportunities

Many clients become clients because, either in their own eyes or in the eyes of others, they are involved in problem situations that they are not handling well. Others come because they feel they are not living as fully as they might. Many come because of a mixture of both. Therefore, clients' problem situations and unused opportunities constitute the starting point of the helping process.

**Problem situations.**  Clients come for help because they have crises, troubles, doubts, difficulties, frustrations, or concerns. These are often called, generically, "problems." However, they are not problems in a mathematical sense because clients' problems in living often cause emotional turmoil and have no clear-cut solutions. It is probably better to say that clients come not with problems but with *problem situations*—that is, with complex and messy *problems in living* that they are not handling well. These problem situations are often poorly defined. Or, if they are well defined, clients still don't know how to handle them. Or they feel that they do not have the resources needed to cope with them adequately. If they have tried solutions, they have not worked.

Problem situations arise in our interactions with ourselves, with others, and with the social settings, organizations, and institutions of life. Clients—whether they are hounded by self-doubt, tortured by unreasonable fears, grappling with the stress that accompanies serious illness, addicted to alcohol or drugs, involved in failing marriages, fired from jobs because of personal behavior, office politics, or disruptions in the economy, confused in their efforts to adapt to a new culture, suffering from a catastrophic loss, jailed because of child abuse, wallowing in a midlife crisis, lonely and out of community with no family or friends, battered by their spouses, or victimized by racism—all face problem situations that move them to seek help or, in some cases, move others to refer them or even send them for help.

People with even devastating problem situations can often, with some help, handle them more effectively. Consider the following example.

Martha S., 58, suffered three devastating losses within six months. One of her four sons, who lived in a different city, died suddenly of a stroke. He was only 32. Shortly after, she lost her job in a downsizing move stemming from the merger of her employer with another company. Finally, her husband, who had been ill for about two years, died of cancer. Though she was not destitute, her financial condition could not be called comfortable, at least not by middle-class North American standards. Two of her other three sons were married with families of their own. One lived in a distant suburb, the other in a different city. The unmarried son, a sales rep for an international company, traveled abroad extensively. After her husband's death, she became agitated, confused, angry, and depressed. She also felt guilty. First, because she believed that she should have done "more" for her husband. Second, because she also felt strangely responsible for her son's early death. Finally, she was deathly afraid of becoming a burden to her children.

At first, retreating into herself, she refused help from anyone. But eventually she responded to the gentle persistence of her church minister. She began attending a support group at the church. A psychologist who worked at a local university provided some direction for the group. Helped by her interactions within the group, she slowly began to accept help from her sons. She began to realize that she was not the only one who was experiencing a sense of loss. Rather, she was part of a grieving family, the members of which needed to help one another cope with the turmoil they were experiencing. She began relating with some of the members of the group outside the group sessions. This filled the social void she experienced when her company laid her off. She had an occasional informal chat with the psychologist who provided services for the group. Through contacts within the group she got another job.

Note that help came from many quarters. Her new-found solidarity with her family, the church support group, the active concern of the minister, the informal chats with the psychologist, and upbeat interactions with her new-found friends helped Martha enormously. Furthermore, since she had always been a resourceful person, the help she received enabled her to tap into her own ingenuity.

It is important to note that none of this "solved" the losses she had experienced. Indeed, the goal of helping is not to "solve" problems but to help the troubled person *manage them* more effectively, or even to transcend them by taking advantage of new posibilities in life. Problems have an up side. They are opportunities for learning. Clients can apply what they learn in the give-and-take of helping sessions first to managing the presenting problem. But what they learn often has wider application. When a client is helped to sort out a difficult relationship, her learning can be applied both to sorting out other problems in relationships and to preventing relationship problems from arising in the first place.

**Missed opportunities and unused potential.**  Some clients come for help not because they are dogged by problems like those listed above, but because they are not living as fully as they would like. And so clients' *missed opportunities* and *unused potential* constitute a second starting point for helping. Most clients have resources they are not using or opportunities they are not developing. People who feel locked in dead-end jobs or bland marriages, who are frustrated because they lack challenging goals, who feel guilty because they are failing to live up to their own values and ideals, who want to do something more constructive with their lives, or who are disappointed with their uneventful interpersonal lives—such clients come to helpers not to manage their problems better, but to live more fully.

In this case, the question is not what is going wrong, but what can be better. Over the years many psychologists have suggested that most of us use only a small fraction of our potential. We are capable of dealing with ourselves, our relationships with others, and our work life much more creatively than we ordinarily do. Consider the following case.

After ten years as a helper in several mental-health centers, Carol was experiencing burnout. In the opening interview with a counselor, she berated herself for not being dedicated enough. Asked when she felt best about herself, she said that it was on those relatively infrequent occasions when she was asked to help provide help for other mental-health centers that were experiencing problems, having growing pains, or reorganizing. The counselor helped her explore her potential as a consultant to human-service organizations, and to make a career adjustment. She enrolled in an organization development program at a local university. Carol stayed in the helping field, but with a new focus and a new set of skills.

In this case, the counselor helped the client manage her problems (burnout, guilt) by helping her explore and develop an opportunity (a new career).

Helping clients identify and develop unused potential and opportunities can be called a "positive-psychology" goal. As guest editors of the special Millennium issue of the *American Psychologist*, Seligman and Csikszentmihalyi (2000) called for a better balance of perspectives in the helping professions. In their minds, too much attention is focused on pathology and too little on what they call "positive psychology": "Our message is to remind our field that psychology is not just the study of pathology, weakness, and damage; it is also the study of strength and virtue. Treatment is not just fixing what is broken; it is nurturing what is best" (p. 7). They and their fellow authors discuss such positive, upbeat topics as:

- subjective well-being, happiness, hope, optimism (see Chang, 2001);
- developing a work ethic and a sense of vocation;
- interpersonal skills, the capacity for love, forgiveness, civility, nurturance, altruism;
- an appreciation of beauty and art;
- responsibility, self-determination, courage, perseverance, moderation;
- future mindedness, originality, creativity, talent;
- a civic sense;
- spirituality, wisdom.

They suggest that counselors would serve their clients better if they were to weave the spirit of these topics into their interactions with them. Seeing problem management as life-enhancing learning, and treating all encounters with clients as opportunity-development sessions are part of the positive psychology approach.

A note of caution: Positive psychology is not an everything's-going-to-be-all-right approach to life and helping. Richard Lazarus (2000) put it well:

> However, it might be worthwhile to note that the danger posed by accentuating the positive is that if a conditional and properly nuanced position is not adopted, positive psychology could remain at a Pollyanna level. Positive psychology could come to be characterized by simplistic, inspirational, and quasi-religious thinking and the message reduced to "positive affect is good and negative affect is bad." I hope that this ambitious and tantalizing effort truly advances what is known about human adaptation, as it should, and that it will not be just another fad that quickly comes and goes.

Positive psychology will be mentioned a number of times in this book, hopefully in the spirit of Lazarus's caution.

## The Primary Goal of Helping

The primary goal of helping can be stated in a simple directive: Help clients manage their problems in living more effectively and develop life-enhancing unused opportunities more fully. Helpers are successful to the degree to which their clients—through client-helper interactions—are in a better position to manage specific problem situations and develop specific unused resources and opportunities more effectively. Notice that I stop short of saying that clients actually end up managing problems and developing opportunities better. Although counselors help clients achieve valued outcomes, they do not control those outcomes directly. In the end, clients can choose to live more effectively or not.

Since helping is a two-way, collaborative process, clients, too, have a primary goal. Clients are successful to the degree that they commit themselves to the helping process and capitalize on what they learn from the helping sessions to manage problem situations more effectively and develop opportunities more fully "out there" in their day-to-day lives.

**The importance of results.**  A corollary to this primary goal is that *helping is about results, outcomes, accomplishments, impact.* For some, this principle is so important that they have developed over the years what is called "solution-focused" therapy (O'Hanlon & Weiner-Davis, 1989; Manthei, 1998; Rowan, O'Hanlon, & O'Hanlon, 1999). Helping is an "-ing" word: It includes a series of activities in which helpers and clients engage. These activities, however, have value only to the degree that they lead to valued outcomes in clients' lives. Indeed, the term *outcomes* appears more and more in the counseling literature (Epstein, Kutash, & Duchnowski, 2005). Ultimately, statements such as "We had a good session," whether spoken by the helper or by the client, must translate into more effective living on the part of the client. If a helper and a client engage in a series of counseling sessions productively, something of value will emerge that makes the sessions worthwhile. For instance, unreasonable fears will disappear or diminish to manageable levels, self-confidence will replace self-doubt, addictions will be conquered, an operation will be faced with a degree of equanimity, a better job will be found, a woman and man will breathe new life into their marriage, a battered wife will find the courage to leave her husband in search of a more decent life, a man embittered by institutional racism will regain his self-respect and take his rightful place in the community. In a word, helping should make a substantive difference in the life of the client. Helping is about constructive change.

**A results-focused case.**  The need for results is seen clearly in the case of a battered woman, Andrea N., outlined by Driscoll (1984, p. 64).

> The mistreatment had caused her to feel that she was worthless even as she developed a secret superiority to those who mistreated her. These attitudes contributed in turn to her continuing passivity and had to be challenged if she was to become assertive about her own rights. Through the helping interactions, she developed a sense of worth and self-confidence. This was the first outcome of the helping

process. As she gained confidence, she became more assertive; she realized that she had the right to take stands, and she chose to challenge those who took advantage of her. She stopped merely resenting them and did something about it. The second outcome was a pattern of assertiveness, however tentative in the beginning, that took the place of a pattern of passivity. When her assertive stands were successful, her rights became established, her social relationships improved, and her confidence in herself increased, thus further altering the original self-defeating pattern. This was a third set of outcomes. As she saw herself becoming more and more an "agent" rather than a "patient" in her everyday life, she found it easier to put aside her resentment and the self-limiting satisfactions of the passive-victim role and to continue asserting herself. This constituted a fourth set of outcomes. The activities in which she engaged, either within the helping sessions or in her day-to-day life, were valuable because they led to these valued outcomes.

Andrea needed much more than "good sessions" with a helper. She needed to work for outcomes that made a difference in her life. Today there is another reason for focusing on outcomes. In the United States many psychological services are offered in managed-care settings. In these settings, more and more third-party payments depend on meaningful treatment plans and the delivery of problem-managing outcomes (Meier & Letsch, 2000). But economics should not force helpers to do what they should be doing anyway in the service of their clients.

## A Secondary Goal

A secondary goal can also be stated simply: Help clients become better at helping themselves in their everyday lives. Clients often are poor problem solvers or whatever problem-solving ability they have tends to disappear in times of crisis. What Miller, Galanter, and Pribram (1960, pp. 171, 174) said many years ago is, unfortunately, probably just as true today.

> In ordinary affairs we usually muddle about, doing what is habitual and customary, being slightly puzzled when it sometimes fails to give the intended outcome, but not stopping to worry much about the failures because there are still too many other things still to do. Then circumstances conspire against us and we find ourselves caught failing where we must succeed – where we cannot withdraw from the field, or lower our self-imposed standards, or ask for help, or throw a tantrum. Then we may begin to suspect that we face a problem . . . An ordinary person almost never approaches a problem systematically and exhaustively unless he or she has been specifically educated to do so.

Most people in our society are not "educated to do so." And if many clients are poor at managing problems in living, they are equally poor in identifying and developing opportunities and unused resources. We have yet to find ways of making sure that our children develop such essential *life skills* as problem management, opportunity identification and development, sensible decision making, and the skills of interpersonal relating.

It is no wonder, then, that clients are often poor problem solvers or that whatever problem-solving ability they have tends to disappear in times of crisis. If the second goal of the helping process is to be achieved—that is, if clients are to go away able to manage their problems in living more effectively and develop opportunities on their own—then sharing some form of the problem-management and opportunity-development process with them is essential.

And so this secondary goal of helping, usually an indirect goal, deals with clients' need (1) to participate actively in the problem-management process during the helping sessions, (2) to apply what they learn to managing immediate problems and opportunities, and (3) to continue to manage their lives more effectively after the period of formal helping is over. Just as doctors want their patients to learn how to prevent illness through exercise, good nutritional habits, and the avoidance of toxic substances and activities, just as dentists want their patients to engage in effective prevention activities, so skilled helpers want to see their clients not just managing the presenting problem situation more effectively, but also becoming more adept at managing subsequent problems in living. That is, helping at its best provides clients with tools to become more effective self-helpers. Therefore, although this book is about a process helpers can use to help clients, more fundamentally, it is also about a problem-management and opportunity-development process that clients can use to help themselves. This process can help clients become more effective learners about problem management and opportunity development, better decision makers, and more responsible "agents of change" in their own lives. Since all of this has a remedial feel to it, teaching young people problem-management skills in the first place is an excellent prevention strategy (Elias & Butler, 2005a, 2005b, 2005c).

## Does Helping Help?

This question must sound strange in a book on helping. However, over the years people have questioned the usefulness of counseling and psychotherapy. There has been an ongoing debate as to the efficacy of both helping in general and different approaches to helping. However, even extreme criticisms should not be dismissed out of hand. What the critics say may be true of some forms of helping, and of some helpers.

There are both good and bad helpers. This is part of the *shadow side* of helping that will be discussed later in this chapter. Helping is a potentially powerful process that is all too easy to mismanage. Helping is not neutral; it is for better or for worse. Inept counseling not only does not help, but sometimes actually makes things worse. Research shows that some helpers are downright incompetent. Helpers who underestimate the severity of clients' problems, fail to establish a solid working relationship, use unproven or faddish approaches to helping— the list is a long one—can make things worse rather than better. Although helping can and often does work, there is plenty of evidence that ineffective helping also abounds. To be fair, research also shows that some of the factors leading to negative outcomes in helping are associated with clients. That is, clients have their shadow side. Clients who are poorly motivated and unwilling to work, or who expect helping to be painless, can become more dysfunctional through counseling.

You should be aware of the ongoing debate concerning the efficacy of helping (see Egan, 2002, Chapter 1), not to be discouraged but to be vigilant for the sake of your clients. Remember that the typical client is not aware of the issues that are being discussed here. Often clients' problems are so pressing that they just want help. No one informs them of the "buyer beware" caution as it applies to helping. Therefore, the burden falls on you, the helper: "Let the practitioner beware." Become competent. Strive continually to become a better helper. Appreciate that, poorly done, helping can actually harm others. Don't overpromise. Develop a solid approach to helping. Remain professionally self-critical. Keep your eye on results—that is, problem-managing and opportunity-developing outcomes for clients. Because helping in some generic sense *works,* do not assume that it will always work with everyone or that you are the best helper for everyone. Build evaluation into the helping process. Do not confuse difficult cases with impossible cases. Rather, appreciate the complexity of the helping process. Develop a helping value system that serves your clients well. This book will help you do all these things.

## Is Helping for Everyone?

Just because in the main, helping works, do not assume that it is for everyone. Helpers see only a fraction of people who have problems. Some don't seek help because they don't know how to get access to helpers. Others feel that they are not ready for change. Some don't acknowledge that they have problems. Some have "aversive expectations" of mental-health services (Kushner & Sher, 1989) and stay clear. Most people muddle through without professional help. They pick up help wherever they find it. I often show a popular series of videotaped counseling sessions in which the client receives help from a number of helpers, each with a different approach to helping. Once the course participants have seen the tapes, I ask, "Does this client need counseling?" They all practically shout, "Yes! " Then I ask, "Well, if this client needs counseling, how many people in the world need counseling?" The question proves to be very sobering because the client in question is struggling with issues we all struggle with. She was one of those some would call "the worried well." The fact that a person might well benefit from counseling does not mean that he or she *needs* counseling. Furthermore, many clients can "get better" in a variety of ways without help.

Working with clients who, for whatever reason, don't want to grapple with their problem situations and develop their unused resources, is a waste of time and money. Of course, it doesn't hurt to try, but the helper should know when to quit. In addition, clients with certain kinds of problems seem to be beyond help, at least at this stage of the helping professions' development. Knowing when to help is important. So is knowing how much to help. Even though the studies mentioned above show that clients who spend a longer time with a helper do reap some benefits, still spending a great deal of time in counseling is often not feasible. Helping is an expensive proposition, both monetarily and psychologically. Even when it is "free," someone is paying for it through tax dollars, insurance premiums, or free-will offerings. Therefore, without rushing nature or your clients, get to the point. Do not assume that you have a client for life.

## An Overview of the Skilled Helper Model

Common sense suggests that problem-solving models, techniques, and skills are important for all of us, since we all must grapple daily with problems in living of greater or lesser severity. Ask parents whether problem-management skills are important for their children, and they say "certainly." But when you ask them where and how their children pick up these skills, they hem and haw. "Sometimes at home, but not always." Parents don't always see themselves as paragons of effective problem solving. "Maybe at school?" Yet, if you review the curricula of our primary, secondary, and tertiary schools, you will find little about problem solving that focuses on problems in living. Some say that formal courses in problem-solving skills are not found in our schools because such skills are picked up through experience. To a certain extent, that's true. However, if problem-management skills are so important, one may well wonder why society leaves the acquisition of these skills to chance. A problem-solving mentality should be second nature to us. The world may be the laboratory for problem solving, but the skills needed to optimize learning in this lab should be taught. They are too important to be left to chance.

Let's move to the helping professions. Institute a search and you will soon discover that there are dozens, if not hundreds, of models or approaches to helping, all of them claiming a high degree of success. Library shelves are filled with books on counseling and psychotherapy—books dealing with theory, research, and practice. All are proposed with equal seriousness and all claim to lead to success. In the face of all this confusion, helpers, especially beginning helpers, need a basic, practical, working model of helping that aids them in making sense of this vast literature and is also effective with their clients.

Since all approaches to helping must eventually help clients manage problem situations and develop unused opportunities, it seems logical to start with a flexible, humanistic, broadly based problem-management and opportunity-development model. Of those who write about problem-solving approaches, some use the term explicitly (Chang, D'Zurilla, & Sanna, 2004; D'Zurilla & Nezu, 1999, 2001; Elias & Tobias, 2002; Nezu, Nezu, Friedman, Faddis, & Houts, 1999), while others use some form of the process but not the name (Hill, 2004). A problem-management model in counseling has the advantage of the vast amount of research that has been done on the problem-solving process itself and on the use of problem-solving approaches to dealing with problems in living (see the extensive research sections in D'Zurilla & Nezu, 1999, and Chang, D'Zurilla, & Sanna, 2004). The model, techniques, and skills outlined in this book tap that extensive research base. Finally, a problem-management approach is cognitive-behavioral in nature, that is, it deals with the ways clients (and the rest of us) get into and out of trouble through both thinking and acting.

All worthwhile approaches to helping ultimately help clients ask and answer the four fundamental questions at the heart of problem management and opportunity development:

- *What's going on?* "What are the problems, issues, concerns, or undeveloped opportunities I need to work on?" This is the client's *present state.*

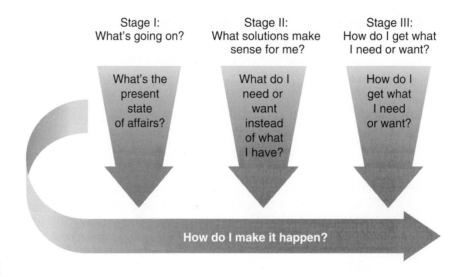

| Stage I:<br>What's going on? | Stage II:<br>What solutions make<br>sense for me? | Stage III:<br>How do I get what<br>I need or want? |

What's the present state of affairs?

What do I need or want instead of what I have?

How do I get what I need or want?

**How do I make it happen?**

FIGURE

**1.1**                                    The Skilled Helper Model

- ***What do I need or want?*** "What do I want my life to look like? What changes would make my life better? What solutions are best for me?" This is the client's *preferred state.*

- ***What do I have to do to get what I need or want?*** "What plan will get me where I want to go?" The plan outlines the *actions* clients need to take to get to the preferred state.

- ***How do I get results?*** "How do I turn thinking, talking, and planning into *solutions, results, outcomes,* or *accomplishments* that make a difference in my life? How do I get going and stay going?" This is about implementation or execution of plans.

These four questions, turned into three logical "stages"and an implementation phase in Figure 1.1, provide the basic framework for the helping process. The term "stage" is somewhat misleading because of its sequential overtones. In practice, the three stages overlap and interact with one another while clients struggle to manage problems and develop opportunities. And, as we shall see, helping, like life itself, is not as logical as the models used to describe it.

Let's take an extended example to bring this process to life. The case, though real, has been disguised and simplified. It is not a session-by-session presentation. Rather, it illustrates ways in which one client was helped to ask and answer the four fundamental questions outlined above. The client, Carlos, is voluntary, verbal, and, for the most part, cooperative. In actual practice, cases do not always flow as easily as this one. The simplification of the case, however, will help you see the main features of the helping process in action.

## Stage I: The Current Picture—Helping Clients Tell Their Stories

The present state spells out the range of difficulties the client is facing. What are the problems, issues, concerns, and undeveloped opportunities with which Carlos needs to grapple? Here is a thumbnail sketch.

> Carlos, a man in his mid-twenties, has been working as a sales rep for a software company for about a year. He is well-educated, though unlike some of his fellow workers, his undergraduate and MBA degrees are not from the "best" schools. He is bright, though practical rather than academic intelligence is his strong suit. He can be quite personable when he wants to. But he doesn't always want to. He comes from a family with traditional Hispanic cultural values. He can speak Spanish fairly well, but prefers not to, even at home.
>
> Trouble started when one of his colleagues, a senior sales rep some twenty years older than Carlos, buttonholed him after a sales meeting one day and said: "I've seen you in action a few times. You know, you're your own worst enemy. You're headed for a fall and don't even know it. If I were you, I'd get some help." With that he walked away and Carlos had no further interaction with him.
>
> Though Carlos was at first bothered by the incident, he sloughed it off. However, a few weeks later, while talking with one of his colleagues who had been at the firm for about three years, Carlos recounted the incident in a joking way. But his colleague didn't laugh. He merely said in a light-hearted manner, "Well, you never know, Carlos, there just might be something there."
>
> Through the company he has access to a couple of "developmental counselors." He is somewhat put off when he finds out that both of them are women, but he makes an appointment with one.

So here is a young man who has received a couple of shots across his bow. While he certainly doesn't think that he needs help, he is unsettled enough to be willing to talk with someone.

As we shall see in the chapters that follow, the work in each stage is divided into three interrelated tasks. These tasks spell out the work to be done in each stage. Like the stages themselves, the tasks are not steps in a mechanistic, now-do-this sense. In Stage I, the three tasks are activities that help clients develop answers to three questions: "What's going on in my life? What am I overlooking? What problems and/or unused opportunities should I work on?"

**Task 1: Help clients tell their *stories* as clearly as possible.**  Though helping is ultimately about solutions, some review of the problem or unused opportunity is called for.

> Elena, Carlos's counselor, listens to him and helps him tell his story. Through their dialogue, she helps him review what is happening in the workplace. Elena knows that if she can help Carlos get an undistorted picture of himself, his problems, and his unused opportunities, he will have a better chance of doing something about them. Her overall goal is to help Carlos manage the interpersonal dimensions of his work life better. His interpersonal style is both problem and opportunity. Carlos gets over his initial reluctance and brings up a number of things that bother him. He feels that he is being discriminated against at work. He doesn't feel that he's on the fast track and he

thinks that he should be. People at work don't understand or appreciate him. Though he lives at home because of financial reasons, he feels distant from his family. "We don't have that much in common any more. They don't want to be mainstream." He and his girl friend are currently at odds. At one point he says, "I think I've moved beyond her anyway."

This, then, is Carlos' story, mostly from his point of view. But, as is often the case, it may not be the full story.

**Task 2: Help clients develop *new perspectives* that help them reframe their stories.** Counselors add great value when they can help their clients identify significant blind spots related to their problems and unused opportunities. Effectively challenged, blind spots yield to new perspectives that help clients think more realistically about problems, opportunities, and solutions.

It does not take Elena long to realize that Carlos has some significant blind spots. For instance, he tends to blame others for his problems: "*They* are keeping me back." In addition, he does not realize how self-centered he is. He gets angry when others stand in his way or don't cater to his needs and wants. Yet he is quite insensitive to anybody else's needs. His arrogant style rubs both colleagues and customers the wrong way. Without being brutal, Elena helps him see himself as others see him. She points out that being Hispanic and coming from the "wrong" school have little to do with his colleagues' hostility toward him. Elena, Hispanic herself, understands both Carlos's struggles and his excuses.

The initial story clients tell may not be the real story or the full story. Effective counselors know how to help clients challenge themselves to put their cards on the table.

**Task 3: Help clients achieve *leverage* by working on issues that make a difference.** If clients have a range of issues, help them gain *leverage* by working on things that will make a real difference in their lives. If a client wants to work only on trivial things or does not want to work at all, then it might be better to defer counseling.

Carlos becomes more cooperative as it becomes clear to him that Elena has both his interests and those of the company in mind. He sees her as "solid"—decent, businesslike, and not overly "psychological." He is surprised that he even likes the fact that she challenges him. He comes to realize that, although he has a number of concerns, he had better work on his interpersonal communication style and his relationships with both colleagues and clients. He also needs to learn something about the "victim" mentality he has developed. With Elena's help he realizes that the flip side of his communication style problem is an enormous opportunity. He quickly sees that becoming a better communicator and relationship builder will help him in every social setting in life. Since this is a work setting and time is limited, Elena does not push him on the problems he's having at home or in his social life outside work. However, she suspects that the changes he makes at work might also apply outside.

These three tasks—the story, the real story, and the right story—are usually intermingled in actual counseling sessions.

### Stage II: The Preferred Picture—Help Clients Identify, Choose, and Shape Problem-Managing Goals

In Stage II the counselors help clients explore and choose possibilities for a better future in which key problem situations are managed and key opportunities developed. "What do you want this future to look like?" asks the helper. The client's answer constitutes his or her "change agenda." Stage II focuses on outcomes in terms of managing problem situations and developing key life-enhancing opportunities.

Elena helps Carlos ask himself such questions as: "What do I want? What do I need, whether I currently want it or not? What would my business life look like if it were more tolerable or—even better—more engaging and fulfilling?"

Unfortunately, some approaches to problem solving skip Stage II. They move from "What's wrong?" (Stage I) to "What do I do about it?" (Stage III). As we shall see, however, helping clients discover what they want has a profound impact on the entire helping process.

Stage II also has three interrelated tasks—that is, three ways of helping clients answer, as creatively as possible, the question "What do I need and want?"

**Task 1: Help clients discover *possibilities for a better future.*** This often helps clients move beyond the problem-and-misery mind-set they bring with them and develop a sense of hope. Brainstorming possibilities for a better future can also help clients understand their problem situations better—"Now that I am beginning to know what I want, I can see my problems and unused opportunities more clearly."

> Elena helps Carlos brainstorm goals that will help him repair some damaged relationships with both colleagues and clients and help him do something about his victim mentality. Carlos declares that he needs to become a "better communicator." Elena, pointing out that "becoming a better communicator" is a rather vague aspiration, asks him, "What do some of the good communicators you know look like?" Carlos comes up with a range of possibilities. "I'd give great presentations like Jeff." "I'd be a good problem solver like Sharon." "Tony listens a lot better than I do." "I'm not patient at all, but I like it when Serena is patient with me." "Roger seems to have good relationships with everyone." And so forth. All of these become possibilities for a better relationship and communication style. Elena also gets him to explore further possibilities by asking him what a "repaired relationship" would look like, both from his perspective and from the perspectives of customers and colleagues he might have alienated.

In actual counseling sessions, both the stages and the tasks within the stages are intermingled. For instance, when clients begin to consider what they want (instead of what they have), they often get a clearer idea of what the real issues in their lives are.

**Task 2: Help clients craft *problem-managing goals.*** Possibilities need to be turned into goals because helping is about solutions and outcomes. A client's goals, then, constitute his or her *agenda for change.* If goals are to be pursued and accomplished, they need to be, ideally, clear, related to the problems and unused opportunities the client has chosen to work on, substantive, realistic, prudent, sustainable, flexible, consistent with the client's values, and set in a reasonable time fame. Effective counselors help clients shape their agendas to meet these requirements.

> Elena helps Carlos sort through some of the possibilities he has come up with. It becomes clear that changes in his interpersonal communication style would help him manage some problems and develop some opportunities at the same time. If he were to communicate well—in terms of both communication skills and the values that support effective relationship-building—he would come across more effectively and repair damaged relationships. But he needs more than skills. He needs to change his self-centered and poor-me attitude. Some of the possibilities Carlos comes up with in his brainstorming session with Elena can be put aside for the present. For instance, things like "becoming a terrific presenter" can wait.
>
> Becoming a better communicator with upbeat interpersonal relationship values and attitudes is a substantial package. It includes Carlos's becoming good at the give-and-take of dialogue and the skills that make it work. These skills include tuning in to others, active listening, thoughtfully processing what he hears, demonstrating an understanding of the key points others are making, getting his own points across clearly, drawing others into the conversation, and the like. They also include embracing the values that make conversations serve relationships—mutual respect, social sensitivity, emotional control, and collaboration.

Counselors need a set of tools to help clients craft a change agenda that is both substantive and realistic. These tools are outlined later.

**Task 3: Help clients find the *incentives to commit* themselves to a better future.** The question clients must ask themselves is: "What am I willing to pay for what I need and want?" Without strong commitment, change agendas end up as no more than a package of nice ideas. This does not mean that clients are not sincere in setting goals. Rather, once they leave the counseling session, they run into the demands of everyday life, and the goals they set for themselves— however useful—face a great deal of competition. Counselors provide an important service when they help clients test their commitment to the better future embedded in the goals they choose.

> For Carlos, becoming a better communicator as a way of repairing and improving relationships is hard work. Elena helps him review the incentives he has for engaging in such work. One very strong one is this: he has to. His current interpersonal communication style will probably get him fired and prevent him from being successful in the future. However, Elena does not dwell on Carlos's bad communication and relationship habits. Rather, she believes that embracing good habits like showing interest in others, listening carefully, and checking his understanding of what they have to say will drive out his bad attitudes and habits. If Carlos does all this work, the upside is enormous. Since communication is at the heart of everything he does, better communication attitudes and skills will serve him well in every dimension of life. Elena does not find it difficult to help Carlos appreciate this attractive package of incentives. Positive psychology wins, at least in theory. In Carlos's case, developing opportunities is the main way of managing problems.

Commitment to change is a key issue. For instance, the work that Carlos has to do to become a better communicator needs to be fitted into a work schedule that is already very demanding.

## Stage III: The Way Forward—Help Clients Develop Strategies and Plans for Accomplishing Goals

Stage III defines the actions that clients need to take to translate goals into problem-managing accomplishments. Stage II answers the question, "How do I get there?" It is about making a plan and choosing strategies that will make the plan work. Stage III, too, has three tasks that intermingle with one another, and with the tasks of the other stages. These tasks, described below, define the work clients need to do to craft a viable goal-accomplishing plan.

**Task 1: Help clients review *possible strategies* to achieve goals.** Help clients see that there are many different ways of achieving their goals. Stimulating clients to think of different ways of achieving their goals is usually an excellent investment of time.

> Elena helps Carlos explore different ways of becoming the competent communicator he wants to be, so he can develop and foster satisfying work relationships and establish a better self-image with colleagues and clients. In order to acquire the skills he needs, he can read books, take courses at local colleges, attend courses for professionals, get a tutor or coach, or come up with his own approach to improving his communication style. Elena helps Carlos brainstorm the possibilities. She also points out where he can get more information, but then lets him do his own research.

Counselors help clients brainstorm possible strategies because, ordinarily speaking, choosing from options makes for better choices. Of course, clients should not leap into action. Hasty and disorganized action is often self-defeating. Complaints such as "I tried this and it didn't work. Then I tried that and it didn't work either!" are often a sign of poor planning rather than of the impossibility of the task.

**Task 2: Help clients choose strategies that *best-fit* their resources.** While Task 1 provides clients with a pool of possible strategies, Task 2 helps them choose the action strategies that best fit their talents, resources, style, temperament, environment, and timetable.

> With Elena's help, Carlos makes some choices. He chooses to attend an interpersonal-communication program for working professionals. While more expensive than university-based programs, there is greater flexibility. This fits better with Carlos's rather hectic travel schedule. Elena helps Carlos see that life is his lab. That is, every conversation is part of the program, because every conversation is an opportunity to practice the skills he will be learning and demonstrating, and to develop the attitudes that foster relationship building. To Carlos's credit he recognizes that he needs to find a way to monitor the degree that the spirit of effective dialogue is permeating his conversations. He decides to get a peer coach—a colleague who has an excellent communication style and excellent relationships with colleagues and clients. He finds a colleague who fits the bill and who is willing to help. He says to his colleague, "Be honest with me. Tell me the way it is." His colleague replies, "Don't worry."

Of course, Carlos could have made his own choices without Elena's help, but in this overview we see the counseling playing a helping role for each task.

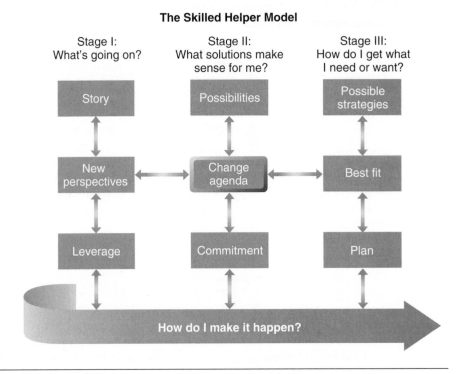

**The Skilled Helper Model**

Stage I:
What's going on?

Stage II:
What solutions make
sense for me?

Stage III:
How do I get what
I need or want?

Story

Possibilities

Possible
strategies

New
perspectives

Change
agenda

Best fit

Leverage

Commitment

Plan

**How do I make it happen?**

FIGURE

**1.2**    The Helping Model Showing Interactive Stages and Steps

**Task 3: Help clients pull strategies together into *a manageable plan.*** Help clients organize the actions they need to take to accomplish their goals. Plans are simply maps clients use to get where they want to go. A plan can be quite simple. Indeed, overly sophisticated plans are often self-defeating.

> Carlos's plan is straightforward. He will begin the interpersonal-communication program within two weeks. He will use every conversation with colleagues and clients as his lab. At the end of each working day he will review the conversations he has had in terms of both skillfulness and values. He creates a checklist for himself that includes such questions as "How effectively did I listen? How clearly did I get my points across? How respectful and collaborative was I? How did I handle sensitive conversations with colleagues and clients who had been turned off by my interpersonal style?" Time and schedules willing, he will meet with his peer coach once a week and with Elena once a month.

It goes without saying that as Carlos comes to grips with his issues, sets some goals, and crafts a plan, he can be engaging in a range of little actions that keep him moving forward.

Figure 1.2 presents the full model in all its stages, together with the tasks specific to each stage, and adds their relationship to the action arrow considered in the next section. It includes two-way arrows between both stages and tasks to suggest the kind of flexibility needed to make the process work. Figure 1.2 is map of the helping process. Maps help you know where you are at any given moment and where your next movement should be.

## The Action Arrow: Help Clients Implement Their Plans and Get Results

All three stages of the helping model sit on the "action arrow," indicating that clients need to act in their own behalf right from the beginning of the helping process. Stages I, II, and III are about planning for change, not engaging in change itself. Planning is not action. Without goal-accomplishing action, talking about problems and opportunities, discussing goals, and figuring out strategies for accomplishing goals is just so much blah, blah, blah. There is nothing magic about change; it is hard work. But, as we shall see in subsequent chapters, each stage and task of the process can promote problem-managing and opportunity-developing action right from the beginning.

> Carlos, like most clients, runs into a number of obstacles as he tries to implement his plan. Right away, his travel schedule keeps conflicting with even the flexible communication-skills program in which he has enrolled. Since he finds the program very useful, he decides to use a tutor from the program to bring him up to speed whenever he cannot fit a session into his travel schedule. Although the company is paying for the program, he has to pay the tutor out of his own pocket. But it's worth it. And, since it is his own money, he becomes more serious about the program. He also finds that he is not very consistent in using the skills he is learning in the so-called lab of life. His progress is much slower that he thought it would be. Sometimes he loses heart. He often fails to do the evening self-evaluation sessions. Getting together with his peer coach proves to be almost impossible. And, while he is doing better with clients, he doesn't see much repair going on in his relationships with his colleagues. So in one of his discussions with Elena, he reviews the ups and downs of his program, and discusses his discouragement about how slowly some of his disenchanted colleagues are warming up to him.
>
> Carlos uses the discussion to reset his program. First, he agrees to see Elena for a half hour every other week. This provides him with an incentive to keep to the program. He wants to give her a good report. Second, he discusses his "bad attitude." Even though she said that the whole program would be a lot of work, Carlos didn't realize just how much work. He finds that he can't do the program well without changing some basic attitudes. It's not just a skills program. It cuts much deeper. He has to recommit himself to a deeper kind of attitudinal change. New attitudes look fine on paper. Developing them is another story. And so Carlos moves on—two small steps forward and a half step backward. He resets his schedule so he can get together with his peer coach. His sessions with Elena and his peer coach both help and Carlos does make progress.

As we shall see, every stage and task of the helping process has the ability to drive problem-managing action on the part of the client. Without action there are no solutions.

## Help Clients Evaluate Their Progress Along the Way: "How Are We Doing?"

In the light of the earlier "Does helping help?" discussion, how do helpers who use the problem-management and opportunity-development framework evaluate what is happening with each client? By making each case a mini-experiment in itself. In many helping models, evaluation is presented as the last step in the model. However, if evaluation occurs only at the end, it is too late. Early detection of what is going wrong in the helping process is needed to prevent failure. The problem-management and opportunity-development model outlined in this chapter fills the bill. Evaluation is a tool to check progress throughout the helping process. As we shall see, it provides criteria for helper effectiveness, for client participation, and for assessing outcomes. Since helpers and clients need to collaborate in this ongoing evaluation process, Elena and Carlos, together, use the helping model as the evaluation framework. Carlos comes to appreciate Elena's skill in using the model. In later chapters, questions designed to help with the evaluation process will be provided for each stage and task.

Carlos begins to use the feedback he gets—from self evaluation, the observations of his peer coach, and from his sessions with Elena—more frequently. Once he takes ongoing evaluation seriously, he begins to make progress. The best form of feedback comes from goal accomplishment. In what ways and to what degree is he becoming a more competent communicator? How effectively are relationships with clients and colleagues being repaired? To what degree is he shedding his self-centered approach to relationships that contribute to his belief that "others are out to get me"?

## Be Flexible in the Use of the Model

There are many reasons why you need to use the helping model flexibly. The main one is this: Helping is for the client. Clients' needs take precedence over any model. That said, a number of points about flexibility need to be made.

First, clients start and proceed differently. Any stage or task of the helping process can be the entry point. For instance, Client A might start by talking about a way he tried to solve a problem that did not work—"I threatened to quit if they didn't give me a leave of absence, but it backfired. They told me to leave." The starting point is a failed Stage III strategy. Client B might start with what she believes she wants but does not have—"I need a boyfriend who will take me as I am. Joe keeps trying to redo me." Stage II, creating the preferred picture, is her entry point. Client C might start with the roots of his problem situation—"I don't think I've ever gotten over being abused by my uncle." Stage I, helping the client tell his story, is the entry point. Client D might announce that she really has no problems but is still vaguely dissatisfied with her life— "I don't know. Everyone tells me I've got a great life, but something's missing." The implication here is that she has not been seizing the kind of opportunities that could make her happy. Opportunity-development, rather than problem-management, is the starting point.

Second, clients engage in each stage and task of the model differently. Take clients' stories. Some clients spill out their stories all at once. Others leak bits and pieces of their story throughout the helping process. Still others tell only those parts that put them in a good light. Most clients talk about problems rather than opportunities. Since clients do not always present all their problems at once in neat packages, it is impossible to work through Stage I completely before moving on to Stages II and III together with the problem-managing actions that flow from each. It is not even advisable to do so. Some clients don't even understand their problems until they begin talking about what they want but don't have. Some clients need to engage in some kind of remedial action before they can adequately define the problem situation. That is, action sometimes precedes understanding. If some supposed problem-solving action is not successful, then the counselor helps the client learn from it and return to the tasks of clarifying the problem or opportunity and then setting some realistic goals. Take the case of Woody.

> Woody, a sophomore in college, came to the student counseling services with a variety of interpersonal and somatic complaints. He felt attracted to a number of women on campus but did very little to become involved with them. After exploring this issue briefly, he said to the counselor, "Well, I just have to go out and do it." Two months later he returned and said that his experiment had been a disaster. He had gone out with a few women, but the chemistry never seemed right. Then he did meet someone he liked quite a bit. They went out a couple of times, but the third time he called, she said that she didn't want to see him any more. When he asked why, she muttered vaguely about his being too preoccupied with himself and ended the conversation. He felt so miserable he returned to the counseling center. He and the counselor took another look at his social life. This time, however, he had some experiences to probe. He wanted to explore this "chemistry thing" and his reaction to being described as "too preoccupied with himself."

Woody put into practice Weick's (1979) dictum that chaotic action is sometimes preferable to orderly inactivity. Once he acted, he learned a few things about himself. Some of this learning proved to be painful, but he now had a better chance of examining his interpersonal style much more concretely.

Third, since the stages and tasks of the model intermingle, helpers will often find themselves moving back and forth in the model. Often two or more tasks or even two stages of the process merge into one another. For instance, clients can name parts of a problem situation, set goals, and develop strategies to achieve them in the same session. New and more substantial concerns arise while goals are being set, and the process moves back to an earlier, exploratory stage. Helping is seldom a linear event. One client, in discussing a troubled relationship with a friend, said something like this:

> Every time I try to be nice to her, she throws it back in my face. So who says being more considerate is the answer? Maybe my problem is that I'm a wimp, not the self-centered jerk she makes me out to be. Maybe I'm being a wimp with you and you're letting me do it. Maybe it's time for me to start looking out for my own interests—you know, my own agenda, rather than trying to make myself fit into everyone else's plans. I need to take a closer look at the person I want to be in my relationships with others.

In these few sentences the client mentions a failed action strategy, questions an established goal, hints at a new problem, suggests a difficulty with the helping relationship itself, offers, at least generically, a different approach to managing his problem, and recasts the problem as an opportunity to develop a more solid interpersonal style. Your challenge is to make sense of clients' entry points and guide them through whatever stage or task that will help them move toward problem-managing and opportunity-developing action.

Flexibility, of course, is not mere randomness or chaos. Focus and direction in helping are also essential. Letting clients wander around in a morass of problem situations under the guise of flexibility leads nowhere. The structure of the helping model is the very foundation for flexibility; it is the underlying system that keeps helping from being a set of random events. A helping model is like a map that helps you locate, at any given moment, where you are with clients and what kinds of interventions would be most useful. In the map metaphor, the stages and tasks of the model are orientation devices. At its best, it is a *shared* map that helps clients participate more fully in the helping process. They, too, need to know where they are going.

## Moving from Smart to Wise: Managing the Shadow Side of Helping

This book outlines a model of helping that is rational, linear, and systematic. What good is that, you well might ask, in a world that is often irrational, nonlinear, and chaotic? One answer is that rational models help clients bring much-needed discipline and order into their chaotic lives. Effective helpers do not apologize for using such models. But they also make sure that their humanity permeates their models.

More than intelligence is needed to apply the model well—smart is not smart enough. The helper who understands and uses the model together with the skills and techniques that make it work might well be smart, but he or she must also be wise. Effective helpers understand the limitations, not only of helping models, but also of helpers, the helping profession, clients, and the environments that affect the helping process. One dimension of wisdom is the ability to understand and manage these limitations which, in sum, constitute what I call the "arational" dimensions, or the "shadow side," of life. The shadow side of helping can be defined as:

> *All those things that adversely affect the helping relationship, process, outcomes, and impact in substantive ways but that are not identified and explored by helper or client, or even by the profession itself.*

Helping models are flawed; helpers are sometimes selfish, lazy, and even predatory, and they are prone to burnout. Clients are sometimes selfish, lazy, and predatory even in the helping relationship. Indeed, the very assumptions on which psychology, including counseling psychology, are based have come into question (Chamberlin, 2005; Slife, Reber, & Richardson, 2005).

Helping models, too, have their shadow side, both in terms of the models themselves and the way they are used with clients. Here are a few of the shadow-side dimensions of helping models.

**No model.**  Some helpers "wing it." They have no consistent, integrated model that has a track record of benefitting clients. Professional training programs often present a wide variety of approaches to helping drawn from the "major brands." If helpers-to-be leave such programs knowing a great deal about different approaches but lacking an integrated approach for themselves, then they need to develop one quickly.

**Fads.**  The helping professions are not immune to fads. A fad is an insight or a technique that would have some merit if it were integrated into some overriding model or framework of helping. Instead it is marketed on its own as the central, if not the only meaningful, intervention needed. A fad need not be something new; it can be the "rediscovery" of a truth or a technique that has not found its proper place in the helping tool kit. Helpers become enamored of these ideas and techniques for a while and then abandon them. There will always be "hot topics" in helping. Note them and integrate them into a comprehensive approach to your clients. Many new approaches to helping make outrageous claims. Don't ignore them, but take the claims with a grain of salt and test the approach. If something works, integrate it into your approach.

**Rigid applications of helping models.**  Some helpers buy into a model early on and then ignore subsequent challenges or alterations to the model. They stop being learners. The so-called purity of the model becomes more important than the needs of clients. Other helpers, especially beginners, apply a useful helping model too rigidly. They drag clients in a linear way through the model even though flexibility would serve clients better. All of this adds up to excessive control. Effective models effectively used are liberating rather than controlling.

**Virtuosity.**  Another problem is virtuosity. Virtuoso helpers are prone to fads. And there are plenty of fads in the helping professions. Virtuoso helpers also tend to specialize in certain techniques and skills—exploring the past, in-depth assessment, focusing on solutions, challenging, and the like. Helpers who specialize like this not only run the risk of ignoring client needs but also often are not very effective even in their chosen specialties. For example, the counselor whose specialty is challenging clients is often an ineffective challenger. The reason is obvious: Challenge must be based on understanding. Challenge, as we shall see later, is just one dimension of an entire helping process.

The antidote to all these shadow-side pitfalls is simple. Helpers need to become radically client-centered. Client-centered helping means that the needs of the client, not the models and methods of the helper, constitute the starting point and guide for helping. Therefore, a flexible model of helping is essential. In the end, helping is about solutions, results, outcomes, and impact driven by a client-serving process. The values that drive client-centered helping are reviewed in Chapter 2.

## EVALUATION QUESTIONS

- What is the primary goal of helping?
- What is the secondary goal?
- What accountability does the client have?
- Why use a problem-management approach to helping?
- Give a brief description of the three stages of the helping process.
- What is the Action Arrow and why is it associated with every stage?
- Describe how evalution fits into the helping process.
- What does flexibility in the use of the model mean?
- What is meant by the shadow side of helping? How does it apply to the helping model?

## EXERCISES ON PERSONALIZING

At the end of each chapter there will be a couple of exercises to help you personalize some of the significant learnings in that chapter.

1a. Think of a problem situation you faced in the past. Jot down its main elements. Describe how you handled the problem situation.

1b. Now walk the problem through the major stages of the helping model outlined in this chapter. How clear was the problem situation? What blind spots did you have? What new perspectives did you develop? What goals did you set? What did you do about the problem? What were the results? How did this affect your life? In the light of the helping model, what would you have done differently?

2a. Think of an unused opportunity you had in the past. Jot down the main features of this opportunity.

2b. Now walk the unused opportunity through the helping model. What did you do? What part of the model did you omit? How did it turn out? What could you have done differently?

3a. Think of someone you know who is currently facing a problem situation. Jot down the main features of the problem situation.

3b. Now, using the stages of the model, jot down what you think this person might do to handle the problem. In the light of the helping model, how might you help this person manage the problem situation more effectively?

4a. Think of a problem situation you are currently facing. Jot down the main features of this problem.

4b. As above, walk the problem through the stages of the helping model. Indicate what you want (your goal) and how you might go about achieving it. What obstacles stand in the way of your acting on the problem?

# CHAPTER 2

# The Helping Relationship and the Values That Drive It

The Helping Relationship

Values As Guides and Drivers

Key Values in Helping: Respect, Empathy, Genuineness, Self-Responsibility, and a Bias toward Action

Respect As a Foundation Value

Empathy As a Primary Orientation Value

Genuineness As a Professional Value

Self-Responsibility As an Outcome Value

A Bias toward Action As a Solution-Focused Value

Shadow-Side Realities in the Helping Relationship

Exercises On Personalizing

## The Helping Relationship

Although theoreticians, researchers, and practitioners alike, not to mention clients, agree that the relationship between client and helper is important, there are significant differences as to how this relationship is to be characterized and played out in the helping process. Some authors stress the *relationship* itself, others highlight the *work* that is done through the relationship, and still others focus on the *outcomes* to be achieved through the relationship. All are important in facilitating change (Murphy & Dillon, 2003; Norcross, 2002; Teyber, 2000).

**The relationship itself.**  Carl Rogers (1951, 1957), one of the great pioneers in the field of counseling, emphasized the quality of the relationship in representing the humanistic-experiential approach to helping. Rogers claimed that the unconditional positive regard, accurate empathy, and genuineness offered by the helper and perceived by the client were both necessary and often *sufficient* conditions for therapeutic progress. Through this highly empathic relationship counselors, in his eyes, helped clients understand themselves, liberate their resources, and manage their lives more effectively. Rogers's work spawned the widely discussed client-centered approach to helping (Rogers, 1965).

**The relationship as a means to an end.**  Others see the helping relationship as very important but still as a means to an end. In this view, a good relationship is practical because it enables client and counselor to do the work called for by whatever helping process is being used. The relationship is instrumental in achieving the goals of the helping process. For them, overstressing the relationship is a mistake because it obscures the ultimate goal of problem management. This goal won't be achieved if the relationship is poor, but if too much focus is placed on the relationship itself, both client and helper can be distracted from the real work to be done.

**The relationship as a working alliance.**  The term *working alliance* can be used to bring together the best of the relationship-in-itself, relationship-as-means, and solution-focused approaches. Bordin (1979) defined the working alliance as the collaboration between the client and the helper based on their agreement on the goals and tasks of counseling. The term "working alliance" has gained credibility in the helping professions and is also the subject of cross-cultural research (Wei & Heppner, 2005). Although there is, predictably, considerable disagreement among practitioners as to what the critical dimensions of the working alliance are, how it operates, and what results it is to produce, it is relatively simple to outline what it means in the context of the problem-management and opportunity-development process outlined and illustrated in this book.

**The collaborative nature of helping.**  In the working alliance, helpers and clients are collaborators. Helping is not something that helpers do to clients; rather, it is a process that helpers and clients work through together. Helpers do not "cure" their patients. Both have work to do in the problem-management and opportunity-development stages of helping, and both have responsibilities related to outcomes. Outcomes depend on the competence and motivation of the helper, on the competence and motivation of the client, and on the quality of their interactions. If either party refuses to play or plays incompetently, then the entire enterprise can fail.

**The relationship as a forum for relearning.**  Even though helpers don't cure their clients, the relationship itself can be therapeutic. In the working alliance, the relationship itself is often a forum for social-emotional relearning. Effective helpers model attitudes and behavior that help clients challenge and change their own attitudes and behavior. It is as if a client were to say to himself (though not in so many words), "She [the helper] obviously cares for and trusts me, so perhaps it is all right for me to care for and trust myself." Or, "He takes the risk of challenging me, so what's so bad about challenge when it's done well?" Or, "I came here frightened to death by relationships and now I'm experiencing a nonexploitative relationship that is very beneficial." Furthermore, protected by the safety of the helping relationship, clients can experiment with different behaviors during the sessions themselves. The shy person can speak up, the secretive person can open up, the aggressive person can back off, the overly sensitive person can ask to be challenged, and so forth.

These learnings can then be transferred to other social settings. It is as if a client might say to himself, "She [the helper] listens to me so carefully and makes sure that she understands my point of view even when she thinks I should reconsider it. My relationships outside would be a lot different if I were to do the same." Or, "I do a lot of stuff in the sessions that would make anyone angry. But she doesn't let herself become a victim of emotions, either her own or mine. And her self-control doesn't diminish her humanity at all. That would make a big difference in my life." The relearning dynamic, however subtle or covert, is often powerful. In sum, necessary changes in both attitudes and behavior often take place within the sessions themselves through the relationship.

**Relationship flexibility.**  The idea that one kind of perfect relationship or alliance fits all clients is a myth. Different clients have different needs, and those needs are best met through different kinds of relationships and different activities within the same relationship. One client may work best with a helper who expresses a great deal of warmth, whereas another may work best with a helper who is more objective and businesslike. Some clients come to counseling with a fear of intimacy. If helpers, right from the beginning, communicate a great deal of empathy and warmth, these clients might be put off. Once the client learns to trust the helper, stronger interventions can be used. Effective helpers use a mix of styles, skills, and techniques tailored to the kind of relationship that is right for each client. And yet they remain themselves while they do so.

## Values As Guides and Drivers

One of the best ways to characterize a helping relationship is through the values that should permeate and drive it. Since it has become increasingly clear that helpers' values influence clients' values over the course of the helping process, it is essential to build a value orientation into the process itself. Helpers not only need to be aware of their own values, but must also come to understand the values of their clients.

**Values and personal culture.** Values are an important part of the *personal culture* of each individual. In society culture refers to the shared assumptions, beliefs, values, and norms that tend to drive patterns of behavior—"the way we do things here." However, counselors don't deal with societies as such but with individuals. So let's apply this basic culture framework to an individual. It goes something like this:

- Over the course of life individuals develop *assumptions and beliefs* about themselves, other people, and the world around them. For instance, Isaiah, a client suffering from posttraumatic stress disorder stemming from a history of gang activity in his neighborhood and a recent brutal attack on his person, has come to believe that the world is a dangerous and heartless place.

- Values are more than good ideas or conceptual ideals. They are also a set of practical criteria for making decisions. As such, they are drivers of behavior. Real values, like honesty or altruism, are not only prized but actually guide and drive behavior. Values, although picked up or inculcated along the path of life, become true values when they are owned and acted upon by the individual. Isaiah, because of a number of ups and downs in his life, has come to value or prize personal security. He steers clear of trouble.

- Assumptions and beliefs, interacting with values, generate *norms,* the "dos and don'ts" we carry around inside ourselves. For Isaiah one of these norms is, "Don't trust people. You'll get hurt."

- These norms drive *patterns of behavior* and these patterns of behavior constitute, as it were, the *bottom line* of personal or individual culture—"the way I live my life." For Isaiah this means not taking chances with people. He's a loner.

Effective helpers come to understand the personal cultures of clients and the impact these individual cultures have, both in everyday life and in helping sessions. Of course, since no individual is an island, personal cultures do not develop in a vacuum. The beliefs, values, and norms people develop are greatly influenced by their environments—the cultures of their families, their local communities, and their societies.

**The helper's values.** Helpers, of course, have their own personal cultures just as clients have theirs. Inevitably, the helper's personal culture interacts with the client's for better or for worse. The way helpers, together with their clients, "do" helping constitutes the culture of helping. Therefore, the Greek dictum "Know Thyself" is essential for helpers. For instance, a helper might say to himself or herself during a session with a difficult client something like this:

> The arrogant, I'm-always-right attitude of this client needs to be challenged. How I challenge her is important, since I don't want to damage our relationship. I value genuineness and openness. Therefore, I can challenge her by describing her behavior

and the impact it has on me and might have on others and I can do so respectfully, that is, without belittling her.

Helpers without a set of working values are adrift. Helpers who don't have an explicit set of values have an implicit or "default" set which may or may not serve the helping process.

Helping-related values, like your other values, cannot be handed to you on a platter, much less shoved down your throat. Therefore, this chapter is meant to stimulate your thinking about the values that should drive helping. In the final analysis, as you sit with your clients, only those beliefs, values, and norms that you have made your own will make a difference in your helping behavior. Therefore, you need to be proactive in your search for the beliefs, values, and norms that will govern your interactions with your clients.

## Key Values in Helping: Respect, Empathy, Genuineness, Self-Responsibility, and a Bias toward Action

Developing a set of values to guide your interactions with clients does not mean that you will invent a set of values different from everyone else's. Tradition is an important part of value formation, and we all learn from the rich tradition of the helping professions. The five major values that form the tradition of the helping professions—respect, empathy, genuineness, empowered self-responsibility, and a bias toward action—are briefly reviewed here and translated into a set of norms. Respect is the *foundation* value; empathy is the value that *orients* helpers in their dialogues with their clients; genuineness is the what-you-see-is-what-you-get *professional* value; *self-responsibility* is at the heart of client empowerment, one of the main goals of helping, and, as such, is an *outcome* value that energizes the entire helping process; a *bias toward action* is the value that underscores the centrality of constructive change in lives of clients. These values, outlined below, are meant to serve as a starting point for your reflection on the values that should drive the helping process. Don't just swallow them. Analyze, reflect on, and debate them.

### Respect As a Foundation Value

Respect for clients is the foundation on which all helping interventions are built. Respect is such a fundamental concept that, like most such concepts, it is difficult to define. The word comes from a Latin root that includes the idea of "seeing" or "viewing." Indeed, respect is a particular way of viewing oneself and others. Respect, if it is to make a difference, cannot remain just an attitude or a way of viewing others. Here are some norms that flow from the interaction between a belief in the dignity of the person and the value of respect.

**Do no harm.** This is the first rule of the physician and the first rule of the helper. Yet some helpers do harm either because they are unprincipled or because they are incompetent. Helping is not a neutral process—it is for better or for worse. In a world in which such things as child abuse, wife battering, and exploitation of workers is much more common than we care to think, it is important to emphasize a nonmanipulative and nonexploitative approach to clients.

**Become competent and committed.** There is no place for the "caring incompetent" in the helping professions. Get good at whatever model of helping you use. Get good at the basic problem-management and opportunity-development framework outlined in this book and the skills that make it work. They are fundamental to almost any approach to helping.

**Make it clear that you are "for" the client.** The way you act with clients will tell them a great deal about your attitude toward them. Your manner should indicate that you are *for* the client, that you care for him or her in a down-to-earth, nonsentimental way. It is as if you are saying to the client, through your attitudes and behavior, "Working with you is worth my time and energy." Respect is both gracious and tough-minded. "Being for" means taking clients' points of view seriously even when they need to be challenged. Respect also involves helping clients place demands on themselves.

**Assume the client's goodwill.** Work on the assumption that clients want to work at living more effectively, at least until that assumption is proved false. The reluctance and resistance of some clients, particularly involuntary clients, are not necessarily evidence of ill will. Respect means entering the world of the clients to understand both their reluctance and their resistance, and having a willingness to help clients work through them.

**Keep the client's agenda in focus.** Helpers should pursue their clients' agendas, not their own. Here are three examples of helpers who lost clients because of lack of appreciation of the clients' agendas. One helper recalled, painfully, that he lost a client because he had become too preoccupied with his theories of depression rather than the client's painful depressive episodes. Another helper, who dismissed as either trivial or irrelevant a client's bereavement over a pet that had died, was dumbfounded and crushed when the client made an attempt on her life. The helper had overlooked the depth of the client's feelings. A third helper, a white male who prided himself on his multicultural focus in counseling, went for counseling himself when a Hispanic client quit therapy saying, perhaps somewhat unfairly, as he was leaving, "I don't think you're interested in me. You're more interested in Anglo-Hispanic politics."

## Empathy As a Primary Orientation Value

Empathy, though a rich concept in the helping professions, has been a confusing one (see Bohart & Greenberg, 1997 and Duan & Hill, 1996 for overviews). For the sake of clarity, empathy as a *value* that should permeate the entire helping process is discussed in this chapter, while empathy as an *interpersonal communication skill* is discussed later. Skilled helpers work hard to understand their clients and then to communicate this understanding to clients. When done well, communicating understanding helps clients better understand themselves, their problem situations, their unused resources and opportunities, and their feelings. Better understanding leads to better problem management.

**Empathy: A rich term.** A number of authors look at empathy from a value point of view and talk about the behaviors that flow from it. Sometimes their language is almost lyrical. For instance, Kohut (1978) said, "Empathy, the accepting, confirming, and understanding human echo evoked by the self, is a psychological nutrient without which human life, as we know and cherish it, could not be sustained" (p. 705). In this view, empathy is a value, a philosophy, or a cause with almost religious overtones. Covey (1989), naming empathic communication one of the "seven habits of highly effective people," said that empathy provides those with whom we are interacting with "psychological air" that helps them breathe more freely in their relationships. Goleman (1995, 1998) puts empathy at the heart of emotional intelligence. It is the individual's "social radar" through which he or she senses others' feelings and perspectives and takes an active interest in their concerns. Rogers (1980) talked passionately about basic empathic listening—being with and understanding the other— even calling it "an unappreciated way of being" (p. 137). He used the word *unappreciated* because, in his view, few people in the general population developed this deep listening ability, and even so-called expert helpers did not give it the attention it deserved. Here is his description of basic empathic listening:

> It means entering the private perceptual world of the other and becoming thoroughly at home in it. It involves being sensitive, moment by moment, to the changing felt meanings which flow in this other person, to the fear or rage or tenderness or confusion or whatever that he or she is experiencing. It means temporarily living in the other's life, moving about in it delicately without making judgments. (p. 142)

Such empathic listening is selfless because helpers must put aside their own concerns to be fully with their clients.

**Empathy: A key helping value.** Empathy as a value is a radical commitment on the part of helpers to understand clients as fully as possible in three different ways. First, empathy is a commitment to work at understanding each client from *his or her point of view,* together with the feelings surrounding this point of view, and to communicate this understanding whenever the helper deems it beneficial. Second, empathy is a commitment to understand individuals in and through the *context* of their lives. The social settings, both large and small, in which they have developed and currently live and move and have their being provide routes to understanding. Third, empathy is also a commitment to understand the *dissonance* between the client's point of view and reality.

**Understand diversity.** While clients have in common their humanity, they differ from one another in a whole host of ways—accent, age, attractiveness, color, developmental stage, abilities, disabilities, economic status, education, ethnicity, fitness, gender, group culture, health, national origin, occupation, personal culture, personality variables, politics, problem type, religion, sexual orientation, social status, to name some of the major categories. We differ from one another in hundreds of ways. And who is to say which differences are key? This presents several challenges for helpers. For one, it is essential that helpers understand clients and their problem situations contextually. For instance, a life-threatening illness might be one kind of reality for a 20-year-old and quite a different reality for an 80-year-old. We know that homelessness is a complex phenomenon. A homeless client with a history of drug abuse who has dropped out of graduate school is far different from a drifter who hates shelters for the homeless and resists every effort to get him to go to them.

Helpers can, over time, come to understand a great deal about the characteristics of the populations with whom they work. For instance, they can and—should—understand the different development tasks and challenges that take place over the life span and, if they work with the elderly, they should grow in their understanding of the challenges, needs, problems, and opportunities of the aged. Still, it is impossible to know everything about every population. This impossibility becomes even more dramatic if the combinations and permutations of individual characteristics are taken into consideration. How could an African American, middle-class, highly educated, younger, urban, Episcopalian, female psychologist possibly understand a poor, unemployed, homeless, middle-aged, uneducated, lapsed-Catholic male, born of migrant workers, his father a Mexican, his mother a Polish immigrant? Indeed, how can anybody fully understand anybody else? If the legitimate principles relating to diversity were to be pushed too far, no one would be able to understand and help anybody else.

**Multicultural diversity.** Dealing knowledgeably and sensitively with that particular form of diversity called *multiculturalism* demands both respect and empathy. There has been an explosion of literature on diversity and multiculturalism, together with related counselor competencies, over the past few years (Diller, 2004; Guadalupe & Lum, 2005; Hogan-Garcia, 2003; Roysircar, Sandhu, & Bibbins, 2003; Slattery, 2004; Sternberg & Grigorenko, 2004). There is both an upside and a downside to this avalanche. One upside is that helpers are forced to take another look at the blind spots they may have about diversity and culture, and to take another look at the world in which we live. One downside is the infighting that goes on among helpers as to the place of multiculturalism in the helping professions, the multicultural competencies helpers should be required to have, and the relationship of multiculturalism to social justice.

Let's look at an example. Sue, a Midwestern American, is married to Lee, an immigrant from Singapore. They are having problems. Many clients come to helpers because they are having difficulties in their relationships with others or because relationship difficulties are part of a larger problem situation. Therefore, understanding clients' different approaches to developing and sustaining relationships is important. Guisinger and Blatt (1994) put this is in a broader multicultural perspective.

Western psychologies have traditionally given greater importance to self-development than to interpersonal relatedness, stressing the development of autonomy, independence, and identity as central factors in the mature personality. In contrast, women, many minority groups, and non-Western societies have generally placed greater emphasis on issues of relatedness. (p. 104)

Sue is deep into the development of autonomy, independence, and her identity as a successful working woman. Lee runs a successful small Web site development business. Guisinger and Blatt go on to point out that both interpersonal relatedness and self-definition are essential for maturity. Helping Sue and Lee, individuals from different cultures, achieve some kind of mutually acceptable balance depends on understanding what "mutually acceptable balance" means in any given culture. Here are a few principles to help you achieve this balance.

**Come to terms with your diversity blind spots.**  Since helpers often differ from their clients in many ways, the helper is often challenged to avoid diversity-related blind spots that can lead to inept interactions and interventions during the helping process. For instance, a physically attractive and extroverted helper might have blinds spots with regard to the social flexibility and self-esteem of a physically unattractive and introverted client. Much of the literature on diversity and multiculturalism targets such blind spots. Counselors would do well to become aware of their own cultural values and biases. They should also make every effort to understand the world views of their clients. Helpers with diversity blind spots are handicapped. Helpers should, as a matter of course, become aware of the key ways in which they differ from their clients and take special care to be sensitive to those differences.

**Tailor your interventions in a diversity-sensitive way.**  Your understanding of diversity needs to be translated into appropriate interventions. The way a Hispanic helper challenges a Hispanic client may be inappropriate if the client is white, and vice versa. The way a younger helper shares his own experience with a younger client might be inappropriate for a client who is older, and vice versa. Client self-disclosure, especially more intimate disclosure, might be relatively easy for a person from one culture, let us say North American culture, but very difficult for a client from another, let us say Asian or British culture. In this case, interventions that call for intimate self-disclosure may be seen as inappropriate by such a client. Even though a client may be from a culture that is more open to self-disclosure, he or she may be frightened to death by self-disclosure. Therefore, with clients who come from a culture that has a different perception of self-disclosure, or with any client who finds self-disclosure difficult, it might be best, after an initial discussion of the problem situation in broad terms, to move to what the client wants instead of what he or she currently has (Stage II), rather than to the more intimate details of the problem situation. Once the helping relationship is on firmer ground, the client can move to the work he or she sees as more intimate or demanding. Although the helping model outlined here is a "human universal," helpers need to apply its stages and tasks with cultural sensitivity.

**Work with individuals.**  The diversity principle is clear: The more helpers understand the broad characteristics, needs, and behaviors of the populations with whom they work—African Americans, Caucasian Americans, diabetics, the elderly, the drug addict, the homeless, you name it—the better positioned they are to adapt these broad parameters and the counseling process itself to the individuals with whom they work. But, whereas multicultural diversity focuses on differences both between and within groups—between cultures and subcultures, helpers interact with clients as individuals. As Satel (1996) pointed out: "Psychotherapy can never be about celebrating racial diversity because it is not about groups; it is about individuals and their infinite complexity" (p. A 14). Your clients are individuals, not cultures, subcultures, or groups. Remember that cultural traits can destroy understanding as well as facilitate it.

Even though individual members have group characteristics, they also have distinct individual differences. There are no homogeneous groups. One of the principal learnings of social psychology is this: *There are as many differences, and sometimes more, within groups as between groups.* This middle-class black male is this individual. This poor Asian woman is this person. In a very real sense, a conversation between identical twins is a cross-cultural event because they are different individuals with differences in personal assumptions, beliefs, values, norms, and patterns of behavior. Genetics and group culture account for commonalities among individuals, but personhood and personal cultures emphasize each person's uniqueness.

Since diversity in all its forms is such an important issue in helping, a brochure called *Skilled Helping Around the World* has been packaged with this book.

## Genuineness As a Professional Value

Like respect, helper genuineness refers to both a set of attitudes and a set of counselor behaviors. Some writers call genuineness "congruence." Genuine people are at home with themselves and, therefore, can comfortably be themselves in all their interactions. Being genuine has both positive and negative implications; it means doing some things and not doing others.

**Do not overemphasize the helping role.**  Genuine helpers do not take refuge in the role of counselor. Ideally, relating at deeper levels to others and helping are part of their lifestyle, not roles they put on or take off at will. This keeps them far away from being patronizing and condescending. Do not use your role as a facade. Do not manipulate or fool clients in any way. Be yourself. Be spontaneous with clients, but don't overwhelm them.

**Avoid defensiveness.**  Genuine helpers are nondefensive. They know their own strengths and deficits and are—presumably—trying to live mature, meaningful lives. When their clients express negative attitudes toward them, they examine the behavior that might cause the client to think negatively, try to understand the clients' points of view, and continue to work with them. Consider the following example:

CLIENT: I don't think I'm really getting anything out of these sessions at all. I still feel drained all the time. Why should I waste my time coming here?

Now consider three possible responses on the part of the helper.

HELPER A: If you were honest with yourself, you'd see that you are the one wasting time. Change is hard and you keep putting it off.

HELPER B: Well, that's your decision.

Counselors A and B are both defensive, though in different ways. It is more likely that the client will react to their defensiveness than that she will move forward.

HELPER C: So from where you're sitting, there's no payoff for being here. Just a lot of dreary work and nothing to show for it.

Counselor C centers on the experience of the client, with a view to resetting the system and helping her explore her responsibility for making the helping process work. Since genuine helpers are at home with themselves, they can allow themselves to examine negative criticism honestly. Counselor C, for instance, would be the most likely of the three to ask himself or herself whether he or she is contributing to the apparent stalemate.

## Self-Responsibility As an Outcome Value

The second, or indirect, goal of helping outlined in the first chapter deals with empowerment—that is, helping clients identify, develop, and use resources that will make them more effective agents of change both within the helping sessions and in their everyday lives. The opposite of empowerment is dependency, deference, and oppression. It is important to understand the dynamics of dependent clients in counseling and dependent patients in health care settings (Bornstein, 2005a, 2005b). And, since helpers are often experienced by clients as relatively powerful people and, since even the most egalitarian and client-centered of helpers do influence clients, it is necessary to come to terms with social influence in the helping process.

**Helping as a social-influence process.** People influence one another every day in every social setting of life. William Crano (2000) suggests that "social influence research has been, and remains, the defining hallmark of social psychology" (p. 68). Parents influence each other and their kids. In turn, they are influenced by their kids. Teachers influence students, and students influence teachers. Bosses influence subordinates, and vice versa. Team leaders influence team members, and members influence both one another and the leader. The world is abuzz with social influence. It could not be otherwise. However, social influence is a form of power and power too often leads to manipulation and oppression.

Helpers can influence clients without robbing them of self-responsibility. Even better, they can exercise their trade in such a way that clients are, to use a bit of current business jargon, *empowered* rather than oppressed, in helping sessions and in everyday life situations as well. With empowerment, of course, comes increased self-responsibility. Imagine a continuum. At one end lies *directing clients' lives* and at the other *leaving clients completely to their own devices.* Somewhere along that continuum is *helping clients make their own decisions and then act on them.* Most forms of helper influence will fall somewhere in between the extremes. Preventing a client from jumping off a bridge moves, understandably, to the controlling end of the continuum. On the other hand, simply accepting and in no way challenging a client's decision to put off dealing with a troubled relationship because he or she is "not ready" moves toward the other end.

**Norms for empowerment and self-responsibility.** Helpers don't self-righteously empower clients. That would be patronizing and condescending. In a classic work, Freire (1970) warned helpers against making helping itself just one more form of oppression for those who are already oppressed. Effective counselors help clients discover, develop, and use the untapped power within themselves. Here, then, is a range of empowerment-based norms, some adapted from the work of Farrelly and Brandsma (1974):

*Start with the premise that clients can change if they choose.* Clients have more resources for managing problems in living and developing opportunities than they—or sometimes their helpers—assume. The helper's basic attitude should be that clients have the resources both to participate collaboratively in the helping process and to manage their lives more effectively. These resources may be blocked in a variety of ways or simply unused. The counselor's job is to help clients identify, free, and cultivate these resources. The counselor also helps clients assess their resources realistically so that their aspirations do not outstrip their resources.

*Do not see clients as victims.* Even when clients have been victimized by institutions or individuals, don't see them as helpless victims. The cult of victimhood is too widespread. Even if victimizing circumstances have diminished a client's degree of freedom—the abused spouse's inability to leave a deadly relationship—work with the freedom that is left. Don't be fooled by appearances. One counselor trainer, in a meeting with his colleagues, dismissed a reserved, self-deprecating trainee with the words, "She'll never make it. She's more like a client than a trainee." Fortunately, his colleagues did not work from the same assumption. The woman went on to become one of the program's best students. She was accepted as an intern at a prestigious mental-health center and was hired by the center after graduation.

*Share the helping process with clients.* Clients, like helpers, can benefit from maps of the helping process. Clients should not have to buy a so-called pig in a poke. Helping should not be a black box for them. Clients have a right to know what they are getting into (see Manthei & Miller, 2000). How to clue clients into the helping process is another matter. Helpers can simply explain what helping is all about. A simple pamphlet outlining the goals and stages of the helping process can be of great help, provided that it is in language that clients can readily understand. Just what kind of detail will help will differ from client to client. Obviously, clients should not be overwhelmed by distracting detail from the beginning. Nor should highly distressed clients be told to contain their anxiety until helpers teach them the helping model. Rather, the details of the model can be shared over a number of sessions. There is no one right way.

*Help clients see counseling sessions as work sessions.* Helping is about client-enhancing change. Therefore, counseling sessions deal with exploring the need for change, the kind of change needed, creating programs of constructive change, experimenting with change-producing pilot projects, and finding ways of dealing with obstacles to change. This is work, pure and simple. Searching for solutions and implementing them can be arduous, even agonizing, but it can also be deeply satisfying, even exhilarating. Helping clients develop the *work ethic* that makes them partners in the helping process can be one of the helper's most formidable challenges.

*Accept helping as a natural, two-way-influence process.* Tyler, Pargament, and Gatz (1983) moved a step beyond the consultant role in what they called the "resource collaborator role." Seeing both helper and client as people with defects, they focused on the give-and-take that should characterize the helping process. In their view, either client or helper can approach the other to originate the helping process. The two have equal status in defining the terms of the relationship, in originating actions within it, and in evaluating both outcomes and the relationship itself. In the best case, positive change occurs in both parties. Helping is a two-way street. Clients and therapists change one another in the helping process. Even a cursory glance at helping reveals that clients can affect helpers in many ways. For instance, Wei-Lian has to correct Timothy, his counselor, a number of times when Timothy tries to share his understanding of what Wei-Lian has said. At one point, when Timothy says, "So you don't like the way your father forces his opinions on you," Wei-Lian replies, "No, my father is my father and I must always respect him. I need to listen to his wisdom." The problem is that Timothy has been inadvertently basing some of his responses on his own cultural assumptions rather than on Wei-Lian's. When Timothy finally realizes what he is doing, he says to Wei-Lian, "When I talk with you, I need to be more of a learner. Chinese culture is quite different from mine. I need your help."

*Focus on learning instead of helping.* Although many see helping as an education process, it is probably better characterized as a learning process. Effective counseling helps clients get on a learning track. Both the helping sessions themselves and the time between sessions involve learning, unlearning, and relearning. Howell (1982) gave us a good description of learning when he said that "learning is incorporated into living to the extent that viable options are increased" (p. 14). In the helping process, learning takes place when viable options that add value to life are opened up, seized, and acted on.

***Do not see clients as overly fragile.***  Neither pampering nor brutalizing clients serves their best interests. However, many clients are less fragile than helpers make them out to be. Helpers who constantly see clients as fragile may well be acting in a self-protective way. Too many helpers shy away from doing much more than listening early in the helping process. The natural deference many clients display early in the helping process—including their reluctance to criticize the therapist, meeting the perceived expectations of the therapist, and showing indebtedness to the therapist—can send the wrong message to helpers. Clients early on may be fearful of making some kind of error, but this does not mean that they are fragile. Reasonable caution in interacting with clients is certainly appropriate, but you can become too cautious. Helping, of course, needs to be tailored to the needs of the individual client, but in many cases helpers can intervene more right from the beginning—for instance, by reasonably challenging the way clients think and act, and by getting them to begin to outline what they want and are willing to work for.

***Appreciate the self-healing nature of clients.***  Helpers should not underestimate the agency of their clients. Bohart and Tallman (1999), in their book *How Clients Make Therapy Work,* lay out the principles of client self-healing, principles which, they say, must be respected if therapy is to be a collaborative enterprise. These principles are part of a positive-psychology approach to helping. I have taken their principles and have reworded and reordered them, while making every attempt to retain their spirit (see pp. 227–235 in their book for a full description of each of their principles):

- Respect clients' ability to act in their own best interest.
- Support clients' capacity to think for themselves.
- Respect clients' wisdom and actively elicit their ideas and intuitions.
- Give clients time to think things through and avoid premature closure.
- Support clients' efforts to generate new ways of thinking about themselves and their problems, and to act on what they discover.
- Believe that clients are capable of learning, and help them amplify small signs of learning and change.
- Work on the assumption that clients prefer positive and proactive, rather than immature and defensive, solutions.
- Help clients in their efforts to understand the constraints that keep them stuck in old ways and to deal with them creatively.
- Remember that clients are agents in their own lives and do not always need our guidance.

While the ideal might be "the client as active self-healer, co-director, and collaborator in therapy" (Bohart & Tallman, p. 141), the authors realize that neither all clients nor all therapists live up to the ideal. But one message is clear—don't sell clients short.

## A Bias toward Action As a Solution-Focused Value

The overall goal of helping clients become more effective in problem-management and opportunity-development was mentioned in Chapter 1. This means helping clients become more effective *agents* in both the helping process and their daily lives—doers rather than mere reactors, preventers rather than fixers, initiators rather than followers.

> Lawrence was liked by his superiors for two reasons. First, he was competent—he got things done. Second, he did whatever they wanted him to do. They moved him from job to job when it suited them. He never complained. However, as he matured and began to think more of his future, he realized that there was a great deal of truth in the adage, "If you're not in charge of your own career, no one is." After a session with a career counselor, he outlined the kind of career he wanted and presented it to his superiors. He pointed out to them how this would serve both the company's interests and his own. At first they were taken aback by Lawrence's assertiveness, but then they agreed. Later, when they seemed to be sidetracking him, he stood up for his rights. Assertiveness was his bias for action.

Doers are more likely to pursue "stretch" rather than merely adaptive goals in managing problems. A stretch goal is one that seems to be a bit beyond the abilities of the client. He or she has to stretch himself or herself—that is, work harder than usual—to achieve the goal. But the benefits of achieving the goal outweigh the added effort. Doers are also more likely to move beyond problem management to opportunity development.

**Action and discretionary change.** If clients are to become more effective agents in their lives, they need to understand the difference between discretionary and nondiscretionary change. Nondiscretionary change is mandated change. If the courts say to a divorced man negotiating visiting rights with his children, "You can't have visiting rights unless you stop drinking," then the change is nondiscretionary. There will be no visiting rights without the change. In contrast, a man and wife having difficulties with their marriage are not under the gun to change the current pattern. Change here is discretionary. "IF you want a more productive relationship, then you must change in the following ways."

The fact of discretionary change is central to mediocrity. If we don't *have* to change, very often we don't. We need merely review the track record of our New Year's resolutions. Unfortunately, in helping situations, clients often see change as discretionary. They may talk about it as if it were nondiscretionary, but deep down the clients' belief that "I don't really have to change" pervades the helping process. "Other people should change; the world should change. But I don't have to." This is not cynical. It's the way things are. The sad track record of discretionary change is not meant to discourage you, but to make you more realistic about the challenges you face as a helper and about the challenges you help your clients face.

A pragmatic bias toward client action on your part—rather than merely talking about action—is a cardinal value. Effective helpers tend to be active with clients, and see no particular value in mere listening and nodding. They engage clients in a dialogue. During that dialogue, they constantly ask themselves, "What can I do to raise the probability that this client will act on her own behalf, intelligently and prudently?" I know a man who years ago went "into therapy" (as "into another world") because, among other things, he was indecisive. Over the years he became engaged several times to different women and each time broke it off. So much for decisiveness.

**Real-life focus.**  If clients are to make progress, they must "do better" (a relative assessment) in their day-to-day lives. The focus of helping, then, is not narrowly on the helping sessions and client-helper interactions themselves, but on clients' behavior in their day-to-day lives. A friend of mine in his early days as a helper exulted in the "solid relationship" he said he was building with a client until in the third interview she stopped, stared at him, and said, "You're really filled with yourself, aren't you? But, you know, we're not getting anywhere." He had become so lost in relationship-building that he forgot about the client's pressing everyday concerns.

## Shadow-Side Realities in the Helping Relationship

There are common flaws in the working alliance that remain in the shadows either because they are not effectively dealt with by the helping professions themselves, or because individual helpers are inept at addressing these flaws with their clients as they arise in the helping relationship.

**Ethical flaws.**  The fact is that at times some helpers shortchange, deceive, manipulate, and exploit their clients. There is a vast literature on ethical responsibilities in the helping professions (Fisher, 2003; Cohen & Cohen, 1999; Corey, Corey, & Callanan, 2003; O'Donohue & Ferguson, 2003; Pack-Brown & Williams, 2003). There is also a growing literature on ways in which helpers violate their ethical responsibilities. An ethics course should be part of every helper training program and should be a theme in every course taken. Since this area is too vast and too important to be given summary treatment here, helpers-to-be are urged to make this part of their professional development program.

**Human tendencies in both helpers and clients.**  Neither helpers nor clients are usually heroic figures. They are human beings with all-too-human tendencies. For instance, helpers find clients attractive or unattractive. There is nothing wrong with this. However, they must be able to manage closeness in therapy in a way that furthers the helping process. They must deal with both positive and negative feelings toward clients, lest they end up engaging in inappropriate behavior with clients. They may have to fight the tendency to be less challenging with attractive clients, or not to listen carefully to unattractive clients. Clients, too, have their tendencies. Some have unrealistic expectations of counseling, while others trip over their own distorted views of their helpers. In such cases, helpers have to manage both the expectations and the relationship. Very often these human tendencies on the part of both client and helper are not center-stage in awareness. Rather, they constitute a subtext within the relationship. Unskilled helpers can get caught up in their own games as well as their clients', causing the working alliance to break down. Skilled helpers, on the other hand, understand and manage the shadow side—their own and their clients'.

**Trouble in the relationship itself.** The helping relationship might be flawed from the beginning. That is, the fit or chemistry between helper and client might not be right. But, for a variety of reasons, it is not easy for a helper to say, "I don't think I'm the one for you." Clients find themselves in the same dilemma. High-level helpers can work with a wide variety of clients. They create their own chemistry. They make the relationship work.

One coach/counselor in a work setting was asked to work with a very bright manager whose interpersonal style left much to be desired. The relationship was troubled from the start. Early on, the coach reported in a supervisory consultation with a colleague that, his client expected him to "say good things" about him to senior managers. The client also had a tendency to play what the coach called "mind" games, saying things like, "I wonder what's going on in your mind right now. I bet you're thinking things about me that you're not telling me." Managing expectations and managing the relationship proved to be hard work. However, the coach knew enough about the company to realize that interpersonal style was an important factor in choosing managers for promotion. Because the client was bright and innovative, promotion was a distinct possibility; but, because of his style, promotion was probably, as is said, his to lose. The coach remained respectful and empathetic, but challenged what he called "the crap." This shocked the client because he perceived himself as always having been able to "win" in his encounters with subordinates, peers, and consultants. He stopped playing games and eventually realized that becoming better at interpersonal relations had only an upside.

Even if the relationship starts off on the right foot, it can deteriorate. In fact, some deterioration is normal. Kivlighan and Shaugnessy (2000) talk about the "tear-and-repair" phenomenon. Many therapeutic relationships start well, get into trouble, and then recover. Experienced helpers are not surprised by this. However, some helping relationships get caught up in self-defeating interactions. Uncomfortable and even hostile interchanges between helpers and clients are common in all treatment models. When impasses and ruptures in the relationship take place, ineffective helpers get bogged down. Many helpers and clients lack both the skill and/or the will for repair. If impasses and ruptures are not addressed, premature termination often takes place. When this happens, helpers predictably blame clients: "She wasn't ready," "He didn't want to work," "She was impossible," and so forth.

**Vague and violated values.** Helpers do not always have a clear idea of what their values are. Or the values they say they hold—that is, their espoused values do not always coincide with their actions. Values too often remain "good ideas" and are not translated into specific norms that drive helping behavior. For instance, even though helpers value self-responsibility in their clients, they see them as helpless, make decisions for them, and direct rather than guide. Often they do so out of frustration. Expediency leads them to compromise their values and then rationalize their compromises. "I blew up at a client today, but he really deserved it. It probably did more good than my unappreciated patience." Fat chance.

**Failure to share the helping process.**  Some counselors are reluctant to let the client know what the helping process is all about. Of course, helpers who "fly by the seat of their pants" can't tell clients what it's all about because they don't know what it's all about themselves. Still others seem to think that knowledge of helping processes is secret or sacred or dangerous and should not be communicated to the client, even though there is little evidence to support such beliefs.

**Flawed contracts.**  There is an implicit contract between helpers and their clients as to what helping is about, what the role of the helper is, what the role of the client is, and what the relationship should look like. Though the contract is usually implicit, at least its fundamental provisions should be explicit (see Egan, 2002) However, there is an extensive shadow side to both explicit and implicit contracts. Even when a contract is written, the contracting parties interpret some of its provisions differently. Over time they forget what they contracted to, and differences become more pronounced. These differences are seldom discussed. In counseling, the helper-client contract has been, traditionally, implicit, even though the need for more explicit structure has been discussed for years. Because of this, the expectations of clients may differ from the expectations of their helpers. Even though implicit contracts are probably the norm, they are often not enough.

 ## EXERCISES ON PERSONALIZING

1. How would you describe yourself in terms of establishing, building, and sustaining relationships?
2. Review the dimensions of the helping alliance, and then describe your strengths and weaknesses in terms of this kind of alliance.
3. Name three of your most important values. What impact do they have on your personal culture?
4. Rate yourself on each of the five helping values reviewed in this chapter. To what degree is each one currently part of your life? What are your strengths and weaknesses with respect to each?
5. How would people who know you well describe you in terms of respect?
6. What role does empathy, as described, currently play in your life?
7. What are your strengths and weaknesses in terms of genuineness in your relationships?
8. Describe yourself in terms of your own self-responsibility.

9. What kind of self-starter are you? How good are you in sustaining momentum in the projects you undertake? How would others rate you in terms of getting things done?

10. Review the main dimensions of your interpersonal style. When you are at your best, which facets of your style would help you develop good working relationships with clients? Explain why this is so. When you are not at your best, which facets of your style might cause difficulty in establishing a working alliance? Why is that?

11. If you were a client in a counseling relationship, which of the values outlined in this chapter would you most prize? Explain. Think of yourself as a client, then ask yourself what other values you would add to the list.

# PART TWO

# The Therapeutic Dialogue

*P*art Two reviews and illustrates the basic communication skills needed to be an effective helper. These skills are integrated under the title The Therapeutic Dialogue. There are less high-sounding names than "therapeutic dialogue"—the helping dialogue, the problem-management dialogue, the opportunity-development dialogue. But dialogue is at the heart of the communication between helper and client.

This section deals with the individual communication skills that go into effective dialogue (Ivey & Ivey, 2003). While individual communication skills are a necessary part of communication competence, dialogue is the integrating mechanism. Chapter 3 reviews the first two basic communication skills. The first is called "attending" or physically and emotionally "tuning in" to clients. This skill focuses on the ways helpers can be fully present to their clients. The second is active listening, the foundation of empathy. Both skills are essential for understanding clients and their concerns. Helpers tune in and actively listen, not only to demonstrate their solidarity with their clients, but also to understand what their clients are saying both directly and indirectly. Chapter 4 deals with a third skill, sharing empathic highlights with clients. Helpers need this skill both to share their understanding with clients and check its accuracy. It is one of the most important ways helpers express the value of empathy in their interactions with their clients. But it is not the only way. Empathy is a value that should permeate all communication with clients. Chapter 5 introduces the skills of probing and summarizing. Chapters 6 and 7 deal with the communication skills needed to help clients challenge the kinds of dysfunctional thinking, emotional expression, and behavior that stand in the way of managing problem situations and identifying and developing opportunities. Effective helpers weave these communication skills together seamlessly in their interactions with clients. Communication skills need to become second nature if they are to serve the helping process and clients' needs.

# CHAPTER 3

# The Basics of Communication: Tuning In to Clients and Active Listening

The Importance of Dialogue in Helping

Tuning In to Clients: The Importance of Empathic Presence
   Nonverbal Behavior as a Channel of Communication
   Helpers' Nonverbal Behavior
   The Skill of Physically Tuning In to Clients

Active Listening: The Foundation of Understanding
   Forms of Inadequate Listening
   Empathic Listening
   Listening to Words: Clients' Stories, Points of View, Decisions, Intentions, and Proposals
   Listening to the Clients' Affect: Feelings, Emotions, and Moods
   Listening to Clients' Nonverbal Messages as Modifiers

Processing What You Hear: The Thoughtful Search for Meaning
   Identifying Key Messages and Feelings
   Understanding Clients Through Context
   Hearing the Slant or Spin: Tough-Minded Listening and Processing
   Identifying What's Missing

Listening to Yourself: The Helper's Internal Conversation

The Shadow Side of Listening to Clients

Exercises on Personalizing

## The Importance of Dialogue in Helping

In keeping with the collaborative nature of helping, conversations between helpers and their clients should be a therapeutic, or helping, *dialogue*. There are four requirements for true dialogue (Egan, 2001):

- *Turning taking.* A dialogue is an interactive conversation. The participants take turns talking and listening. In counseling this means that, generally speaking, monologues on the part of either client or helper don't add value. Turn taking makes mutual learning possible. Through the give-and-take of the dialogue, helpers learn about their clients and base their interventions on what they come to understand. Through dialogue, clients come to understand themselves and their concerns more fully.

- *Connecting.* Turn taking is not enough. Helper and client need to engage each other if their relationship is to be productive. Dialogue involves building on what the other person has said. Both helper and client need to listen carefully in order to move forward in the helping process.

- *Mutual influencing.* In true dialogue, each person is open to being influenced by what the other person has to say. This echoes the social-influence dimension of counseling discussed in Chapter 2. Helpers certainly influence their clients, but the best helpers learn from, and are influenced by, their clients.

- *Co-creating outcomes.* Good dialogue leads to outcomes that are created by, and benefit, both parties. As we have seen, counseling is about results, accomplishments, outcomes. Through dialogue, client and helper are co-creating the way forward. The counselor's job is to act as a catalyst for the kind of problem-managing dialogue that helps clients find their own answers. Only clients can change themselves, but the helping dialogue makes these changes richer and more effective. At least, that is the ideal.

Very often clients will not have the communication skills outlined in these pages. What then? How can they engage in the kind of dialogue described here? The answer is simple. Counselors with the communication skills described here can help clients hold up their side of the dialogue. That is, by using such skills as probing, summarizing, and challenging, they can help clients engage in the give-and-take of dialogue. For instance, if a client tends to ramble or engage in monologues, the helper can gently intervene by asking questions to provide focus, summarizing for clarity, or challenging the client on the implications of what he or she is saying. Edgar, a client dissatisfied with his relationship with his father in the family business, tends not to listen. Rather, he comes up with example after example of the ways his father treats him unfairly. At one point, Susan, his counselor, summarizes what Edgar has been saying, and asks if she has captured the main points he has been making. "That pretty well covers it," Edgar says, and then launches into another example. Susan gently interrupts him and says, "If your father were sitting here, what are some of the things he might be saying about you? This may give us both sides of the picture." Edgar is taken aback for a moment but then, with Susan's help, begins to explore his own contributions to the impasse between himself and his father.

> EDGAR: Well, for starters, I think he resents the way I keep pointing out how he violates the principles I'm learning in business school. I really get on his case.

> SUSAN: So you point out some of the mistakes you think he is making in the business. Well . . . How does he react to that?

> EDGAR: He either ignores me or says that all I know about business is what I read in books.

> SUSAN: So you thrust and he counters, or he thrusts and you counter.

> EDGAR: Yes, there's a lot of that. I guess if you taped it, it would sound silly.

So Susan uses her skills to promote a dialogue that moves beyond Edgar's recitation of complaints. At one point she asks, "From your point of view, what would a truce between the two of you look like?" Edgar pauses for a while and then asks, "Do you mean a real truce?" This leads to a completely different, and much more productive, conversation. Examples of how to help clients engage in dialogue are scattered throughout this book.

The skills involved in dialogue are not special skills peculiar to helping. Rather, they are extensions of the kinds of skills all of us need in our everyday interpersonal transactions. Ideally, helpers-to-be would enter training programs with this basic set of interpersonal communication skills in place, and training would simply help them adapt the skills to the helping process. Unfortunately, this is often not the case. While interpersonal communication competence is critical for effective everyday living—it is the principal enabling skill for just about everything we do—it is, in my view, one of the skills that is forgotten by society. In that respect, it suffers a fate shared by a number of essential "life skills" such as problem-solving, opportunity-development, parenting, and managing (knowing how to make some system work), to name a few.

Let me make my point. In lecturing, I have often asked audiences to answer two questions. The first question goes something like this:

> "Given the importance of effective human relationships in just about every area of life, how important is it for your kids to develop a solid set of interpersonal communication skills? On a scale from 1-100, how high would you rate the importance?"

Inevitably, the ratings are at the high end, always near 100. The second question goes something like this:

> Given the importance of these skills, where do your kids pick them up? How does society make sure that they acquire them? In what forums do they learn them?

Then the hemming and hawing begins. "Well, I guess they get some of them at home. That is, if they find good role models at home." Or, "Life itself is the best teacher of these skills. They learn them on the run." The members of the audience go on like this for a while, until I say:

Let me summarize what I've been hearing. And, by the way, it's no different from what I hear every place else. Although most parents rate the importance of these sets of skills very highly, we live in a society that leaves their development to chance. Nothing is done systematically to make sure that our kids learn these skills. And, by the way, there is no assurance that they will pick them up on the run.

Children learn a bit from their parents, they might get a dash in school, perhaps a soupçon of TV helps. But, in the main, they are more often exposed to poor communicators than good ones.

Since trainees don't ordinarily arrive equipped with the requisite set of interpersonal skills, they need time to come up to speed in communication competence. This, in a strange way, creates its own problem. Some helper-training programs focus almost exclusively on interpersonal communication skills. As a result, trainees know how to communicate but not necessarily how to help. Furthermore, most adults feel that they are "pretty good" at these skills. What they actually mean is they see themselves as good as others. This is not good enough.

## Tuning In to Clients: The Importance of Empathic Presence

At some of the more dramatic moments of life, simply being with another person is extremely important. If a friend of yours is in the hospital, just your being there can make a difference, even if conversation is impossible. Similarly, being with a friend who has just lost his wife can be very comforting to him, even if little is said. Your empathic presence is comforting. But tuning in is also important in the give-and-take of everyday life. Most people appreciate it when others pay attention to them. They feel respected. By the same token, being ignored is often painful: The averted face is too often a sign of the averted heart. Given how sensitive most of us are to others' attention or inattention, it is paradoxical how insensitive we can be at times about our own failure to tune in.

Helping and other deep interpersonal transactions demand a certain robustness, or intensity of presence. Tuning in to others, both physically and psychologically, contributes to this presence. Tuning in as an expression of empathy that tells clients that you are with them. Tuning in also puts you in a position and frame of mind to listen carefully to their concerns. Your attention can be manifested in a number of ways. Let's start by briefly exploring nonverbal behavior as a channel of communication.

## Nonverbal Behavior as a Channel of Communication

Over the years, both researchers and practitioners have come to appreciate the importance of nonverbal behavior in counseling (for a wealth of information about nonverbal behavior see the following Internet site: http://www3.usal.es/~nonverbal/introduction.htm). Highlen and Hill (1984) suggested that nonverbal behaviors regulate conversations, communicate emotions, modify verbal messages, provide important messages about the helping relationship, give insights into self-perceptions, and provide clues that clients (or counselors) are not saying what they are thinking. This area has taken on even more importance because of the multicultural nature of helping and the different meanings attributed to different nonverbal behaviors in different cultures.

The face and body are extremely communicative. We know from experience that even when people are together in silence, the atmosphere can be filled with messages. Sometimes the facial expressions, bodily motions, voice quality, and physiological responses of clients communicate more than their words do. The following factors, on the part of both helpers and clients, play an important role in the therapeutic dialogue:

- *bodily behavior,* such as posture, body movements, and gestures
- *eye behavior,* such as eye contact, staring, eye movement
- *facial expressions,* such as smiles, frowns, raised eyebrows, and twisted lips
- *voice-related behavior,* such as tone of voice, pitch, volume, intensity, inflection, spacing of words, emphases, pauses, silences, and fluency
- *observable autonomic physiological responses,* such as quickened breathing, blushing, paleness, and pupil dilation
- *physical characteristics,* such as fitness, height, weight, and complexion
- *space,* how close or far a person chooses to be during a conversation
- *general appearance,* such as grooming and dress

People constantly "speak" to one another through their nonverbal behavior. Effective helpers learn this "language" and how to use it effectively in their interactions with their clients. They also learn how to "read" relevant messages embedded in the nonverbal behavior of their clients.

## Helpers' Nonverbal Behavior

Before you begin interpreting the nonverbal behavior of your clients, take a look at yourself. You speak to your clients through all the nonverbal categories outlined above. At times your nonverbal behavior is as important as, or even more important than, your words. Your nonverbal behavior influences clients for better or for worse. Clients read in your nonverbal behavior cues that indicate the quality of your presence. Attentive presence can invite or encourage them to trust you, open up, and explore the significant dimensions of their problem situations. Half-hearted presence can promote distrust and lead to clients' reluctance to reveal themselves to you. Clients may misinterpret your nonverbal behavior. For instance, you may be comfortable with the space between you and your client, but it is too close for the client. Or remaining silent might, in your mind, mean giving a client time to think, but the client might feel embarrassed. Part of listening, then, is being sensitive to clients' reactions to your nonverbal behavior.

Effective helpers are mindful of the stream of nonverbal messages they send to clients. Reading your own bodily reactions is an important first step. For instance, if you feel your muscles tensing as the client talks to you, you can say to yourself, "I'm getting anxious here. What's going on? And what nonverbal messages indicating my discomfort am I sending to the client?" Again, you probably would not use these words. Rather, you would read the signals your body is sending you without letting them distract you from your client.

You can also use your body to censor instinctive or impulsive messages that you feel are inappropriate. For instance, if the client says something that instinctively angers you, you can control the external expression of the anger (for instance, a sour look) to give yourself time to reflect. Such self-control is not phony because your respect for your client takes precedence over your instinctive reactions. Not dumping your annoyance or anger on your clients through nonverbal behavior is not the same as denying it. Becoming aware of it is the first step in dealing with it.

In a more positive vein, you can "punctuate" what you say with the nonverbal messages. For instance, Denise, a social worker, is especially attentive when Jennie talks about actions she could take to do something about the unruliness of her children. Denise leans forward, nods, and says "uh-huh." She uses nonverbal behavior to reinforce Jennie's intention to act.

On the other hand, don't become preoccupied with your body and the qualities of your voice as a source of communication. Rather, learn to use your body instinctively as a means of communication. Being aware of, and at home with, nonverbal communication can reflect being at peace with yourself, with the helping process, and with your clients. Your nonverbal behavior should enhance, rather than stand in the way of, your working alliance with your clients.

Although the skills of tuning in can be learned, they will be phony if they are not driven by the attitudes and values such as respect and empathy discussed earlier. Your mind-set—and what's in your heart—is as important as your visible presence. If you are not actively interested in the welfare of your client, or if you resent working with a client, subtle or not-so-subtle nonverbal clues will likely color your behavior. I once mentioned to a doctor my concerns about an invasive diagnostic procedure he intended to use. The doctor said the right words to reassure me, but his physical presence and the way he rushed his words said, "I've heard this dozens of times. I really don't have time for your concerns. Let's get on with this." His words were right but the real message was in the nonverbal messages that accompanied his words.

## The Skill of Physically Tuning In to Clients

There are certain key nonverbal skills you can use to tune in to clients. These skills can be summarized in the acronym SOLER. Since communication skills are particularly sensitive to cultural differences, care should be taken in adapting what follows to different cultures. What follows is only a framework.

**S: Face the client *Squarely;*** that is, adopt a posture that indicates involvement. In North American culture, facing another person squarely is often considered a basic posture of involvement. It usually says, "I'm here with you; I'm available to you." Turning your body away from another person while you talk to him or her can lessen your degree of contact with that person. Even when people are seated in a circle, they usually try in some way to turn toward the individuals to whom they are speaking. The word "squarely" here should not be taken too literally; it is not a military term. The point is that your bodily orientation should convey the message that you are involved with the client. If, for any reason, facing the person squarely is too threatening, then an angled position may be more helpful. The point is not inches and angles but the quality of your presence. Your body sends out messages whether you like it or not. Make them congruent with what you are trying to do.

**O: Adopt an *Open* posture.**   Crossed arms and crossed legs can be signs of lessened involvement with, or availability to, others. An open posture can be a sign that you're open to the client and to what he or she has to say. In North American culture, an open posture is generally seen as a nondefensive posture. Again, the word "open" can be taken literally or metaphorically. If your legs are crossed, this does not mean that you are not involved with the client. But it is important to ask yourself, "To what degree does my present posture communicate openness and availability to the client?" If you are empathic and open-minded, let your posture mirror what is in your heart.

**L: Remember that it is possible at times to *Lean* toward the other.**   Watch two people in a restaurant who are intimately engaged in conversation. Very often they are both leaning forward over the table as a natural sign of their involvement. The main thing is to remember that the upper part of your body is on a hinge. It can move toward a person and back away. In North American culture, a slight inclination toward a person is often seen as saying, "I'm with you, I'm interested in you and in what you have to say." Leaning back (the severest form of which is a slouch) can be a way of saying, "I'm not entirely with you" or "I'm bored." Leaning too far forward, however, or doing so too soon, may frighten a client. It can be seen as a way of placing a demand on the other for some kind of closeness or intimacy. In a wider sense, the word "lean" can refer to a kind of bodily flexibility or responsiveness that enhances your communication with a client. And bodily flexibility can mirror mental flexibility.

**E: Maintain good *Eye* contact.**   In North American culture, fairly steady eye contact is not unnatural for people deep in conversation. It is not the same as staring. Again, watch two people deep in conversation. You may be amazed at the amount of direct eye contact. Maintaining good eye contact with a client is another way of saying, "I'm with you; I'm interested; I want to hear what you have to say." Obviously, this principle is not violated if you occasionally look away. Indeed, you have to, if you don't want to stare. But if you catch yourself looking away frequently, your behavior may give you a hint about some kind of reluctance to be with this person or to get involved with him or her. Or, it may say something about your own discomfort. In other cultures, however, too much eye contact, especially with someone in a position of authority is out of order. I have learned much about the cultural meaning of eye contact from my Asian students and clients.

**R: Try to be relatively *Relaxed*, or natural in these behaviors.** Being relaxed means two things. First, it means not fidgeting nervously or making distracting facial expressions. The client may wonder what's making you nervous. Second, it means becoming comfortable with using your body as a vehicle of personal contact and expression. Your being natural in the use of these skills helps put the client at ease.

A counselor trained in the *Skilled Helper* model was teaching counseling to visually impaired students in the Royal National College for the Blind in the UK. Most of her clients were visually impaired. However, she wrote this about SOLER:

> In counseling students who are blind or visually impaired, eye contact has little or no relevance. However, attention on voice direction is extremely important, and people with a visual impairment will tell you how insulted they feel when sighted people are talking to them while looking somewhere else.
>
> I teach SOLER as part of listening and attending skills and can adapt each letter of the acronym [to my visually impaired students] with the exception of the E.... After much thought, I would like to change your acronym to SOLAR, the A being for "Aim," that is, aim your head and body in the direction of your client so that when they hear your voice, be it linguistically or paralinguistically, they know that you are attending directly to what they are saying (private communication).

This underscores the fact that people are more sensitive to how you orient yourself to them nonverbally than you might imagine. Anything that distracts from your "being there" can harm the dialogue. The point to be stressed is that a respectful, empathic, genuine, and caring mind-set might well lose its impact if the client does not see these internal attitudes reflected in your external behaviors. In the beginning, you may become overly self-conscious about the way you tune in, especially if you are not used to being attentive. Still, the guidelines just presented are just that—guidelines. They should not be taken as absolute rules to be applied rigidly in all cases. Box 3.1 summarizes, in question form, the main points related to being tuned in to clients.

## Active Listening: The Foundation of Understanding

Tuning in is not, of course, an end in itself. We tune in both physically and psychologically in order to listen to the stories, points of view, intentions/proposals, and decisions of our clients. Listening carefully to a client's concerns seems to be a concept so simple to grasp and so easy to do that one may wonder why it is given such explicit treatment here. Nonetheless, it is amazing how often people fail to listen to one another. Active listening means listening fully, listening accurately, and listening for meaning. Listening is not merely a skill. It is a rich metaphor for the helping relationship itself. The rest of this section will attempt to tap some of that richness.

## Questions to Ask Yourself: How Physically and Psychologically Are You Tuned In to Your Clients?

- What are my attitudes toward this client?
- How would I rate the quality of my presence to this client?
- To what degree does my nonverbal behavior indicate a willingness to work with this client?
- What attitudes am I expressing in my nonverbal behavior?
- What attitudes am I expressing in my verbal behavior?
- To what degree does this client experience me as effectively present and working with him or her?
- To what degree does my nonverbal behavior reinforce my internal attitudes?
- In what ways am I distracted from giving my full attention to this client? What am I doing to handle these distractions? How might I be more effectively present to this person?

The following case can help you develop a better behavioral feel for both tuning in and listening.

> Jennie, an African American college senior, was raped by a "friend" on a date. She received some immediate counseling from the university Student Development Center and some ongoing support during the subsequent investigation. But, even though she was raped, it turned out that it was impossible for her to prove her case. The entire experience—both the rape and the investigation that followed—left her shaken, unsure of herself, angry, and mistrustful of institutions she had assumed would be on her side (especially the university and the legal system). When Denise, a middle-aged and middle-class African American social worker who was a counselor for a health maintenance organization (HMO), first saw her a couple of years after the incident, Jennie was plagued by a number of somatic complaints, including headaches and gastric problems. At work, she engaged in angry outbursts whenever she felt that someone was taking advantage of her. Otherwise, she had become quite passive and chronically depressed. She saw herself as a woman victimized by society and was slowly giving up on herself.

I will refer back to the interactions between Denise and Jennie to illustrate some of the main points about listening.

## Forms of Inadequate Listening

Effective listening is not a state of mind, like being happy or relaxed. It's not something that "just happens." It's an activity. In other words, effective listening requires work. Let's first take a look at the opposite of active listening. All of us have been, at one time or another, both perpetrators and victims of the following forms of inactive or inadequate listening.

**Non-Listening.** Sometimes we go through the motions of listening, but are not really engaged. At times we get away with it. Sometimes we are caught. "What would you do?" Jennifer asks her colleague, Kieran, after outlining a problem the company is having with the structure of the sales group. Embarrassed, Kieran replies, "I'm not sure." Staring him down, Jennifer says, "You haven't been listening to a word I've said." For whatever reason, he had tuned her out. Obviously, no helper sets out not to listen, but even the best can let their minds wander as they listen to the same kinds of stories over and over again, forgetting that the story is unique to *this* client.

**Partial listening.** This is listening that skims the surface. The helper picks up bits and pieces, but not necessarily the essential points the client is making. For instance, Janice's client, Dean, is talking to her about a date that went terribly wrong. Janice only half listens. It seems that Dean is not that interesting. Dean stops talking and looks rather dejected. Janice tries to pull together the pieces of the story she did listen to. Her attempt to express understanding has a hollow ring to it. Dean pauses and then switches to a different topic. Inadequate listening is totally out of place in counseling relationships.

**Tape-recorder listening.** What clients look for from listening is not the helper's ability to repeat their words. Any recording device could do that perfectly. People want more than physical presence in human communication; they want the other person to be present psychologically, socially, and emotionally. Sometimes helpers fail to demonstrate that tuning in and listening means that they are totally present. The client picks up some signals that the helper is not listening very well. How many times have you heard someone exclaim, "You're not listening to what I'm saying!" When the person accused of not listening answers, almost predictably, "I *am*, too, listening; I can repeat everything you've said," the accuser is not comforted. Clients are usually too polite or cowed or preoccupied with their own concerns to say anything when they find themselves in that situation. But it is a shame if your auditory equipment is in order, but you are elsewhere. Your clients want **you**, a live counselor, not a tape recorder.

**Rehearsing.** Picture Sid, a novice counselor, sitting with Casey, a client. At one point in the conversation, when Casey talks about some "wild dreams" he is having, Sid says to himself, "I really don't believe in dream analysis; I wonder what I'm going to say?" Sid stops listening and begins rehearsing what he's going to say. Even when experienced helpers begin to mull over the perfect response to what their clients are saying, they stop listening. Effective helpers listen intently to clients and to the themes and core messages embedded in what they are saying. They are never at a loss in responding. They don't need to rehearse. And their responses are much more likely to help clients move forward in the problem-management process. When the client stops speaking, they often pause, too, to reflect on what he or she just said and before speaking. Pausing says, "I'm still mulling over what you've just said. I want to respond thoughtfully." They pause because they have listened.

## Empathic Listening

The opposite of inactive or inadequate listening is empathic listening, listening driven by the value of empathy. Empathic listening centers on the kind of attending, observing, and listening—the kind of "being with"— needed to develop an understanding of clients and their worlds. Although it might be metaphysically impossible to actually get *inside* the world of another person and experience that world as he or she does, it is possible to approximate this.

Carl Rogers (1980), one of the great deans of counseling, talked passionately about basic empathic listening—being with, and understanding, the other—even calling it "an unappreciated way of being" (p. 137). He used the word "unappreciated" because, in his view, few people in the general population developed this deep listening ability, and even so-called expert helpers did not give it the attention it deserved. Here is his description of empathic listening or being with:

> It means entering the private perceptual world of the other and becoming thoroughly at home in it. It involves being sensitive, moment by moment, to the changing felt meanings which flow in this other person, to the fear or rage or tenderness or confusion or whatever that he or she is experiencing. It means temporarily living in the other's life, moving about in it delicately without making judgments. (p. 142)

Such empathic listening is selfless because helpers must put aside their own concerns to be fully with their clients. Of course, Rogers pointed out that this deeper understanding of clients remains sterile unless it is somehow communicated to them. Although clients can appreciate how intensely they are attended and listened to, helpers still need to communicate their understanding. Empathic listening leads to empathic understanding, the basis of empathic responding. If your help is based on an incorrect or invalid understanding of the client, then your helping may lead him or her astray. If your understanding is valid but superficial, then you might miss the central issues of the client's life.

## Listening to Words: Clients' Stories, Points of View, Decisions, Intentions, and Proposals

Clients use words to convey many things. For instance, clients tell *stories*, share *points of view*, state *decisions*, reveal *intentions*, and offer *proposals*. Recognizing these modes of client discourse can help you organize your listening because they can help you listen for the client's key points.

**Listening to Clients' Stories.**   Most immediately, helpers listen to clients' *stories*, that is, their accounts of their problem situations and unused opportunities.

Stories tend to be mixtures of clients' experiences, behaviors, and emotions. Traditionally, human activity has been divided into three parts—thinking, feeling, and acting. A slightly different approach is taken in the rest of this section. Clients talk about their **experiences**, that is, what happens to them. If a client tells you that she was fired from her job, she is talking about her problem situation as an experience. Jennie, as you recall, talked about being raped, belittled, and ignored. Clients talk about their **behavior**—what they do, or refrain from doing. If a client tells you that he smokes and drinks a lot, he is talking about his external behavior. If a different client says that she spends a great deal of time daydreaming, she is talking about her internal behavior. Jennie talked about pulling away from her family and friends after the rape investigation. Clients talk about their **affect**—the feelings, emotions, and moods that arise from, or are associated with, their experiences and behavior. If a client tells you how depressed she gets after fights with her fiancé, she is talking about the mood associated with her experiences and behavior. Jennie talked about her shame, her feelings of betrayal, and her anger. Since experiences, actions, and emotions are interrelated in the day-to-day lives of clients, they mix them together in telling their stories. Consider this example:

> A client says to a counselor in the personnel department of a large company, "I had one of the lousiest days of my life yesterday." At this point, the counselor knows that something went wrong and that the client feels bad about it, but she knows nothing about the specific experiences and behaviors that made the day such a horror for the client. However, the client continues, "Toward the end of the day, my boss yelled at me in front of some of my colleagues for not landing an order from a new customer [an experience]. I lost my temper [emotion] and yelled right back at him [behavior]. He blew up and fired me on the spot [an experience for the client]. And now I feel awful [emotion] and am trying to find out if I really have been fired and, if so, if I can get my job back [behavior]."

Problem situations are much clearer when they are spelled out as specific experiences, behaviors, and feelings related to specific situations. Since clients spend so much time telling their stories, a few words about each of these elements are in order.

*Experiences.* Most clients spend a fair amount of time, sometimes too much time, talking about what happens *to* them.

- "My wife doesn't understand me."
- "My ulcer acts up when family members argue."
- "My boss doesn't give me feedback."
- "I get headaches a lot."

It is of paramount importance to listen to, and understand, clients' experiences. However, since experiences often focus on what other people do or fail to do, experience-focused stories can indicate passivity or even a sense of helplessness on the part of the client. At times, the implication is that others—or the world in general—are to blame for the client's problems.

- "She doesn't do anything all day. The house is always a mess when I come home. No wonder I can't concentrate at work."

- "He tells his little jokes, and I'm always the butt of them. He makes me feel bad about myself most of the time."

Some clients talk about experiences that are internal and out of their control.

- "These feelings of depression come from nowhere and seem to suffocate me."
- "I just can't stop thinking of him."

The last statement sounds like an action, but it is expressed as an experience. It is something happening *to* the client, at least to the client's way of thinking. One reason that some clients fail to manage the problem situations of their lives is that they are too passive, or see themselves as victims adversely affected by other people, by such immediate social settings of life as the family, by society in its larger organizations and institutions such as government or the workplace, by cultural prescriptions, or even by internal forces. They feel that they are no longer in control of their lives, or of some dimension of life. Therefore, they talk extensively about these experiences.

- "Company policy discriminates against women. It's that simple."
- "The economy is booming but the kind of jobs I want are already taken."
- "No innovative teacher gets very far around here."

Of course, some clients *are* treated unfairly; they are victimized by the behaviors of others in the social and institutional settings of their lives. Although they can be helped to cope with victimization, full management of their problem situations demands changes in the social settings themselves. One client was helped to cope with a brutal husband but, ultimately, the courts had to intervene to keep him at bay.

For other clients, talking constantly about experiences is a way of avoiding responsibility: "It's not my fault. After all, these things are happening to me." People who are too coddled or comfortable can become "whiners" when problems arise. Counselors must be able to distinguish "whiners" from those who are truly being victimized.

*Behaviors.* All of us do things that get us into trouble and fail to do things that will help us get out of trouble, or to develop opportunities. Clients are no different:

- "When he ignores me, I think of ways of getting back at him."
- "Whenever anyone gets on my case for having a father in jail, I let him have it. I'm not taking that kind of crap from anyone."
- "Even though I feel the depression coming on, I don't take the pills the doctor gave me."
- "When I get bored, I find some friends and go get drunk."
- "I have a lot of sexual partners and have unprotected sex whenever my partner will let me."

Some clients talk freely about their experiences, what happens to them, but seem more reluctant to talk about their behaviors. One reason for this is that they can't talk about behaviors without bringing up issues of personal responsibility.

**Listening to Clients' Points of View.** As clients tell their stories, explore possibilities for a better future, set goals, make plans, and review obstacles to accomplishing these plans, they often share their points of view. A point of view is a client's personal estimation of something. A full point of view includes the point of view itself, the reasons for it, an illustration to bring it to life, and some indication how open the client might be to modifying it. There is no expectation that anyone else need adopt the point of view. That would be a form of selling or persuasion. But, realistically, the implication often is: "I think this way. Why don't other people think this way?" For instance, Aurora, an 80-year-old woman, is talking to a counselor about the various challenges of old age. At one point she says,

> My sister in Florida [85-years-old]—Sis, we call her—is sick. She's probably dying, but she wants to stay at home. She's asked me to come down there and take care of her. I think that's asking too much. I could use some help myself these days. But she keeps calling.

Aurora's point of view is that her sister's request is not realistic because she herself needs some help to get by. But her sister persists in trying to persuade her to change her mind. Points of view reveal clients' beliefs, values, attitudes, and convictions. Clients may communicate many points of view. The ones that are relevant to their problem situations or undeveloped opportunities need to be listened to and understood. Let's return briefly to Jennie and Denise.

> **JENNIE:** You just can't trust the system. They're not going to help. They take the easy way out. I don't care which system it might be. Church, government, the community, sometimes even the family. They're not going to give you much help.

Denise listens carefully to Jennie's point of view and realizes how much it is influencing her behavior. In Jennie's case, it's easy to see where the point of view comes from. But Denise also knows that, at some point, she needs to challenge Jennie's point of view because it may be one of the things that is keeping her locked in her misery. Points of view have power.

**Listening to Clients' Decisions.** From time to time, we all tell others about decisions we are making or that we've made. A client might say, "I've decided to stop drinking. Cold turkey." Or, "I'm tired of being alone. I'm going to join a dating service." Decisions usually have implications for the decision maker and for others. The client who has decided to go cold turkey on drink has his work cut out for him, but there are implications for his spouse. For instance, she's used to coping with a drunk, but now she may have to learn how to cope with this "new person" in the house. Commands, instructions, and even hints are, in a way, decisions about other people's behavior. A client might say to her helper, "Don't bring up my ex-husband any more. I'm finished with him."

Sharing a decision fully means spelling out the decision itself, the reasons for the decision, the implications for self and others, and some indication as to whether the decision—or any part of it—is open to review. For instance, Jennie, in talking with Denise about future employment, says in a rather languid tone of voice, "I'm not going to get any kind of job where I have to fight the race thing. Or the woman thing. I'm tired of fighting. I only get hurt. I know that this limits my opportunities, but I can live with that." Note that this is more than a point of view. Jennie is more or less saying, "I've made up my mind." She notes the implication for herself—a limitation of job opportunities. But is she also saying, "So that's the end of it. Don't try to convince me otherwise." This is something Denise might explore. Denise does believe that some of Jennie's decisions, expressed or implied, need challenging. Decisions can be tricky. Often enough, *how* the client delivers the decision says a great deal about the decision itself. Given the rather languid way in which Jennie delivers her decision, Denise thinks that it might not be Jennie's final decision. This is something that has to be checked out. A dialogue with Jennie about the reasons for her decision and a review of its implications are possible routes for a challenge.

**Listening to Clients' Intentions and Proposals.**  Finally, clients state intentions, offer proposals, or make a case for certain courses of action. Consider Lydia. She is a single parent of two young children who is a member of the *working poor*. Her wages don't cover her expenses. She has no insurance. The father of her children has long disappeared. She says to a social worker:

> I think I should quit my job. I'm making the minimum wage, and with travel expenses and all I just can't make ends meet. I spend too much time traveling and don't see enough of my kids. Friends look after them when I'm gone, but that's hit-or-miss and puts a burden on them. You know, if I go on welfare I could make almost as much. And then I could pick up jobs that would pay me cash. I've got friends who do this. I believe I could make ends meet. My kids and I would be better off. And I wouldn't be hassled as much.

Lydia is making a case, not announcing a decision. The case includes what she might do (quit her job and move into the "alternate" work economy), the reasons for doing it (the inadequacy of her current work situation, the need to make ends meet), and the implications for herself and her children (she'd be less hassled and her kids would see more of her and be better off).

Of course, when clients talk about their concerns, they mix all these forms of discourse together. Here's an example. What follows came out through dialogue in one of Jennie's sessions with Denise. For the sake of illustration, it is presented here in summary form in Jennie's words.

> A couple of weeks ago I met a woman at work who has a story similar to mine. We talked for a while and got along so well that we decided to meet outside of work. I had dinner with her last night. She went into her story in more depth. I was amazed. At times I thought I was listening to myself! Because she had been hurt she was narrowing her world down into a little patch so that she could control everything and not get hurt any more. I saw right away that I'm trying to do my own version of the same thing. I know you've been telling me that, but I haven't been listening very well.

Here's a woman with lots going for her and she's hiding out. As I came back from dinner I said to myself, "you've got to change." So, Denise, I want to revisit two areas we've talked about—my work life and my social life. I don't want to live in the hole I've dug for myself. I could see clearly some of the things she should do. So here's what I want to do. I want to engage in some little experiments in broadening my social life. Starting with my family. And I want to discuss the kind of work I want without putting all the limitations on it. I want to start coming out of the hole I'm in. And I want to help my new friend do the same.

Everything is here. Or elements of everything—a story about her new friend, including experiences, actions, and feelings; points of view about her new friend; decisions about where she wants her life to go; proposals about experiments in her social life and her relationship with her friend. The point is this: Developing frameworks for listening can help you zero in on the key messages your clients are communicating and help you identify and understand the feelings, emotions, and moods that go with them.

**"Hearing" Opportunities and Resources.**  If you listen only for problems, you will end up talking mainly about problems and, in doing so, may well short-change your clients. Every client has something going for him or her. Your job is to spot clients' resources and help them invest these resources in managing problem situations and opportunities. If people generally use only a fraction of their potential, then there is much to be tapped. Consider this case.

A counselor is working with a young Eritrean emigre who is depressed because he has failed to get into an MBA program. The young man works in the maintenance department of a large office building in a major Midwestern city. It's a decent job, but the young man has higher aspirations. Failure to get into graduate school reawakens feelings of being a second-class citizen. After a couple of sessions, the counselor realizes that he is talking with a person who has a great deal going for him. He is intelligent, personable, hard-working, reasonably ambitious, thoughtful, assertive, and reliable. He has loads of resources. But these are not necessarily the resources needed to pass the hurdles for getting into an MBA program. The problem is that the young man sees the MBA as the only way forward. Another way forward is helping him review the range of jobs or vocations or careers that fit his strengths once they have been identified and discussed.

"Hearing"— that is, identifying—clients' strengths and not just their concerns almost inevitably opens up new possibilities.

## Listening to Clients' Affect: Feelings, Emotions, and Moods

Feelings, emotions, and moods constitute a river that continually runs through us—peaceful, meandering, turbulent, or raging—often beneficial, sometimes dangerous, seldom neutral. They are certainly an important part of clients' problem situations and undeveloped opportunities (Plutchik, 2001). Some have complained that psychologists, both researchers and practitioners, don't take emotions seriously enough. For instance, anger is an ubiquitous and extremely important emotion, but "it has been oddly neglected by the clinical community" (Norcross & Kobayashi, 1999, p. 275—this is the opening article of a series of articles on anger in a special section of the *Journal of Clinical Psychology, 55,* March, 1999). But there are some signs that things are changing. In 2001 a new American Psychological Association journal, *Emotion,* entered the scene because of the recognition, some would say belated recognition, that emotion is fundamental to so much of human life and certainly to problems in living. The journal includes articles ranging from the so-called "softer" side of psychology through hard-nosed molecular science. Of course, popular self-help books written by professionals have proved useful to both clients and practitioners for years (McKay & Dinkmeyer, 1994; McKay, Davis, & Fanning, 1997). These very practical books tend to take a positive psychology approach to the experiencing, regulation, and use of emotion in everyday life.

Recognizing key feelings, emotions, and moods (or the lack thereof) is very important for at least three reasons. First of all, they pervade our lives. There is an emotional tone to just about everything we do. Feelings, emotions, and moods pervade clients' stories, points of view, decisions, and intentions or proposals. Second, they greatly affect the quality of our lives. A bout of depression can stop us in our tracks. A depressed client is an unhappy client. A client who gets out from under the burden of self-doubt breathes more freely. Third, feelings, emotions, and moods are drivers of our behavior. Clients driven by anger can do desperate things. On the other hand, enthusiastic clients can accomplish more than they ever thought they could. The good news is we can help clients learn how to tune in to and regulate their emotions at the service of a fuller life.

Therefore, understanding the role of feelings, emotions, and moods in client's problem situations and their desire to identify and develop opportunities is central to the helping process. Emotions highlight learning opportunities.

"I've been feeling pretty sorry for myself ever since he left me." In counseling sessions, this client learns how much her self-pity constricts her world and limits problem-managing action.

"I yelled at my mother last night and now I feel very ashamed of myself." Shame proves to be a wake-up call in this clients relationships with his mother.

"I've been anxious for the past few weeks, but I don't know why. I wake up feeling scared and then it goes away but comes back again several times during the day." Anxiety has become a bad habit for this client. It is self-perpetuating. What can the client do to break through this vicious circle?

"I finally finished the term paper that I've been putting off for weeks and I feel great!" Here a positive emotion becomes a tool in this client's struggle against procrastination.

The last item in this list brings up an important point. In the psychological literature, negative emotions tend to receive more attention than positive emotions. Now work is under way to study positive emotions and their beneficial effects. There are indications that we can use positive emotions to promote both physical and psychological well-being (Salovey, Rothman, Detweiler, & Steward, 2000). Emotions can free up psychological resources, act as opportunities for learning, and promote health-related behaviors. For instance, in managing problems and developing opportunities, social support plays a key role. As Salovey and his colleagues note, clients are more likely to elicit social support if they manifest a positive attitude toward life. Potential supporters tend to shun clients who let their negative emotions get the best of them

Of course, clients often express feelings without talking about them. When a client says, "My boss gave me a raise and I didn't even ask for one!" you can feel the emotion in her voice. A client who is talking listlessly and staring down at the floor may not say, in so many words, "I feel depressed." A dying person may express feelings of fear, anger, and depression without talking about them. Other clients feel deeply about things but do their best to hold their feelings back. Effective helpers can usually find clues or hints, whether verbal or nonverbal, that indicate the feelings and emotions rumbling inside.

Clients' stories, points of view, decisions, and expressed intentions or proposals for action are permeated by feelings, emotions, and moods. Your job is to listen carefully to the ways in which such affect color and give meaning to words they are using. The meaning is not just in the words. It's in the full package.

## Listening to Clients' Nonverbal Messages as Modifiers

Recall what was said about nonverbal behavior in earlier in this chapter. Clients send messages through their nonverbal behavior. Nonverbal behavior has a way of "leaking" messages about what clients really mean. The very spontaneity of nonverbal behaviors contributes to this leakage even in the case of highly defensive clients. It is not easy for clients to fake nonverbal behavior (Wahlsten, 1991). The real messages still tend to leak out. Helpers need to learn how to read clients' nonverbal messages without distorting or overinterpreting them. For instance, when Denise says to Jennie, "It seems that it's hard talking about yourself," Jennie says, "No, I don't mind at all." But the real answer is probably in her nonverbal behavior, for she speaks hesitatingly while looking away and frowning. Reading such cues helps Denise understand Jennie better.

Besides being a channel of communication in itself, such nonverbal behavior as facial expressions, bodily motions, and voice quality often modify and punctuate verbal messages in much the same way that periods, question marks, exclamation points, and underlining punctuate written language. All the kinds of nonverbal behavior mentioned earlier can punctuate or modify verbal communication in the following ways:

**Confirming or repeating.** Nonverbal behavior can confirm or repeat what is being said verbally. For instance, once when Denise responds to Jennie with just the right degree of understanding—she hits the mark—not only does Jennie say, "That's right!" but also her eyes light up (facial expression), she leans forward a bit (bodily motion), and her voice is very animated (voice quality). Her nonverbal behavior confirms her verbal message.

**Denying or confusing.** Nonverbal behavior can deny or confuse what is being said verbally. When challenged by Denise, Jennie denies that she is upset, but her voice falters a bit (voice quality) and her upper lip quivers (facial expression). Her nonverbal behavior carries the real message.

**Strengthening or emphasizing.** Nonverbal behavior can strengthen or emphasize what is being said. When Denise suggests to Jennie that she ask her boss what he means by her "erratic behavior," Jennie says in a startled voice, "Oh, I don't think I could do that!" while slouching down and putting her face in her hands. Her nonverbal behavior underscores her verbal message.

**Adding intensity.** Nonverbal behavior often adds emotional color or intensity to verbal messages. When Jennie tells Denise that she doesn't like to be confronted without first being understood and then stares at her fixedly and silently with a frown on her face, Jennie's nonverbal behavior tells Denise something about the intensity of her feelings.

**Controlling or regulating.** Nonverbal cues are often used in conversation to regulate or control what is happening. Let's say that in a group counseling session Nina looks at Tom and gives every indication that she is going to speak to him. But he looks away. Nina hesitates and then decides not to say anything. Tom has used a nonverbal gesture to control her behavior.

In reading nonverbal behavior—"reading" is used here instead of "interpreting"—caution is a must. We listen to clients in order understand them, not to dissect them. Merely reading about nonverbal behavior is not enough. Identifying relevant clues in videotaped interactions can help a great deal. Once you develop a working knowledge of nonverbal behavior and its possible meanings, you must learn through practice and experience to be sensitive to it and read its meaning in any given situation.

Richmond and McCroskey (2000, pp. 2–3) spell out the shadow side of nonverbal behavior in terms of commonly-held myths (in italics):

*Nonverbal communication is nonsense.* All communication involves language. Therefore, all communication is verbal. This myth is disappearing. It does not stand up under the scrutiny of common sense.

*Nonverbal behavior accounts for most of the communication in human interaction.* Early studies tried to "prove" this, but they were biased. Studies were aimed at dispelling the previous myth and overstepped their boundaries.

*You can read a person like a book.* Some people, even some professionals, would like to think so. Nonverbal behavior and verbal behavior form a package. And understanding clients is not the same as reading them like a book. The former is hard work; the latter is a myth.

*If a person does not look you in the eye while talking to you, he or she is not telling the truth.* Tell this to liars! The same nonverbal behavior can mean many different things.

*Although nonverbal behavior differs from person to person, most nonverbal behaviors are natural to all people.* Cross-cultural studies give the lie to this. But it isn't true within the same culture.

*Nonverbal behavior stimulates the same meaning in different situations. Too often the context is the key.* Yet some practitioners buy the myth and base interpretive systems on it.

Since nonverbal behaviors can often mean a number of things, how can you tell which meaning is the real one? The key is the context in which they take place. Effective helpers listen to the entire context of the helping interview and do not become overly fixated on details of behavior. They are aware of and use the nonverbal communication system, but they are not seduced or overwhelmed by it. Sometimes novice helpers will fasten selectively on this or that bit of non-verbal behavior. For example, they will make too much of a half-smile or a frown on the face of a client. They will seize upon the smile or the frown and, in over-interpreting it, lose the person. There is no need to go overboard on listening. Remember that you are a human being listening to a human being, not a vacuum cleaner indiscriminately sweeping up every scrap. Quality, not quantity.

## Processing What You Hear: The Thoughtful Search for Meaning

As we listen, we are processing what we hear. The trick is to become a thoughtful processor. What does thoughtful processing look like?

### Identifying Key Messages and Feelings

Denise listens to what Jennie has to say early on about her past and present experiences, actions, and emotions. She listens to Jennie's points of view and the decisions she had made, or is in the process of making. She listens to Jennie's intentions and proposals. For instance, Jennie tells Denise about an intention gone awry, and the emotions that went with it: "When the investigation began, I had every intention of pushing my case, because I knew that some of the men on campus were getting away with murder. But then it began to dawn on me that people were not taking me seriously because I was an African American woman. First I was angry, but then I just got numb. . . ." Later, Jennie says, "I get headaches a lot now. I don't like taking pills, so I try to tough it out. I've also become very sensitive to any kind of injustice, even in movies or on television. But I've stopped being any kind of crusader. That got me nowhere." As Denise listens to Jennie speak, questions based on the listening frameworks outlined here arise in the back of her mind:

- "What are the main points here?"
- "What experiences and actions are most important?"
- "What themes are coming through?"
- "What is Jennie's point of view?"
- "What is most important to her?"
- "What does she want me to understand?"
- "What decisions are implied in what she's saying?"
- "What, if anything, is she proposing to do?"

Processing means sorting. If you think that everything your client says is key, then nothing is key. In the end, helpers make a clinical judgment as to what is key. Questions like those outlined above help you do the sorting. Denise doesn't distract herself from Jennie by directly asking herself the above questions; they are woven into her active listening skills and are another indication of her interest in Jennie's world.

## Understanding Clients Through Context

People are more than the sum of their verbal and nonverbal messages. Listening in its deepest sense means listening to clients insofar as they are influenced by the contexts in which they live. Earlier, it was pointed out how important it is to interpret a client's nonverbal behavior in the context of the entire helping session. It is also essential to understand clients' stories, points of view, and messages through the wider context of their lives. All the things that make people different—culture, personality, personal style, ethnicity, key life experiences, religious attitudes, education, travel, economic status, and the others forms of diversity discussed in Chapter 2—provide the context for the client's problems and unused opportunities. Key elements of this context become part of the client's story whether they are mentioned directly or not. Effective helpers listen through this wider context without being overwhelmed by the details of it.

McAuliffe and Eriksen (1999), for example, offer a where-when-how-what model for helping helpers think about their clients in context. There are four questions that go something like this:

- What circumstances surround the client, and how do these circumstances affect the way the client understands and deals with her problems and opportunities?

- What age-related psychosocial tasks and challenges is the client currently facing, and how does the way he goes about these tasks affect the problem situation or opportunity?

- How does the client go about constructing meaning, including such things as determining what is important and what is right?

- How does the client's personality style and temperament affect his understanding of himself and his approach to the world?

These questions, of course, constitute but one framework. You need to discover for yourself the contextual frameworks (Ashford, Lecroy, & Lorti, 2006; Hutchinson, 2003; Conyne & Cook, 2004; Walsh, Galassi, Murphy, & Park-Taylor, 2002) that can help you understand your clients as "people-in-systems" (Egan & Cowan, 1979).

Denise tries to understand Jennie's verbal and nonverbal messages, especially the core messages, in the context of Jennie's life. As she listens to Jennie's story, Denise says to herself, right from the start, something like this:

Here is an intelligent African American woman from a conservative Catholic background. She has been very loyal to the church because it proved to be a refuge in the inner city. It was a gathering place for her family and friends. It provided her with a decent primary—and secondary—school education and a shot at college. She did very well in her studies. Initially, college was a shock. It was her first venture into a predominantly white and secular culture. But she chose her friends carefully and carved out a niche for herself. Since studies were much more demanding, she had to come to grips with the fact that, in this larger environment, she was, academically, closer to average. The rape and investigation put a great deal of stress on what proved to be a rather fragile social network. Her life began to unravel. She pulled away from her family, her church, and the small circle of friends she had at college. At a time when she needed support the most, she cut it off. After graduation she continued to stay out of community. Now she is underemployed as a secretary in a small company. This does little for her sense of personal worth.

The helping context is also important. Denise needs to be sensitive about how Jennie might feel, talking to a woman who is quite different from her.

In sum, Denise tries to pull together the themes she sees emerging in Jennie's story, and tries to see these themes in context. She listens to Jennie's discussion of her headaches (experiences), her self-imposed social isolation (behaviors), and her chronic depression (feelings) against the background of her social history—the pressures of being religious in a secular society at school, and the problems associated with being an upwardly-mobile African American woman in a predominantly white male society. Denise sees the rape and investigation as social, not merely personal, events. She listens actively and carefully, because she knows that her ability to help depends, in part, on not distorting what she hears. She does not focus narrowly on Jennie's inner world, as if Jennie could be separated from the social context of her life. Finally, although Denise listens to Jennie through the context of Jennie's life, she does not get lost in it. She uses context to understand Jennie and to help her manage her problems and develop her opportunities more fully.

## Hearing the Slant or Spin: Tough-Minded Listening and Processing

This is the kind of listening necessary to help clients explore issues more deeply and to pinpoint blind spots that need to be challenged. Skilled helpers not only listen to the client's stories, points of view, decisions, intentions, and proposals, but also to any slant or spin that clients might give their stories. Although clients' views of, and feelings about, themselves, others, and the world are real and need to be understood, their perceptions of themselves and their worlds are sometimes distorted. For instance, if a client sees herself as ugly when in reality she is beautiful, her experience of herself as ugly is real and needs to be listened to and understood. But her experience of herself does not square with the facts. This, too, must be listened to and understood. If a client sees himself as above average in his ability to communicate with others when, in reality, he is below average, his experience of himself needs to be listened to and understood, but reality cannot be ignored. Tough-minded listening includes detecting the gaps, distortions, and dissonance that are part of the client's experienced reality.

Denise realizes from the beginning that some of Jennie's understandings of herself and her world are not accurate. For instance, in reflecting on all that has happened, Jennie remarks that she probably got what she deserved. When Denise asks her what she means, she says, "My ambitions were too high. I was getting beyond my place in life." This is the slant, or spin, Jennie gives to her career aspirations. It is one thing to understand Jennie's interpretation of what has happened; it is another to assume that such an interpretation reflects reality. To be fully client-centered, helpers must first be reality-centered.

Of course, helpers need not challenge clients as soon as they hear any kind of distortion. Rather, they can note gaps and distortions, choose key examples, and challenge those perceptions when it is appropriate to do so (see Chapters 6 and 7).

## Musing on What's Missing

Clients often leave key elements out when talking about problems and opportunities. Frameworks for listening can help you spot key things that are missing. For instance, clients tell their stories but leave out key experiences, behaviors, or feelings. They offer points of view but say nothing about what's behind them or their implications. They deliver decisions but don't give the reasons for them or spell out the implications. They propose courses of action but don't say why they want to head in a particular direction, what the implications are for themselves or others, what resources they might need, or how flexible they are. As you listen, it's important to note what your clients put in and what they leave out. For instance, when it comes to stories, clients often leave out their own behavior or their feelings. Jennie says, "I got a call from an old girlfriend last week. I'm not sure how she tracked me down. We must have chatted away for 20 minutes. You know, catching up." Since Jennie says this in a rather matter-of-fact way, it's not clear how she felt about it at the time, or how she feels now. Nor is there any indication of what she might want to do about it—for instance, stay in touch.

In another session Jennie says, "I was talking with my brother the other day. He runs a small business. He asked me to come and work for him. I told him no. . . . By the way, I have to change the time of our next appointment. I forgot I've got a doctor's appointment." Denise notes the experience (being offered a job) and Jennie's behavior or reaction (a decision conveyed to her brother refusing the offer). But Jennie leaves out the reasons for her refusal or the implications for herself or her brother or their relationship, and moves to another topic.

Note that this is not a search for the "hidden stuff" that clients are leaving unsaid. Rather, Denise, understanding what full versions of stories, points of view, and messages look like, notes which parts are missing. She then uses her clinical judgment—a large part of which is common sense—to determine whether to ask about the missing parts. For instance, when she asks Jennie why she immediately refused her brother's offer, Jennie says, "Well, he's a good guy and I'd probably like the work, but this is no time to be getting mixed up with family." Jennie's response is one more indication of how restricted she has allowed her social life to become. It may be that she has determined that support from her family is out of bounds. Chapter 4 outlines and illustrates ways of helping clients fill out their stories with essential, but missing, detail.

# Listening to Yourself: The Helper's Internal Conversation

The conversation helpers have with themselves during helping sessions is the "internal conversation." To be an effective helper, you need to listen not only to the client, but also to yourself. Granted, you don't want to become self-preoccupied, but listening to yourself on a virtual *second channel* can help you identify both what you might do to be of further help to the client, and what might be standing in the way of your being with and listening to the client. This kind of internal listening is a positive form of self-consciousness.

Some years ago, this second channel did not work very well for me. A friend of mine who had been in and out of mental hospitals for a few years, and whom I had not seen for over six months, showed up unannounced one evening at my apartment. He was in a highly excited state. A torrent of ideas, some outlandish, some brilliant, flowed nonstop from him. I sincerely wanted to be with him as best I could, but I was very uncomfortable. I started by more or less naturally following the guidelines of tuning in, but I kept catching myself at the other end of the couch on which we were both sitting with my arms and legs crossed. I think that I was defending myself from the torrent of ideas. When I discovered myself almost literally tied up in knots, I would untwist my arms and legs, only to find them crossed again a few minutes later. It was hard work being with him. In retrospect, I realize I was concerned for my own security. I have since learned to listen to myself on the second channel a little better. When I listen to my nonverbal behavior as well as my internal dialogue, my interactions with clients are better.

Helpers can use this second channel to listen to what they are *saying* to themselves— their nonverbal behavior, and their feelings and emotions. These messages can refer to the helper, the client, or the relationship.

- "I'm letting the client get under my skin. I had better do something to reset the dialogue."

- "My mind has been wandering. I'm preoccupied with what I have to do tomorrow. I had better put that out of my mind."

- "Here's a client who has had a tough time of it, but her self-pity is standing in the way of her doing anything about it. My instinct is to be sympathetic. I need to talk to her about her self-pity, but I had better go slow."

- "It's not clear that this client is interested in changing. It's time to test the waters."

The point is that this internal conversation goes on all the time. It can be a distraction or it can be another tool for helping. The client, too, is having his or her internal conversation. One intriguing study (Hill, Thompson, Cogar, & Denman, 1993) suggested that both client and therapist are more or less aware of the other's "covert processes." However, this study showed that helpers, even though they knew that clients were having their own internal conversations and were leaving things unsaid, were not very good at determining what those things were. At times there are verbal or nonverbal hints as to what the client's

## Questions on Active Listening

1. What forms of inadequate listening am I prone to?
2. How well do I listen to people's stories in terms of their experiences? Their feelings? Their behavior?
3. How sensitive am I to the fact that, when people talk, they talk about their points of view? Their decisions? Their intentions? Their plans?
4. When I listen to people, how good am I at hearing their strengths and resources?
5. How well do I pick up nonverbal cues without overinterpreting them?
6. How well do I listen to people contextually?
7. What kind of "tough-minded" listener am I?
8. How aware am I of my internal conversations, and how well do I use them to listen meaningfully to others?
9. Which kinds of distorted listening do I have to be careful of?

internal dialogue might be. Helping clients move key points from their internal conversations into the helping dialogue is a key task and will be discussed in later chapters.

Box 3.2 asks questions that will help you assess your strengths and weaknesses in listening.

## The Shadow Side of Listening to Clients

Listening as described here is not as easy as it sounds. Obstacles and distractions abound. Some relate to listening generally. Others relate more specifically to listening to, and interpreting, the nonverbal behavior of clients. The following kinds of distorted listening, as you will see from your own experience, permeate human communication. They also insinuate themselves at times into the helping dialogue. They are part of the shadow side because helpers never intend to engage in distorted listening. Rather, helpers fall into them at times without even realizing that they are doing so. Sometimes more than one kind of distortion contaminates the helping dialogue.

**Filtered listening.**  It is impossible to listen to other people in a completely unbiased way. Through socialization we develop a variety of filters through which we listen to ourselves, others, and the world around us. As Hall (1977) noted:

"One of the functions of culture is to provide a highly selective screen between man and the outside world. In its many forms, culture therefore designates what we pay attention to and what we ignore. This screening provides structure for the world" (p. 85). We need filters to provide structure for ourselves as we interact with the world. But personal, familial, sociological, and cultural filters introduce various forms of bias into our listening and do so without our being aware of it.

The stronger the cultural filters, the greater the likelihood of bias. For instance, a white, middle-class helper probably tends to use white, middle-class filters in listening to others. Perhaps this makes little difference if the client is also white and middle class, but if the helper is listening to an Asian client who is well-to-do and has high social status in his community, to an African American mother from an urban ghetto, or to a poor white subsistence farmer, then the helper's cultural filters might introduce bias. Prejudices, whether conscious or not, distort understanding. Like everyone else, helpers are tempted to pigeonhole clients because of gender, race, sexual orientation, nationality, social status, religious persuasion, political preferences, lifestyle, and the like. Self-knowledge on the part of helpers is essential. This includes ferreting out the biases and prejudices that distort listening.

**Evaluative listening.** Most people, even when they listen attentively, listen evaluatively. That is, as they listen, they are judging what the other person is saying as good/bad, right/wrong, acceptable/unacceptable, likable/unlikable, relevant/irrelevant, and so forth. Helpers are not exempt from this universal tendency. The following interchange took place between Jennie and a friend of hers. Jennie recounts it to Denise as part of her story.

> JENNIE: Well, the rape and the investigation are not dead, at least not in my mind. They are not as vivid as they used to be, but they are there.

> FRIEND: That's the problem, isn't it? Why don't you do yourself a favor and forget about it? Get on with life, for God's sake!

Evaluative listening gives way to advice giving. It might well be sound advice, but the point here is that Jennie's friend listens and responds evaluatively. Clients should first be understood, then, if necessary, challenged or helped to challenge themselves. Evaluative listening, translated into advice-giving, usually puts clients off. Indeed, a judgment that a client's point of view, once understood, needs to be expanded or transcended or that a pattern of behavior, once listened to and understood, needs to be altered can be quite useful. That is, there are productive forms of evaluative listening. It is practically impossible to suspend judgment completely. Nevertheless, it is possible to set one's judgment aside for the time being in the interest of understanding clients, their worlds, their stories, their points of view, and their decisions *from the inside.*

**Stereotyped-based listening.** I remember my reaction to hearing a doctor refer to me as the "hernia in 304." We don't like to be stereotyped, even when the stereotype has some validity. The very labels we learn in our training—paranoid, neurotic, sexual disorder, borderline—can militate against empathic understanding. Books on personality theories and abnormal psychology provide us with stereotypes: "He's a perfectionist." We even pigeonhole ourselves: "I'm a Type A personality" —though, in this case, the stereotype is often used as an excuse. In psychotherapy, diagnostic categories can take precedence over the clients who are being diagnosed. Helpers forget at times that their labels are interpretations rather than understandings of their clients. You can be "correct" in your diagnosis and still lose the person. In short, what you learn as you study psychology may help you to organize what you hear, but it may also distort your listening. To use terms borrowed from Gestalt psychology, make sure that your client remains *figure*—in the forefront of your attention—and that models and theories about clients remain *ground*—knowledge that remains in the background and is used only in the interest of understanding and helping this unique client.

**Fact-centered rather than person-centered listening.** Some helpers ask clients many informational questions, as if clients would be cured if enough facts about them were known. It's entirely possible to collect facts but miss the person. The antidote is to listen to clients contextually, trying to focus on themes and key messages. Denise, as she listens to Jennie, picks up what is called a "pessimistic explanatory style" theme (Peterson, Seligman, & Vaillant, 1988). When talking about unfortunate events, clients with this style tend to say—directly or indirectly—"It will never go away;" "It affects everything I do;" and "It's my fault." Denise knows that research indicates that people who fall victim to this style tend to end up with poorer health than those who do not. There may be a link, she hypothesizes, between Jennie's somatic complaints (headaches, gastric problems) and this explanatory style. This is a theme worth exploring.

**Sympathetic listening.** Since most clients are experiencing some kind of misery, and since some have been victimized by others or by society itself, there is a tendency on the part of helpers to feel sympathy for them. Sometimes these feelings are strong enough to distort the stories that clients are telling. Consider this case.

> Liz was counseling Ben, a man who had lost his wife and daughter to a tornado. Liz had recently lost her husband to cancer. As Ben talked about his own tragedy during their first meeting, she wanted to hold him. Later that day she took a long walk and realized how her sympathy for Ben had distorted what she heard. She heard the depth of his loss but, reminded of her own loss, only half heard the implication that his loss now excused him from getting on with his life.

Sympathy has an unmistakable place in human relationships, but its "use," if that does not sound too inhuman, is limited in helping. In a sense, when I sympathize with someone, I become his or her accomplice. If I sympathize with my client as she tells me how awful her husband is, I take sides without knowing what the complete story is. Expressing sympathy can reinforce self-pity in a client. And self-pity has a way of driving out problem-managing action.

**Interrupting.** I am reluctant to add "interrupting," as some do, to this list of shadow-side obstacles to effective listening. Certainly, when helpers interrupt their clients, they, by definition, stop listening. And interrupters often say things that they have been rehearsing, which means that they have been only partially listening. My reluctance, however, comes from the conviction that the helping conversation should be a dialogue. There are benign and malignant forms of interrupting. The helper who cuts the client off in mid-thought because he has something important to say is using a malignant form. But the case is different when a helper interrupts a monologue with some gentle gesture and a comment such as, "You've made several points. I want to make sure that I've understood them." If interrupting promotes the kind of dialogue that serves the problem-management process, then it is useful. Still, care must be taken to factor in cultural differences in storytelling.

One possible reason counselors fall prey to the kinds of shadow-side listening described above is the unexamined assumption that listening with an open mind is the same as approving what the client is saying. This, of course, is not the case. Rather, listening with an open mind helps you learn and understand. Whatever the reason for shadow-side listening, the outcome can be devastating because of a truth philosophers learned long ago: A small error in the beginning can lead to huge errors down the road. If the foundation of a building is out of kilter, it is hard to notice with the naked eye. But, by the time construction reaches the ninth floor, it begins to look like the leaning Tower of Pisa. Tuning in to clients and listening both actively and with an open mind are foundation counseling skills. Ignore them, and dialogue is impossible.

 **EXERCISES ON PERSONALIZING**

1. Review the four behaviors that make up dialogue. Then, on a scale from 1–7, rate yourself on each of the four. Given these ratings, how effective do you see yourself in both engaging in dialogue, and helping others engage in dialogue in your conversations? What do you think you need to do to become more effective in dialogue? What do you need to stop doing? What do you need to start doing? What do you need to keep on doing?

2. Review the elements of active listening. Which of these do you do best? What mistakes do you tend to make? In what situations are you at your best? In what situations do you make the most mistakes? What changes would produce the most improvement in your active listening style?

# CHAPTER 4

# Communicating Empathy and Checking Understanding

The Three Dimensions of Responding Skills: Perceptiveness, Know-How, and Assertiveness

Sharing Empathic Highlights: Communicating Understanding to Clients

> The Basic Formula

> Respond Accurately to Clients' Feelings, Emotions, and Moods

> Respond Accurately to the Key Experiences and Behaviors in Clients' Stories

> Respond with Highlights to Clients' Points of View, Intentions, Proposals, and Decisions

Principles for Sharing Highlights

Tactics for Communicating Highlights

The Shadow Side of Sharing Empathic Highlights

The Importance of Empathic Relationships

Exercises on Personalizing

Helpers listen to clients both to understand them and their concerns, and to respond to them in constructive ways. The logic of listening includes, as we have seen, tuning in to clients; listening actively; processing what is heard contextually; identifying the key ideas, messages, or points of view the client is trying to communicate—all at the service of understanding clients. Listening, then, is a very active process that serves understanding. But helpers don't just listen; they also respond to clients in a variety of ways. They respond by sharing their understanding, checking to make sure that they've got things right, asking questions, probing for clarity, summarizing the issues being discussed, and challenging clients in a variety of ways. In this chapter, the focus is on communicating empathic understanding to clients and checking to see if that understanding is accurate. When helpers both understand and accurately communicate their understanding to clients, they help their clients understand themselves, their problem situations, and their unused opportunities more fully—a process that helps them move to problem-managing action.

## The Three Dimensions of Responding Skills: Perceptiveness, Know-How, and Assertiveness

The communication skills involved in responding to clients have three dimensions: perceptiveness, know-how, and assertiveness.

**Perceptiveness.**  Feeling empathy for others is not helpful if the helper's perceptions are not accurate. Ickes defined "empathic accuracy" as "the ability to accurately infer the specific content of another person's thought and feelings" (1993, p. 588). According to Ickes, this ability is a component of success in many walks of life.

> Empathically accurate perceivers are those who are consistently good at "reading" other people's thoughts and feelings. All else being equal, they are likely to be the most tactful advisors, the most diplomatic officials, the most effective negotiators, the most electable politicians, the most productive salespersons, the most successful teachers, and the most insightful therapists. (Ickes, 1997, p. 2)

The assumption is, of course, that such people are not only accurate perceivers but they can also weave their perceptions into their dialogues with their constituents, customers, students, and clients. Helpers do this by sharing empathic highlights with their clients.

Your responding skills are only as good as the accuracy of the perceptions on which they are based. Consider the difference between these two examples:

> Beth is counseling Ivan in a community mental-health center. Ivan is scared to talk about an "ethical blunder" he made at work. Beth senses his discomfort but thinks that he is angry rather than scared. She says, "Ivan, I'm wondering what's making you so angry right now." Since Ivan does not feel angry, he says nothing. In fact, he's startled by what she says and feels even more insecure. Beth takes his silence as a confirmation of his "anger." She tries to get him to talk about it.

Beth's perception is wrong, and therefore disrupts the helping process. She misreads Ivan's emotional state and tries to engage in a dialogue based on her flawed perception. Contrast this to what happens in the following example.

> Mario, a manager, is counseling Enrique, a relatively new member of his team. During the past week, Enrique has made a significant contribution to a major project, but he has also made one rather costly mistake. Enrique's mind is on his blunder, not his success. Mario, sensing Enrique's discomfort, says, "Your ideas in the meeting last Monday helped us reconceptualize and reset the entire project. It was a great contribution. That kind of outside-the-box thinking is very valuable here." He pauses. "I'd also like to talk to you about Wednesday. Your conversation with Acme's purchasing agent on Wednesday made him quite angry." He pauses briefly once more. "Something tells me that you might be more worried about Wednesday's mistake than delighted with Monday's contribution. I just wanted to let you know that I'm not." Enrique is greatly relieved. They go on to have a useful dialogue about what made Monday so good and what could be learned from Wednesday's blunder.

Mario's perceptiveness and his ability to defuse a tense situation lays the foundation for an upbeat dialogue.

The kind of perceptiveness needed to be a good helper comes from basic intelligence, social intelligence, experience, reflecting on your experience, developing wisdom and, more immediately, tuning in to clients, listening carefully to what they have to say, and objectively processing what they say. Perceptiveness is part of social-emotional maturity.

**Know-how.** Once you are aware of what kind of response is called for, you need to be able to deliver it. For instance, if you are aware that a client is anxious and confused because this is his first visit to a helper, it does little good if you don't know how to translate your perceptions and your understanding into words. Let's return to Ivan and Beth for a moment.

> Ivan and Beth end up arguing about his "anger." Ivan finally gets up and leaves. Beth, unfortunately, takes this as a sign that she was right in the first place. The next day Ivan goes to see his minister. The minister sees quite clearly that Ivan is scared and confused. His perceptions are right. He says something like this: "Ivan, you seem to be very uncomfortable. It may be that whatever is on your mind might be difficult to talk about. But I'd be glad to listen to it, whatever it is. But I don't want to push you into anything." Ivan blurts out, "But I've done something terrible." The minister pauses and then says, "Well, let's see what kind of sense we can make of it." Ivan hesitates a bit, then leans back into his chair, takes a deep breath, and launches into his story.

The minister is not only perceptive but also knows how to address Ivan's anxiety and hesitation. The minister says to himself, "Here's a man who is almost exploding with the need to tell his story, but fear or shame or something like that is paralyzing him. How can I put him at ease, let him know that he won't get hurt here? I need to recognize his anxiety and offer an opening." He does not use these words, of course, but these are the kinds of sentiments that instinctively run through his mind. This chapter is designed to help you develop know-how.

**Assertiveness.** Accurate perceptions and excellent know-how are meaningless if they remain locked up inside you. They need to become part of the therapeutic dialogue. For instance, if you see that self-doubt is a theme that weaves itself throughout a client's story about her frustrating search for a better relationship with her estranged brother but fail to share your hunch with her, you do not pass the assertiveness test. Consider this example:

> Nina, a young counselor in the Center for Student Development, is in the middle of the first session with Antonio, a graduate student. During the session, he briefly mentions a very helpful session he had the previous year with Carl, a middle-aged counselor on the staff. Carl has accepted an academic position at the university and is no longer involved with the Center. Nina realizes that Antonio is disappointed that he couldn't see Carl, and might have some misgivings about being helped by a new counselor— a younger woman. She has faced sensitive issues like this before and would not take it amiss if Antonio were to choose a different counselor. During a lull in the conversation, she says something like this: "Antonio, could we take a time-out here for a moment? I think you might be a bit disappointed to find out that Carl is no longer here. Or, at least I probably would be if I were in your shoes. You were just more or less assigned to me and I'm not sure the fit is right. Maybe you can give that a bit of thought. Then, if you think I can be of help, you can schedule another meeting with me. But you're certainly free to review who is on staff and choose whomever you want."

In this case, perceptiveness, know-how, and assertiveness all come together. This is not to suggest that assertiveness is an overriding value in and of itself. To be assertive without perceptiveness and know-how is to court disaster.

## Sharing Empathic Highlights: Communicating Understanding to Clients

Although many people may "feel empathy" for others—that is, are motivated in many different ways by the value of empathy described in Chapter 2—the truth is that few know how to put empathic understanding into words. And so, sharing empathic highlights as a way of showing understanding during conversations remains an improbable event in everyday life. Perhaps that's why it is so powerful in helping settings. When clients are asked what they find helpful in counseling sessions, being understood gets top ratings. There is such an unfulfilled need to be understood.

In some ways this section is a kind of anatomy lesson. That is, we are going to take the process of sharing empathic highlights apart and look at the pieces. Further on, we'll put them back together again.

### The Basic Formula

Basic empathic understanding can be expressed in the following stylized formula:

> *You feel* . . . [here name the correct emotion expressed by the client] . . .
> *because* . . . [here indicate the correct thoughts, experiences, and behaviors that give rise to the feelings] . . .

For instance, Leonardo is talking with a helper about his arthritis and all its attendant ills. There is pain, of course but, more to the point, he can't get around the way he used to. At one point the helper says:

> You feel bad, not so much because of the pain, but because your ability to get around—your freedom—has been curtailed.

Leonardo replies:

> That's just it. I can manage the pain. But not being able to get around is killing me! It's like being in jail.

They go on to discuss ways in which Leonardo, with the help of family and friends, can get out of "jail," that is, become more mobile and have more ways of coping with the time he is in "jail."

The formula "You feel . . . because . . ." is a beginner's tool to get used to the concept of sharing highlights. It focuses on the key points of clients' stories, points of view, intentions, proposals, and decisions, together with the feelings, emotions, and moods associated with them. The formula is used in the following examples. For the moment, ignore how stylized it sounds. Ordinary human language will be substituted later. In the first example, a divorced mother with two young children is talking to a social worker about her ex-husband. She has been talking about the ways he has let her and their kids down. She ends by saying:

> CLIENT:  I could kill him! He failed to take the kids again last weekend. This is three times out of the last six weeks.

> HELPER:  You feel furious because he keeps failing to hold up his part of the bargain.

> CLIENT:  I just have to find some way to get him to do what he promised to do. What he told the court he would do.

His not taking the kids according to their agreement [an experience for the client] infuriates her [an emotion]. The helper captures both the emotion and the reason for it. And the client moves forward in terms of thinking about possible actions she could take.

In the next example, a woman who has been having a great deal of gastric and intestinal distress is going to have a colonoscopy. She is talking with a hospital counselor the night before the procedure. She says:

> PATIENT:  God knows what they'll find when they go in. I keep asking questions, but they keep giving me vague answers.

> HELPER:  You feel troubled because you believe that you're being left in the dark.

> PATIENT:  In the dark about my body, my life! If they'd only tell me! Then I could prepare myself better.

They go on to discuss what she needs to do to get the kind of information she wants. The accuracy of the helper's response does not solve the woman's problems, but the patient does move a bit. She gets a chance to vent her concerns, she receives a bit of understanding, and says why she wants more information. This, perhaps, puts her in a better position to ask for a more open relationship with her doctors.

The key elements of an empathic highlight are the same as the key elements of Jennie's story discussed in Chapter 3—that is, the thoughts, experiences, behaviors, and feelings that make up the client's story. Here are some guidelines.

## Respond Accurately to Clients' Feelings, Emotions, and Moods

The importance of feelings, emotions, and moods in our lives was discussed in Chapter 3. Helpers need to respond to clients' emotions in such a way as to move the helping process forward. This means identifying key emotions that the client either expresses or discusses (helper perceptiveness), and weaving them into the dialogue (helper know-how) even when they are sensitive or part of a messy situation (helper courage or assertiveness). Do you remember the last time you, as a consumer, got a problem resolved with a good customer service representative? She might have said something like this to you: "I know you're angry right now because the package didn't arrive, and you have every right to be. After all, we did make you a promise. Here's what we can do to make it right for you . . ." Rather than ignoring the customer's emotions, good customer service reps face up to them as helpfully as possible.

**Use the right family of emotions and the right intensity.** In the basic highlight formula, "You feel . . ." should be followed by the correct family of emotions and the correct intensity.

*Family.* The statements "You feel hurt," "You feel relieved," and "You feel enthusiastic" specify different families of emotion.

*Intensity.* The statements "You feel annoyed," "You feel angry," and "You're furious" specify different degrees of intensity in the same family (anger).

The words sad, mad, bad, and glad refer to four of the main families of emotion, whereas "content," "quite happy," and "overjoyed" refer to different intensities within the "glad" family.

**Distinguish between expressed and discussed feelings.** Clients both *express* emotions they are feeling during the session, and *talk about* emotions they felt at the time of some incident. For instance, consider this interchange between a client involved in a child custody proceeding and a counselor. She is talking about her husband.

CLIENT (calmly): I get furious with him when he says things, little snide things, that suggest that I don't take good care of the kids.

HELPER: You feel especially angry when he intimates that you're not a good mother.

The client isn't angry right now. Rather, she is talking about the anger. In the following example, a woman is talking about one of her colleagues at work. This scenario deals with expressed, rather than discussed, feelings.

CLIENT (enthusiastically): I threw caution to the wind and confronted him about his sarcasm and it actually worked. He not only apologized, but behaved himself the rest of the trip.

HELPER: You feel great because you took a chance and it paid off.

Clients don't always name their feelings and emotions. However, if they express emotion, it is part of the message and needs to be identified and understood.

**Read and respond to feelings and emotions embedded in clients' nonverbal behavior.** Often, helpers have to read their clients' emotions—both the family and the intensity—in their nonverbal behavior. In the following example, a North American student comes to a helper, sits down, looks at the floor, hunches over, and haltingly says,

CLIENT: I don't even know where to start. (He falls silent).

HELPER: It's pretty clear that you're feeling miserable. Can we talk about why?

CLIENT (after a pause): Well, let me tell you what happened . . .

The helper sees that he is depressed, and his nonverbal behavior indicates that the feelings are quite intense. His nonverbal behavior reveals the broad family ("You feel bad") and the intensity ("You feel very bad"). Of course, the helper does not yet know the experiences and behaviors that give rise to these emotions.

**Be sensitive in naming emotions.** Naming and discussing feelings and emotions threaten some clients. If this is the case, it might be better to focus on thoughts, experiences, and behaviors and proceed only gradually to a discussion of feelings. The following client, an unmarried man in his mid-30s who has come to talk about "certain dissatisfactions" in his life, has shown some reluctance to express, or even to talk about, feelings.

CLIENT (in a pleasant, relaxed voice): My mother is always trying to make a little kid out of me. And I'm 35! Last week, in front of a group of my friends, she brought out my rubber boots and an umbrella and gave me a little talk on how to dress for bad weather (laughs).

COUNSELOR A: It might be hard to admit it, but I get the feeling that down deep you were furious.

CLIENT: Well, I don't know about that. Anyway, at work. . . .

Counselor A pushes the emotion issue and is met with some resistance. The client changes the topic.

COUNSELOR B (in a somewhat lighthearted way): So she's still playing the mother role—to the hilt, it would seem.

CLIENT (with more of a bite in his voice): And the hilt includes not wanting me to grow up. But I am grown up . . . well, pretty grown up. But I don't always act grown up around her.

Counselor B, choosing to respond to the strong-mother issue rather than the more sensitive being-kept-a-kid-and-feeling-really-lousy-about-it issue, gives the client more room to move. This works, for the client himself moves toward the more sensitive issue—his playing the child, at least at times, when he's with his mother.

Some clients are hesitant to talk about certain emotions. One client might find it relatively easy to talk about his anger but not his hurt. The following client is talking about his disappointment at not being chosen for a special team at work.

> **CLIENT:** I worked as hard as anyone else to get the project up and running. In fact, I was at the meeting where we came up with the idea in the first place . . . And now they've dropped me.

> **COUNSELOR A:** So you feel really hurt—left out of your own project.

> **CLIENT** (hesitating): Hmm . . . I'm really ticked off. Why shouldn't I be!

Here is a client with lots of ego. He doesn't like the idea that he has been "hurt." Counselor B takes a different tact.

> **COUNSELOR B:** So, it's more than annoying to be left out of what, in many ways, is your own project.

> **CLIENT:** How could they do that?—It is more than annoying. It's . . . well . . . humiliating!

Counselor B, factoring in the client's ego, sticks to the anger, allowing the client himself to name the more sensitive emotion. Contextual listening—in this case listening to the client's emotions through the context of the pride he takes in himself and his accomplishments—is part of social intelligence. However, being sensitive to clients' sensitive emotions should not rob counseling of its robustness. Too much tiptoeing around clients' sensitivities does not serve them well. Remember what was said earlier. Clients are not as fragile as we sometimes make them out to be.

**Use different ways to share highlights about feelings and emotions.** Since clients express feelings in a number of different ways, helpers can communicate an understanding of feelings in a variety of ways.

- *By single words.* You feel good. You're depressed. You feel abandoned. You're delighted. You feel trapped. You're angry.
- *By different kinds of phrases.* You're sitting on top of the world. You feel down in the dumps. You feel left in the lurch. Your back's up against the wall. You're really on a roll.
- *By what is implied in behavioral statements.* You feel like giving up (implied emotion: despair). You feel like hugging him (implied emotion: joy). Now that you see what he's been doing to you, you almost feel like throwing up (implied emotion: disgust).
- *By what is implied in experiences that are revealed.* You feel you're being dumped on (implied feeling: victimized). You feel you're being stereotyped

(implied feeling: resentment). You feel you're at the top of her list (implied feeling: elation). You feel you're going to get caught (implied feeling: fear). Note that the implication of each could be spelled out: You feel angry because you're being dumped on. You resent the fact that you're being stereotyped. You feel great because it seems that you're at the top of her list.

Since, ultimately, you must discard formulas and use your own language—words that are you, rather than words from a textbook—it helps to develop a variety of ways to communicate your understanding of clients' feelings and emotions. Using your own language keeps you from being wooden in your responses. Consider this example: The client tells you that she has just been given the kind of job she has been looking for the past two years. Here are some possible responses to her emotion.

> *Single word.* You're really happy.
>
> *A phrase.* You're on cloud nine.
>
> *Experiential statement.* You feel you finally got what you deserve.
>
> *Behavioral statement.* You feel like going out and celebrating.

With experience, you can extend your range of expression at the service of your clients. Providing variety will become second nature.

**Neither overemphasize nor underemphasize feelings, emotions, and moods.** Some counselors take an overly rational approach to helping and almost ignore clients' feelings. Others become too preoccupied with clients' emotions and moods. They pepper clients with questions about feelings, and at times extort answers. To say that feelings, emotions, and moods are important is not to say that they are everything. The best defense against either extreme is to link feelings, emotions, and moods to the experiences and behaviors that give rise to them.

## Respond Accurately to the Key Experiences and Behaviors in Clients' Stories

Key experiences and behaviors give rise to clients' feelings, emotions, and moods. The "because . . ." in the empathic-highlight formula is to be followed by an indication of the experiences and behaviors that underlie the client's feelings. In the following example, the client, a graduate student in law school, is venting his frustration.

> CLIENT (heatedly): You know why he got an "A"? He took my notes and disappeared. I didn't get a chance to study them. And I never even confronted him about it.
>
> HELPER: You feel doubly angry because, not only did he steal your notes, but you let him get away with it.

The response specifies both the client's experience (the theft) and his behavior (in this case, a failure to act) that give rise to his distress. His anger is directed not only at his classmate, but also himself.

In the following example, a mugging victim has been talking to a social worker to help cope with his fears of going out. Before the mugging he had given no thought to urban problems. Now he tends to see menace everywhere.

CLIENT: This gradual approach of getting back in the swing seems to be working. Last night I went out without a companion. First time. I have to admit that I was scared. But I think I've learned how to be careful. Last night was important. I feel I can begin to move around again.

HELPER: You feel comfortable with the one-step-at-a-time approach you've been taking. And it paid off last night when you regained a big chunk of your freedom.

CLIENT: That's it! I know I'm going to be free again . . . Here's what I've been thinking of doing . . .

The client is talking about the success of the action phase of the program. The helper's response recognizes the client's satisfaction with the success of the program, and how important it is for the client to feel both safe and free. The client moves on to describe the next phase of his change program.

Another client is working to rebuild her life after a devastating car accident. After a few sessions spread out over six months, she's back at her job and has been working with her husband at rebuilding their marriage. She says something like this about the progress she is making:

CLIENT (talking in an animated way): I really think that things couldn't be going better. I'm doing very well at my new job, and my husband isn't just putting up with it. He thinks it's great. He and I are getting along better than ever, even sexually, and I never expected that. We're both working at our marriage. I guess I'm just waiting for the bubble to burst.

HELPER: You feel great because things have been going better than you ever expected—and it seems almost too good to be true.

CLIENT: Well, a "bubble bursting" might be the wrong image. I think there's a difference between being cautious and waiting for disaster to strike. I'll always be cautious, but I'm finding out that I can make things come true instead of sitting around waiting for them to happen as I usually do. I guess I've got to keep making my own luck.

This client talks about her experiences and behaviors and expresses feelings, the flavor of which is captured in the helper's highlight. The response, capturing as it does both the client's enthusiasm and her lingering fears, is quite useful because the client makes an important distinction between reasonable caution and expecting the worst the happen. She moves on to her need to make things happen, to become more of an agent in her life.

## Respond with Highlights to Clients' Points of View, Intentions, Proposals, and Decisions

Sharing highlights communicates to clients that you are working hard at understanding them at the service of constructive change. This means not only understanding the key elements of the stories they tell, but also the key elements of anything they share with you. Here are some examples that relate to clients' points of view, decisions, intentions, plans, and proposals, and the feelings and emotions that permeate them.

**Communicating your understanding of clients' points of view.** In the following example, the client, a forty-five-year-old man, is a construction worker, married with four children between the ages of nine and sixteen. He has been expressing concerns about his children. At one point he says:

> CLIENT: I don't consider myself old-fashioned, but I think kids these days suffer from over-indulgence. We keep giving them things. We let them do what they want. I fall into the same trap myself. It's just not good for them. I don't think we're preparing them for what the world is really like. People assume that the economy will keep booming. Everyone keeps shouting, "Free lunch!" This isn't doing kids any good.

> COUNSELOR: So you see the "do-what-you-want" and "free-lunch" messages as a lot of hogwash. It's going to backfire and your kids could end up getting hurt.

> CLIENT: Right. . . But I'm not in control. My kids can get one set of messages from me and then get a flood of contradictory messages outside and from TV and the Internet. . . I don't want to be a tyrant. Or come across as a kill-joy. That doesn't work, anyway. At work I see problems and I take care of them. But this has got me stymied.

> COUNSELOR: So the whole picture seems pretty gloomy right now. You're not exactly sure what to do about it. You handle problems at work, but it's a lot harder to do something about societal problems that could hurt your kids. What have you been doing so far to handle all of this?

Once the counselor communicates understanding of the client's point of view, he (the client) moves on to share his sense of helplessness. The helper realizes, however, that the client probably has tried some approaches to managing this problem situation. He is probably not as helpless as he makes himself out to be.

**Communicating your understanding of intentions or proposals.** In the next example, the client, who is hearing impaired, has been discussing ways of becoming, in her words, "a full-fledged member of my extended family." The discussion between client and helper takes place through a combination of lipreading and signing.

> CLIENT (enthusiastically): Let me tell you what I'm thinking of doing. . . . . First of all, I'm going to stop fading into the background in family and friends conversation groups. I'll be the best listener there. And I'll get my thoughts across even if I have to use props. That's how I really am . . . inside, you know, in my mind.

> HELPER: Sounds exciting. You're thinking of getting right into the middle of things— where you think you belong. You might even try a bit of drama.

> CLIENT: And I think that, well, socially, I'm pretty smart. So I'm not talking about being melodramatic or anything. I can do all this with finesse, not just barge in.

HELPER: You'll make it all natural. Draw me a couple of pictures of what this would look like.

The client comes up with a proposal for a course of action that will help her take her "rightful place" in conversations with family and friends, setting her agenda (Stage III). The helper's response recognizes her enthusiasm and sense of determination. They go on to have a dialogue about practical tactics.

**Communicating your understanding of clients' decisions.** When clients announce key decisions or express their resolve to do something, it's important to recognize the core of what they are saying. In the following example, a client being treated for social phobia, has benefitted greatly from cognitive-behavioral therapy. For instance, in uncomfortable social situations, he has learned to block self-defeating thoughts and to keep his attention focused externally—on the social situation itself and on the agenda of the people involved—instead of turning in on himself. At one point the client says:

CLIENT (emphatically): I'm not going to turn back. I've had to fight to get where I am now. But I can see how easy it could be to slide back into my old habits. I bet a lot of people do. I see it all around me. People make resolutions and then they peter out.

HELPER: Even though it's possible for you to give up your hard-earned gains, you're not going to do it. You're just not.

CLIENT: But what can I do to make sure that I won't? I'm convinced I won't, but . . .

HELPER: You need some rachets. They're the things that keep roller-coaster cars from sliding back. You hear them going click, click, click on the way up.

CLIENT: Ah, right! But I need psychological ones . . .

HELPER: And social ones . . . What's kept you from sliding back so far?

In a positive psychology mode, the counselor focuses on past successes. They go on to discuss the kind of "rachets" he needs to stay on track.

## Principles for Sharing Highlights

Here are a number of principles that can guide you as you share highlights. Remember that these guidelines are principles, not formulas to be followed slavishly.

**Use empathic highlights at every stage and in every task of the helping process.** Communicating and checking understanding is always helpful. Here are some examples of helpers sharing highlights at different stages and tasks of the helping process:

*Stage I: Problem clarification and opportunity identification.* A teenager in his third year of high school who has just found out that he is moving with his family to a different city. A school counselor responds,

"You're miserable because you have to leave all your friends. But it sounds like you may even feel a bit betrayed. You didn't see this coming at all." The counselor realizes that he has to help his client pick up the pieces and move on, but sharing his understanding helps build a foundation to do so. The teen goes on to talk in positive terms about the large city they will be moving to and the opportunities it will offer. At one point the school counselor responds, "So there's an upside to all this. Big cities are filled with things to do. You like theater and there's loads there. That's something to look forward to."

*Stage II: Evaluating goal options.*  A woman who has been discussing the trade-offs between marriage and career: At one point her helper says, "There's some ambivalence here. If you marry Jim, you might not be able to have the kind of career you'd like. Or, did I hear you half-say that it might be possible to put both together? Sort of get the best of both worlds." The client goes on the explore the possibilities around "getting the best of both worlds." It helps her greatly in preparing for her next conversation with Jim.

*Stage III: Choosing actions to accomplish goals.*  A man has been discussing his desire to control his cholesterol level without taking a medicine whose side effects worry him. He says that his plan might work. The counselor responds, "It's a relief to know that sticking to the diet and exercise might mean that you won't have to take any medicine . . . Hmm . . . Let's explore the 'might' part. I'm not exactly sure what your doctor said." The helper recognizes an important part of the client's message, but then seeks further clarification.

*The Action Arrow: Implementation issues.*  A married couple have been struggling to put into practice a few strategies to improve their communication with each other. They've both called their attempts a "disaster." The counselor replies: "OK, so you're annoyed with yourselves for not accomplishing even the simple active-listening goals you set for yourselves . . . Well, let's see what we can learn from the 'disaster'," he says somewhat light-heartedly. The counselor communicates understanding of their disappointment in not implementing their plan but, in the spirit of positive psychology, focuses on what they can learn from the failure.

Sharing empathic highlights is a mode of human contact, a relationship builder, a conversational lubricant, a perception-checking intervention, and a mild form of social influence. It is always useful. Driscoll (1984), in his common-sense way, referred to highlights as "nickel-and-dime interventions" which each contribute only a smidgen of therapeutic movement, but without which the course of therapeutic progress would be markedly slower (p. 90). Since sharing highlights provides a continual trickle of understanding, it is a way of providing support for clients throughout the helping process. It is never wrong to let clients know that you are trying to understand them from their frame of reference. Of course, thoughtful listening and processing lead to highlights that are much more than "nickel-and-dime" interventions. Clients who feel they are being understood participate more effectively and more fully in the helping process. Since sharing highlights helps build trust, it paves the way for stronger interventions on the part of the helper, such as challenging.

**Respond selectively to core client messages.** It is impossible to respond with highlights to everything a client says. Therefore, as you listen to clients, make every attempt to identify and respond to what you believe are core messages—that is, the heart of what the client is saying and expressing, especially if the client speaks at any length. Sometimes this selectivity means paying particular attention to one or two messages, even though the client communicates many. For instance, a young woman, in discussing her doubts about marrying her companion, says at one point during a session that she is tired of his sloppy habits, is not really interested in his friends, wonders about his lack of intellectual curiosity, is dismayed at his relatively low level of career aspirations, and vehemently resents the fact that he faults her for being highly ambitious.

COUNSELOR: The whole picture doesn't look very promising, but the mismatch in career expectations is especially troubling.

CLIENT: You know, I'm beginning to think that Jim and I would be pretty good friends, even *because* we're so different. But partners? Maybe that's pushing it.

In this example, the counselor's empathic highlights helps the client herself to identify what is core. The counselor follows her lead. His summary highlight at the end allows her to question the direction in which she and her friend are headed. Of course, since clients are not always so obliging, helpers must continually ask themselves as they listen, "What is key? What is most important here?" and then find ways of checking it out with the client. This helps clients sort out things that are not clear in their own minds.

Responding selectively sometimes means focusing on experiences or actions or feelings rather than all three. Consider the following example of a client who is experiencing stress because of his wife's poor health and concerns at work.

CLIENT: This week I tried to get my wife to see the doctor, but she refused, even though she fainted a couple of times. The kids had no school, so they were underfoot almost constantly. I haven't been able to finish a report my boss expects from me next Monday.

HELPER: It's been a lousy, overwhelming week all the way around.

CLIENT: As bad as they come. When things are lousy both at home and at work, there's no place for me to relax. I just want to get the hell out of the house and find some place forget it all . . . Almost run away . . . But I can't . . . I mean, I won't.

Here the counselor chooses to emphasize the feelings of the client, because she believes that his feelings of frustration and irritation are uppermost in his consciousness right now. This helps him move deeper into the problem situation—and then find a bit of resolve at the bottom of the pit.

At another time, or with another client, the emphasis might be quite different. In the next example, a young woman is talking about her problems with her father.

CLIENT: My dad yelled at me all the time last year about how I dress. But just last week I heard him telling someone how nice I looked. He yells at my sister about the same things he ignores when my younger brother does them. Sometimes he's really nice with my mother, and other times, too much of the time, he's just awful—demanding, grouchy, sarcastic.

HELPER: The inconsistency is killing you.

CLIENT: Absolutely! It's hard for all of us to know where we stand. I hate coming home when I'm not sure which dad will be there. Sometimes I come late to avoid all this. But that makes him even madder.

In this response, the counselor emphasizes the client's experience of her father's inconsistency. Her intervention hits the mark, and she explores the problem situation further.

**Respond to the context, not just the words.**  A good empathic response is not based just on the client's immediate words and nonverbal behavior, but also takes into account the context of what is said, everything that surrounds and permeates a client's statement. This client may be in crisis. That client may be doing a more leisurely taking stock of where he is in life. You are listening to clients in the context of their lives. The context modifies everything the client says.

Consider this case. Jeff, a white teenager, is accused of beating a black youth whose car stalled in a white neighborhood. The beaten youth is in a coma. When Jeff talks to a court-appointed counselor, the counselor listens to what Jeff says in light of Jeff's upbringing and environment. The context includes the racist attitudes of many people in his blue-collar neighborhood, the sporadic violence there, the fact that his father died when Jeff was in primary school, a somewhat indulgent mother with a history of alcoholism, and easy access to drugs. The following interchange takes place.

JEFF: I don't know why I did it. I just did it, me and these other guys. We'd been drinking a bit and smoking up a bit—but not too much. It was just the whole thing.

HELPER: Looking back, it's almost like it's something that happened to you rather than something you did, and yet you know, somewhat bitterly, that you actually did it.

JEFF: More than bitter! I've screwed up the rest of my life. It's not like I got up that morning saying that I was going to bash someone that day.

The counselor's response is in no way an attempt to excuse Jeff's behavior, but it does factor in some of the environmental realities. Later on he will challenge Jeff to decide whether he is to remain a victim of his environment in terms of the prejudices he has acquired, gang membership, family history, and the like, or whether he is going to take charge of his life.

**Use highlights as a mild social-influence process.**  Because helpers cannot respond with highlights to everything their clients say, they are always searching for core messages. They are forced into a selection process that influences the course of the therapeutic dialogue. So, even sharing highlights can be part of the social-influence dimension of counseling mentioned in Chapter 2. Helpers believe that the messages they select for attention are core primarily because they are core for the client. But helpers also believe, at some level, that certain messages *should* be important for the client. In the following example, an incest victim turned incest perpetrator is in jail awaiting trial. In a session with a counselor he is trying to exonerate himself by blaming what happened to him in the past. He has been talking so quickly that the helper finds it difficult to interrupt. Finally, the helper, who has a pretty good working relationship with the client, breaks in.

> HELPER: You've used some strong language to describe yourself. Let me see if I have some of them right. You said something about being "structurally deformed." I believe you also used the term "automatic reactions." You describe yourself as "haunted" and "driven."

> CLIENTS: Well . . . I guess it's strong language . . . Makes me sound like a psychological freak. Which I'm not.

The helper wants the client to listen to himself. So his let-me-get-this-straight response is a kind of empathic highlight form of challenge. It hits the mark because the client pulls himself up short. Of course, helpers need to be careful not to put words in a client's mouth.

**Use highlights to stimulate movement in the helping process.** Sharing highlights is an excellent tool for building the helping relationship. But the tool also needs to serve the goals of the entire helping process. Therefore, sharing highlights is useful to the degree that it helps the client move forward. What does "move forward" mean? That depends on the stage or task in focus. For instance, sharing highlights helps clients move forward in Stage I if it helps them explore a problem situation or an undeveloped opportunity more realistically. It helps clients move forward in Stage II to the degree that it helps them identify and explore possibilities for a better future, craft a change agenda, or discuss commitment to that agenda. Moving forward in Stage III means clarifying action strategies, choosing specific things to do, and setting up a plan. In the action phase, moving forward means identifying obstacles to action, overcoming them, and accomplishing goals.

In the following example, a young woman visits the student services center at her college to discuss an unwanted pregnancy.

> CLIENT: And so here I am, two months pregnant. I don't want to be pregnant. I'm not married, and I don't even love the father. To tell the truth, I don't even think I like him. Oh, Lord, this is something that happens to other people, not me! I wake up thinking this whole thing is unreal. Now people are trying to push me toward abortion.

> HELPER: You're still so amazed that it's almost impossible to accept that it's true. To make things worse, people are telling you what to do.

> CLIENT: Amazed? I'm stupefied! Mainly, at my own stupidity for getting myself into this. I've never had such an expensive lesson in my life. But I've decided one thing. No one, no one is going to tell me what to do now. I'll make my own decisions.

After the helper's highlight, self-recrimination over lack her lack of self-responsibility helps the client make a stand. She says she wants to capitalize on a very expensive mistake. It often happens that sharing highlights that hit the mark puts pressure on clients to move forward. So sharing highlights, even though it is a communication of understanding, is also part of the social-influence process.

**Recover from inaccurate understanding.** Although helpers should strive to be accurate in the understanding they communicate, all helpers can be inaccurate at times. You may think you understand the client and what he or she has said only to find out, when you share your understanding, that you were off the mark. Therefore, sharing highlights is a perception-checking tool. If the helper's response is accurate, the client often tends to confirm its accuracy in two ways. The first is some kind of verbal or nonverbal indication that the helper is right. That is, the client nods or gives some other nonverbal cue or uses some assenting word or phrase such as "that's right" or "exactly." This happens in the following example in which a client, who has been arrested for selling drugs, is talking to his probation officer.

HELPER: So your neighborhood makes it easy to do things that can get you into trouble.

CLIENT: You bet it does! For instance, everyone's selling drugs. You not only end up using them, but you begin to think about pushing them. It's just too easy.

The second and more substantive way in which clients acknowledge the accuracy of the helper's response is by moving forward in the helping process—for instance, by more fully clarifying the problem situation or possibilities for a better future. In the preceding example, the client not only verbally acknowledges the accuracy of the helper's empathy—"you bet it does"—but, more important, also outlines the problem situation in greater detail. If the helper again responds by sharing a highlight, this leads to the next cycle. The next cycle may mean further clarification of the problem or opportunity or it may mean moving on to goal-setting or some kind of problem-managing action.

On the other hand, when a response is inaccurate, the client often lets the counselor know in different ways. He or she may stop dead, fumble around, go off on a different tangent, tell the counselor "That's not exactly what I meant," or even try to get the helper back on track. Helpers need to be sensitive to all these cues. In the following example, Ben, a man who lost his wife and daughter in a train crash, has been talking about the changes that have taken place since the accident.

HELPER: So you don't want to do a lot of the things you used to do before the accident. For instance, you don't want to socialize much anymore.

BEN (pausing a long time): Well, I'm not sure that it's a question of wanting to or not. I mean that it takes much more energy to do a lot of things. It takes so much energy for me just to phone others to get together. It takes so much energy sometimes being with others that I just don't try.

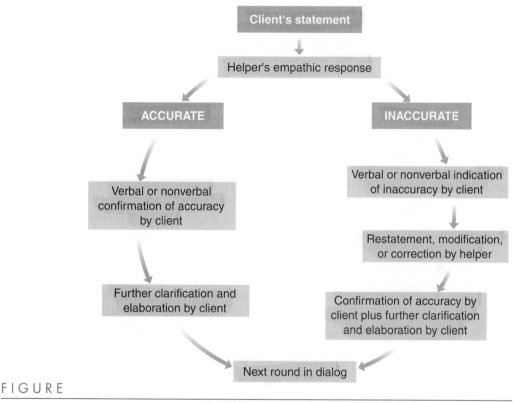

FIGURE

**4.1**    The Movement Caused by Accurate and Inaccurate Highlights

HELPER: It's like a movie of a man in slow motion—it's so hard to do almost anything.

BEN: I'm in low gear, grinding away. And I don't know how to get out of it.

Ben says that it is not a question of motivation, but of energy. The difference is important to him. By picking up on it, the helper gets the interview back on track. Ben wants to regain his old energy but he doesn't know how. His "lack of energy" is most likely some form of depression. And there are a number of ways to help clients deal with depression. This provides an opening for moving the helping process forward.

If you are intent on understanding your clients, they will not be put off by occasional inaccuracies on your part. Figure 4-1 indicates two different paths— one where helpers hit the mark in sharing highlights the first time, the other where they are inaccurate and then recover.

**Use empathic highlights as a way of bridging diversity gaps.** This princi- ple is a corollary of the preceding two. Highlights based on effective tuning in and listening constitute one of the most important tools you have in interacting with clients who differ from you in significant ways. Sharing highlights is one way of telling clients that you are a learner, Scott and Borodovsky (1990) referred to empathic listening as "cultural role taking." They could have said "diversity role taking." In the following example, a younger white male counselor is talking with an elderly African American woman who has recently lost her husband. She is in the hospital with a broken leg.

> CLIENT: I hear they try to get you out of these places as quick as possible. But I seem to be lying around here doing nothing. Jimmy [her late husband] wouldn't even recognize me.
>
> HELPER: It's pretty depressing to have this happen so close to losing your husband.
>
> CLIENT: Oh, I'm not depressed. I just want to get out of here and get back to doing things at home. Jimmy's gone, but there's plenty of people around there to help me take care of myself.
>
> HELPER: Getting back into the swing of things is the best medicine for you.
>
> CLIENT: Now you got it right. What I need right now is to know when I can go home and what I need to do for my leg once I get there. I've got to get things in order. That's what I do best.

The helper makes assumptions that might be true for him and some people in his culture, but they miss the mark with the client. She's taking her problems in stride and counting on her social system and a return to everyday household life to keep her going. The helper's second response hits the mark and she, in Stage-II fashion, outlines some of things she wants.

## Tactics for Communicating Highlights

The principles outlined in the previous section provide strategies for sharing empathic highlights. Here are a few tactics to help you improve the quality of your responses.

**Give yourself time to think.** Beginners sometimes jump in too quickly with an empathic response when the client pauses. They do not give themselves enough time to reflect on what the client has just said, which is what they need to identify the core message being communicated. Watch some experts on tape. They often pause and allow themselves to assimilate what the client is saying.

**Use short responses.** I find that the helping process goes best when I engage the client in a dialogue rather than give speeches or allow the client to ramble. In a dialogue, the helper's responses can be relatively frequent, but lean and trim. In trying to be accurate, the novice helper is often long-winded, especially if he or she waits too long to respond. Again, the question "What is the core of what this person is saying to me?" can help you make your responses short, concrete, and accurate.

**Gear your response to the client, but remain yourself.** If a client speaks animatedly, telling you how he finally got his partner to listen to his point of view about a new venture, and you reply accurately but in a flat, dull voice, your response is not fully empathic. This does not mean that you should mimic your clients, go overboard, or not be yourself. It means that part of being with the client is sharing, in a reasonable way, in his or her emotional tone. Consider this example:

> TWELVE-YEAR-OLD CLIENT: My teacher started picking on me from the first day of class. I don't fool around more than anyone else in class, but she gets me anytime I do. I think she's picking on me because she doesn't like me. She doesn't yell at Bill Smith, and he acts funnier than I do.

> COUNSELOR A: This is a bit perplexing. You wonder why she singles you out for so much discipline.

Counselor A's language is stilted, not in tune with the way a 12-year-old speaks. Here's a different approach.

> COUNSELOR B: You're mad because the way she picks on you seems unfair.

On the other hand, helpers should not adopt a language that is not their own, just to be on the client's wavelength. An older counselor using "hip" language or slang with a young client sounds ludicrous.

## The Shadow Side of Sharing Empathic Highlights

Some helpers are poor communicators without even realizing it. Many responses that novice or inept helpers make are really poor substitutes for sharing accurate empathic highlights. The example that follows illustrates a range of such responses.

Robin is a young woman who has just started a career in law. This is her second visit to a counselor in private practice. In the first session, she said she wanted to "talk through" some issues relating to the "transition" from school to business life. She appeared quite self-confident. In this session, after talking about a number of transition issues, she begins speaking in a rather strained voice and avoids eye contact with the counselor.

> ROBIN: Something else is bothering me a bit . . . . Maybe it shouldn't. After all, I've got the kind of career that a lot of women would die for. Well—I'm glad that none of my feminist colleagues is around—I don't like the way I look. I'm neither fat nor thin, but I don't really like the shape of my body. And I'm uncomfortable with some of my facial features. Maybe this is a strange time of life to start thinking about this. In two years I'll be thirty . . . I bet I seem like an affluent, self-centered yuppie . . .

Robin pauses and looks at a piece of art on the wall. What would you do or say? Here are some possibilities that are better avoided.

**No response.** It can be a mistake to say nothing, though cultures differ widely in how they deal with silence (Sue, 1990). In North American culture, generally speaking, if the client says something significant, respond to it, however briefly. Otherwise, the client may think that what he or she has just said doesn't merit a response. Don't leave Robin sitting there stewing in her own juices. A skilled helper would realize that a person's nonacceptance of his or her body could generalize to other aspects of life and therefore should not be treated as just a "vanity" problem.

**Distracting questions.** Some helpers, like many people in everyday life, cannot stop themselves from asking questions. Instead of responding with an empathic highlight, a counselor might ask something like, "Is this something new now that you've started working?" This response ignores what Robin has said and the feelings she has expressed and focuses rather on the helper's mistaken agenda to get more information.

**Cliches.** A counselor might say, "The workplace is competitive. It's not uncommon for issues like this to come up." This is cliche-talk. It turns the helper into an insensitive instructor and probably sounds dismissive to the client. Cliches are hollow. The helper is saying, in effect, "You don't really have a problem at all, at least not a serious one." Cliches are not a substitute for understanding.

**Interpretations.** For some helpers, interpretive responses based on their theories of helping seem more important than expressing understanding. Such a counselor might say something like this: "Robin, my bet is that your body-image concerns are probably just a symptom. I've got a hunch that you're not really accepting yourself. That's the real problem." The counselor fails to respond to the client's feelings and distorts the content of the client's communication. The response implies that the client is totally unaware of what the real issues are.

**Advice.** In everyday life, giving unsolicited advice is extremely common. It happens in counseling, too. For instance, a counselor might say to Robin, "Hey, don't let this worry you. You'll be so involved with work issues that these concerns will disappear." Advice-giving at this stage is out of order and, to make things worse, the advice given has a cliche flavor to it. Furthermore, advice giving robs clients of self-responsibility.

**Parroting.** Sharing a highlight does not mean merely repeating what the client has said. Such parroting is a parody of sharing empathic highlights. Review what Robin said about herself at the beginning of this section. Then evaluate the following empathic response.

COUNSELOR: So, Robin, even though you have a great job, one that many people would envy, it's your feelings about your body that bother you. The feminist in you recoils a bit from this news. But there are things you don't like—your body shape, some facial features. You're wondering why this is hitting you now. You also seem to be ashamed of these thoughts. "Maybe I'm just self-centered," is what you're saying to yourself.

Most of this is accurate, but it sounds awful. Mere repetition or restatement or paraphrasing carries no sense of real understanding of, no sense of being with, the client. Real understanding, since it passes through you, should convey some part of you. Parroting doesn't. To avoid parroting, tap into the processing you were doing as you listened, come at what the client has said from a slightly different angle, use your own words, change the order, refer to an expressed but unnamed emotion—in a word, do whatever you can to let the client know that you are working at understanding.

**Sympathy and agreement.** Being empathic is not the same as agreeing with the client or being sympathetic. An expression of sympathy has much more in common with pity, compassion, commiseration, and condolence than with empathic understanding. Although these are fully human traits, they are not particularly useful in counseling. Sympathy denotes agreement, whereas empathy denotes understanding and acceptance of the person, of the client. At its worst, sympathy is a form of collusion with the client. Note the difference between Counselor A's response to Robin and Counselor B's response.

> COUNSELOR A: This is not an easy thing to struggle with. It's even harder to talk about. It's even worse for someone who is as self-confident as you usually are.

> ROBIN: I guess so.

Note that Robin does not respond very enthusiastically to collusion-talk. She is interested in managing her problem. The helping process does not move forward. Let's see a different approach.

> COUNSELOR B: You've got some misgivings about how you look, yet you wonder whether you're even justified talking about it.

> ROBIN: I know. It's like I'm ashamed of my being ashamed. What's worse, I get so preoccupied with my body that I stop thinking of myself as a person. It blinds me to the fact that I more or less like the person I am.

Counselor B's response gives Robin the opportunity to deal with her immediate anxiety and then to explore her problem situation more fully.

**Faking it.** Clients are sometimes confused, distracted, and in a highly emotional state. All these conditions affect the clarity of what they are saying about themselves. Helpers may fail to pick up what the client is saying because of the client's confusion or because clients are not stating their messages clearly. Or, the helpers themselves have become distracted in one way or another. In any case, it's a mistake to feign understanding. Genuine helpers admit that they are lost and then work to get back on track again. A statement like "I think I've lost you. Could we go over that once more?" indicates that you think it important to stay with the client. It is a sign of respect. Admitting that you're lost is infinitely preferable to such cliches as "uh-huh," "um," and "I understand." On the other hand, if you often catch yourself saying that you don't understand, then you'd better find out what's going on. In any case, faking it is never a substitute for competence.

## The Importance of Empathic Relationships

In day-to-day conversations, sharing empathic highlights is a tool of civility. Making an effort to get in touch with your conversational partner's frame of reference sends a message of respect. Therefore, sharing highlights plays an important part in building relationships. However, the communication skills, as practiced in helping settings, don't automatically transfer to the ordinary social settings of everyday life. In everyday life, understanding does not necessarily have to be put into words. Given enough time, people establish empathic relationships with one another in which understanding is communicated in a variety of rich and subtle ways without necessarily being put into words. A simple glance across a room as one spouse sees the other trapped in conversation with someone he or she does not want to be with can communicate worlds of understanding. The glance says, "I know you feel caught. I know you don't want to hurt the other person's feelings. I can feel the struggles going on inside you. And I also know that you'd like me to rescue you, as soon as I can do so tactfully."

People with empathic relationships often express empathy in actions. An arm around the shoulders of someone who has just suffered a defeat expresses both empathy and support. I was in the home of a poor family when the father came bursting through the front door shouting, "I got the job!" His wife, without saying a word, went to the refrigerator, got a bottle of beer with a makeshift label on which "Champagne" had been written, and offered it to her husband. Beer never tasted so good.

On the other hand, some people enter caringly into the world of their relatives, friends, and colleagues and are certainly "with" them but don't know how to communicate understanding through words. When a wife complains, "I don't know whether he really understands," she is not necessarily saying that her relationship with her husband is not mutually empathic. She is more likely saying that she would appreciate it if he were to put his understanding into words more often. In general, more frequent use of empathic highlights in everyday life, especially when relationships are not going as well as they might, is highly desirable. Sharing highlights plays an important role in developing empathic relationships.

 ## Evaluation Questions for Sharing Empathic Highlights

### How effectively do you do the following?

- Remember that empathy is a value, a way of being, that should permeate all communication skills.
- Tune in carefully, both physically and psychologically, and listen actively to the client's point of view.
- Make every effort to set your judgments and biases aside for the moment and walk in the shoes of the client.

- As the client speaks, listen carefully for core messages.
- Listen to both verbal and nonverbal messages, and their context.
- Respond with highlights fairly frequently, but briefly, to the client's core messages.
- Be flexible and tentative enough that the client does not feel pinned down.
- Use highlights to keep the client focused on important issues.
- Move gradually toward the exploration of sensitive topics and feelings.
- After sharing a highlight, attend carefully to cues that either confirm or deny the accuracy of your response.
- Determine whether your highlights are helping the client remain focused, and stimulate the clarification of key issues.
- Note signs of client stress or resistance; try to judge whether these arise because you are inaccurate, or because you are too accurate, in your responses.
- Keep in mind that the communication skill of sharing empathic highlights, however important, is just one tool to help clients see themselves and their problem situations more clearly, with a view to managing them more effectively.
- Take special care when the client's personal culture differs considerably from your own.

## EXERCISES ON PERSONALIZING

1. To what degree is sharing empathic highlights and checking understanding part of your interpersonal communication style?

2. How do you react—either internally or externally—when people ignore what you say, or fail to understand what you are trying to communicate? Give some examples.

3. What's it like, talking with people who have little ability to indicate that they have understood what you have said to them?

4. Describe someone you know who approaches the ideal outlined in this chapter. What's it like, talking with this person?

# CHAPTER 5

# The Art of Probing and Summarizing

Using Prompts and Probes Effectively

  Verbal and Nonverbal Prompts

  Different Forms of Probes

  Using Questions Effectively

  Guidelines for Using Probes

  The Relationship Between Sharing Highlights and Using Probes

The Art of Summarizing: Providing Focus and Direction

The Shadow Side of Communication Skills

Exercises on Personalizing

In most of the examples used in the discussion of sharing empathic highlights, clients have demonstrated a willingness to explore themselves and their behavior relatively freely. Obviously, this is not always the case. Although it is essential that helpers respond with highlights when their clients do reveal themselves, it is also necessary at times to encourage, prompt, and help clients to explore their concerns when they fail to do so spontaneously. Therefore, the ability to use prompts and probes well is another important communication skill. If sharing highlights is the lubricant of dialogue, then probes provide often-needed nudges.

## Using Prompts and Probes Effectively

Prompts and probes are verbal and nonverbal tactics for helping clients talk more freely and concretely about any issue at any stage of the helping process. For instance, counselors can use probes to help clients identify and explore opportunities they have been overlooking, to clear up blind spots, to translate dreams into realistic goals, to come up with plans for accomplishing goals, and for working through obstacles to action. Probes, judiciously used, provide focus and direction for the entire helping process. Let's start with prompts.

### Verbal and Nonverbal Prompts

Prompts are brief verbal or nonverbal interventions designed to let clients know that you are with them or to encourage clients to talk further or even to reconsider what they have just said.

**Nonverbal prompts.** Nonverbal behaviors on the part of counselors can act as prompts. For example, a client who has been talking about how difficult it is to make a peace overture to a neighbor she is at odds with says, "I just can't do it!" The helper says nothing but, rather, simply leans forward attentively and waits. The client pauses and then says, "Well, you know what I mean. It would be very hard for me to take the first step. It would be like giving in. You know, weakness." They go on to explore how such an overture, properly done, could be a sign of strength rather than weakness. Such things as bodily movements, gestures, nods, eye movement, and the like can be used as nonverbal prompts.

**Vocal and verbal prompts.** You can use such responses as "um," "uh-huh," "sure," "yes," "I see," "ah," "okay," and "oh" as prompts, provided you use them intentionally and they are not simply a sign that the your attention is flagging, that you don't know what else to do, or are on automatic pilot. In the following example, the client, a 33-year-old married woman, is struggling with perfectionism both at work and at home.

> CLIENT (hesitatingly): I don't know whether I can kick the habit, you know, just let some trivial things go at work and at home. I know I've made a contract with myself. I'm not sure that I can keep it.

HELPER: (The helper utters a brief "oh" and then remains silent.)

CLIENT (pauses, then laughs): Here I am deep into perfectionism and I hear myself saying that I can't do something! How ironic. Of course I can. I mean it's not going to be easy, at least at first.

The helper's "oh" prompts the client to reconsider what she has just said. Prompts should never be the main course. They are part of the therapeutic dialogue only as a condiment.

## Different Forms of Probes

Probes, used judiciously, help clients name, take notice of, explore, clarify, or further define any issue at any stage of the helping process. They are designed to provide clarity and to move things forward. Probes take different forms.

**Statements.**  One form of probe is a statement indicating the need for further clarity. For instance, a helper is talking to a client who is having problems with his 25-year-old daughter, who is still living at home. The helper says, "It's still not clear to me whether you want to challenge her to leave the nest or not." The client replies, "Well, I want to but I don't know how to do it without alienating her. I don't want it to sound like I don't care about her and that I'm just trying to get rid of her." Probes in the form of statements often take the form of the helper's confessing that he or she is in the dark in some way. "I'm not sure I understand how you intend . . ." "I guess I'm still confused about . . ." This kind of request puts the responsibility on clients without accusing them of failing to cough up the truth.

**Requests.**  Probes can take the form of direct requests for further information or more clarity. A counselor, talking to a woman living with her husband and her mother-in-law, says, "Tell me what you mean when you say that three's a crowd at home." She answers, "I get along fine with my husband, I get along fine with my mother-in-law. But the chemistry among the *three* of us is very unsettling." This is helpful new information. Obviously, requests should not sound like commands: "Come on, just tell me what you thinking." Tone of voice and other paralinguistic and nonverbal cues help to soften requests.

**Questions.**  Direct questions are perhaps the most common type of probe. "How do you react when he flies off the handle?" "What keeps you from making a decision?" "Now that the indirect approach to letting him know what your needs are is not working, what might Plan B look like?" Consider this case: A client has come for help in controlling her anger. With the help of a counselor, she comes up with a solid program. In the next session, the client gives signs of backtracking. The counselor says, "You seemed enthusiastic about the program last week. But now, unless I'm mistaken, I hear a bit of hesitancy in your voice. What's going on?" The client responds, "Well, after taking a second look at the program, I'm afraid it will make me look like a wimp. My fellow workers could get the wrong idea and begin pushing me around." The counselor says,

"So there's something about yourself you don't want to lose. What might that be?" The client hesitates for a moment and then says, "Spunk!" The counselor replies, "Well, let's see how you can keep your spunk and still get rid of the outbursts that get you in trouble." They go on to discuss the difference between assertiveness and aggression.

**Single words or phrases used as questions or requests.** Sometimes single words or simple phrases are, in effect, probes. A client talking about a difficult relationship with her sister at one juncture says, "I really hate her." The helper responds simply and unemotionally, " 'Hate'." The client responds, "Well, I know that 'hate' is too strong a term. What I mean is that things are getting worse and worse." This kind of clarity helps. Another client, troubled with irrational fears, says. "I've had it. I just can't go on like this. No matter what, I'm going to move forward." The counselor replies, "Move forward to . . .?" The client says, "Well . . . to not indulging myself with my fears. That's what they are, a form of self-indulgence. From our talks I've learned that it's a bad habit. A very bad habit." They go on to discuss ways of controlling such thoughts.

Whatever form probes take, they are often, directly or indirectly, questions of some sort. Therefore, a few words about the use of questions is in order.

## Using Questions Effectively

Helpers, especially novices or inept counselors, tend to ask too many questions. When in doubt about what to say or do, they ask questions that often add no value, as if gathering information were the goal of the helping session. Questions, judiciously used, can be an important part of your interactions with clients. Here are two guidelines.

**Do not ask too many questions.** When clients are asked too many questions, they feel grilled, and that does little for the helping relationship. Furthermore, many clients instinctively know when questions are just filler, used because the helper does not have anything better to say. I have caught myself asking questions, the answers to which I didn't even want to know. Let's look at an example in which the helper is working with Rolin, an inmate in a state facility for young offenders. He is doing time for burglary and drug use. He is difficult to work with, especially because he blames his dysfunctional family for all his woes. Out of frustration, the counselor ends up asking a whole series of questions:

"When did you first feel caught in the messiness of your family?"

"What did you do to try to get away from their influence?"

"What could you do different?"

"What kind of friends did you have?"

Rolin is an expert in evading questions like this. He has been grilled by professionals. When he tires of being questioned, he just clams up. Such questions constitute a random search for information, the value of which is questionable.

Helping that turns into question-and-answer sessions tends to go nowhere.

**Ask open-ended questions.** As a general rule, ask open-ended questions—that is, questions that require more than a simple yes or no or similar one-word answer. Not, "Now that you've decided to take early retirement, do you have any plans?" but, "Now that you've decided to take early retirement, how do you see the future?" Counselors who ask closed questions find themselves asking more and more questions. One closed question begets another. Of course, if a specific piece of information is needed, then a closed question may be used. A career counselor might ask, "How many jobs have you had in the past two years?" The information is relevant to helping the client draw up a resume and a job-search strategy. Of course, occasionally, a sharp closed question can have the right impact. For instance, a client has been outlining what he was going to do to get back at his "ungrateful" son. The counselor asks, "Is getting back at him what you really want?" Rhetorical questions like this are a form of challenge. On the other hand, open-ended questions in moderation can help clients fill in what is missing at every stage of the helping process.

## Guidelines for Using Probes

Here, then, are some suggestions that can guide you in the use of all probes, whatever form they may take.

**Use probes to help clients engage as fully as possible in the therapeutic dialogue.** As noted earlier, many clients do not have all the communication skills needed to engage in the problem-managing and opportunity-developing dialogue. Probes are the principal tools needed to help all clients engage in the give-and-take of the helping dialogue. The following exchange takes place between a counselor at a church parish center and a parishioner who has been struggling to tell her story about her attempts to get her insurance company to respond to the claim she filed after a car accident.

> CLIENT:  They just won't do anything. I call and get the cold shoulder. They ignore me, and I don't like it!
>
> HELPER:  You're angry with the way you're being treated. And you want to get to the bottom of it . . . I'd like a brief overview of what you've done so far.
>
> CLIENT:  Well, they sent me forms that I didn't understand very well. I did the best I could. I think they were trying to show that it was my fault. I even kept copies. I've got them with me.
>
> HELPER:  You're not sure you can trust them . . . Let's see what the forms look like . . .

The forms turn out to be standard claims forms. The fact that this is the client's first encounter with an insurance company, and that she has poor communication skills, gives the counselor some insight into what the phone conversations between her and the insurance company must be like. By sharing highlights and using probes, he gets her to see that her experience might well be normal. The outcome is that the client gets help from a fellow parishioner who has completed insurance forms a number of times.

**Use probes to help clients achieve concreteness and clarity.** Probes can help clients turn what is abstract and vague into something concrete and clear—something you can get your hands on and work with. In the next example, a man is talking about an intimate relationship that has turned sour.

CLIENT: She treats me badly, and I don't like it!

HELPER: Tell me what she actually does.

CLIENT: She talks about me behind my back. I know she does. Others tell me what she says. She also cancels dates when something more interesting comes up.

HELPER: That's pretty demeaning . . . How have you been reacting to all this?

CLIENT: Well, I think she knows that I have an idea of what's going on. But we haven't talked about it much.

In this example, the helper's probe leads to a clearer statement of the client's experience and behavior. By sharing highlights and using probes, the helper discovers that the client puts up with a great deal because he is afraid of losing her. He goes on to help the client deal with the psychological "economics" of such a one-sided relationship.

In the next example, a man who is dissatisfied with living a somewhat impoverished social life is telling his story. A simple probe leads to a significant revelation.

CLIENT: I do funny things that make me feel good.

HELPER: What kinds of things?

CLIENT: Well, I daydream about being a hero, a kind of tragic hero. In my daydreams, I save the lives of people whom I like but who don't seem to know I exist. And then they come running to me but I turn my back on them. I choose to be alone! I come up with all sorts of variations of this theme.

HELPER: So in your daydreams you play a character who wants to be liked or loved, but who gets some kind of satisfaction from rejecting those who haven't loved him back . . . I'm not sure I've got that right.

CLIENT: Well . . . yeah . . . I sort of contradict myself. . . I do want to be loved but I guess I don't do very much to get a real social life. It's all in my head.

The helper's probe leads to a clearer statement of what's going on in the client's head. Helping the client to explore his fantasy life could be a first step toward finding out what he really wants from relationships and what he needs to do to get it.

The next client has become the breadwinner since her husband suffered a stroke. Someone takes care of her husband during the day.

> CLIENT: Since my husband had his stroke, coming home at night is rather difficult for me. I just . . . Well, I don't know.

> HELPER: It really gets you down . . . What's it like?

> CLIENT: When I see him sitting immobile in the chair, I'm filled with pity for him and the next thing I know it's pity for myself and it's mixed with anger or even rage, but I don't know what or whom to be angry at. I don't know how to focus my anger. Good God, he's only forty-two and I'm only forty!

In this case, the helper's probe leads to a fuller description of the intensity of the client's feelings and emotions, her sense of desperation.

In each of these cases, the client's story gets more specific. Of course, the goal is not to get more and more detail. Rather, it is to get the kind of detail that makes the problem or unused opportunity clear enough to see what can be done about it.

**Use probes to explore and clarify clients' points of view, intentions, proposals, and decisions.** Clients often fail to clarify their points of view, intentions, proposals, and decisions. For instance, a client might announce some decision she has made. But the decision itself is unclear and the reasons behind it and the implications for the client and others are not spelled out. In the following case, the client, driving under the influence, has had a bad automobile accident. Luckily, he was the only one hurt. He is recovering physically, but his psychological recovery has been slow. The accident opened a Pandora's box of psychological problems which were not being handled—for instance, a lack of self-responsibility. A counselor has been helping him work through some of these issues. During an early session the client says,

> CLIENT: I don't think that the laws around driving under the influence should be as tough as they are. I'm scared to death of what might happen to me if I ever had an accident again.

> COUNSELOR: So you feel you're in jeopardy . . . What makes you think that the laws are too tough?

> CLIENT: Well, they bully us. One little mistake and bingo! Your freedom goes out the window. Laws should make people free.

> COUNSELOR: Well, let's explore a little. Hmm . . . let's say all laws on driving under the influence were dropped. What then?

The counselor knows that the client is running away from taking responsibility for his actions. Using probes to get him to spell out the implications of his point of view on DUI laws and their implications is the beginning of an attempt to help the client face up to himself.

In a later session, the client talks about the legal ramifications of the accident; he has to go to court.

CLIENT: I've been thinking about this. I'm going to get me a really good lawyer and fight this thing. I talked with a friend and he thinks he knows someone who can get me off. I need a break. It might cost me a bundle. After all, I messed up someone's property a bit, but I didn't hurt anyone.

COUNSELOR: What's the best thing that could happen in court?

CLIENT: I'd get off scot-free. Well, maybe a slap on the wrist of some kind. A warning.

COUNSELOR: And what's the worst thing that could happen?

CLIENT: I haven't given that a lot of thought. I don't really know much about the laws or the courts or how tough they might be. That sort of stuff. But with the right lawyer . . .

COUNSELOR: Hmm. . . . I'm trying to put myself in your shoes . . . I think I'd try to find out what the most likely deal in court might be before I made up my mind about lawyers and things . . . What do you think?

The counselor is using probes to help the client explore the implications of a decision he's making.

The state has very tough DUI laws. In the end the client, because of the alcohol level in his blood, has his license suspended for six months, is fined heavily, and has to spend a month in jail. All of this is very sobering. The counselor visits him in jail and they talk about the future.

CLIENT: I feel like I've been hit by a train.

COUNSELOR: You had no idea that it would be this bad.

CLIENT: Right. No idea . . . I know you tried to warn me in your own way, but I wasn't ready to listen . . . Now I have to begin to put my life back together. Though I don't feel like it.

COUNSELOR: But now you've had the wake-up call, a horrible wake-up call . . . So what does the future hold . . . even if you don't feel up to looking at it?

CLIENT: I've been thinking. One thing I want to do is to make some sort of apology to my family. They're hurting as bad as I am. I feel so awkward. I know how to act in cocky mode. Humble mode I'm not used to. Do I write a long letter? Do I wait and just apologize through my actions? Do I take each one of them aside? I don't know, but I've just got to do it.

COUNSELOR: Somehow you have to make things right with them. Just how, well that's another matter. Maybe we could start by finding out what you want to accomplish through an apology, however it's done.

Here we find a much more sober and cooperative client. He proposes, roughly, a course of action. The counselor supports his need to move beyond past stupidities and present misery. It's about the future, not the past. The counselor's last statement is a probe aimed at giving substance and order to the client's proposal: What do you want to accomplish?

**Use probes to help clients fill in missing pieces of the picture.** Probes further the therapeutic dialogue by helping clients identify missing pieces of the puzzle—thoughts, experiences, behaviors, and feelings that would help both clients and helpers get a better fix on the problem situation, possibilities for a better future, or drawing up a plan of action. The client in the following example is at odds with his wife over his mother-in-law's upcoming visit.

> **HELPER:** I realize now that you often get angry when your mother-in-law stays for more than a day. But I'm still not sure what she does that makes you angry.

> **CLIENT:** First of all, she throws our household schedule out and puts in her own. Then she provides a steady stream of advice on how to raise the kids. My wife sees this as an "inconvenience." For me it's a total family disruption. When she leaves, there's a lot of emotional cleaning up to be done.

Just what the client's mother-in-law does to get him going has been missing. Once the behavior has been spelled out in some detail, it is easier to help him come up with some remedies. Still missing, however, is what *he* does in face of his mother-in-law's behavior. The helper continues:

> **HELPER:** So, when she takes over, everything gets turned upside down. . . . How do you react in the face of all this turmoil?

> **CLIENT:** Well . . . well . . . I guess I go silent. Or I just get out of there, go somewhere, and fume. After she's gone I take it out on my wife, who still doesn't see why I'm making such a fuss.

So now it's clear that the client does little to change things. It is also obvious that he is a little taken aback by being asked how he handles the situation.

In the next example, a divorced woman is talking about the turmoil that takes place when her ex-husband visits the children. This situation has some similarities with the case we've just seen.

> **HELPER:** The Sundays your ex-husband exercises his visiting rights with the children end in his taking verbal potshots at you, and you get these headaches. I've got a fairly clear picture of what he does when he comes over, and how it gets to you, but it might help if you could describe what you do.

> **CLIENT:** Well, I do nothing.

> **HELPER:** So last Sunday he just began letting you have it for no particular reason. Or just to make you feel bad.

> **CLIENT:** Well . . . not exactly. I asked him about increasing the amount of the child-support payments. And I asked him why he's dragging his feet about getting a better job. He's so stupid. He can't even take a bit of sound advice.

Through probes, the counselor helps the client fill in a missing part of the picture—her own behavior. She keeps describing herself as total victim and her ex-husband as total aggressor. That doesn't seem to be the full story. Her behavior contributes to the drama.

**Use probes to help clients get a balanced view of problem situations and opportunities.** Clients, in their eagerness to discuss an issue or make a point, often describe one side of a picture or one viewpoint. Probes can be used to help them fill out the picture. In the following example, the client, a manager who has been saddled with a bright, highly ambitious, and aggressive young woman who plays politics to further her own interests, has been agonizing over his plight.

COUNSELOR: I've been wondering whether you see any upside to this. Any hidden opportunities.

CLIENT: I'm not sure what you mean. It's just a disaster.

COUNSELOR: Well, you strike me as a pretty bright guy. I'm wondering if there are any lessons for you hidden in all this.

CLIENT (pausing): Oh, well, you know I tend to ignore politics around here, but now it's in my face. Where there are people there are politics, I suppose. I think she's being political to serve her own career. But I don't want to play her game. There must be some other kind of game or something that would let me keep my integrity. The days of avoiding all of this are probably over.

The problem situation has a flip side. It is an opportunity for rethinking and learning. As such, problems are incentives for constructive change. The client can learn something through all this. It's an opportunity to come to grips with the male-female dynamics of the workplace and a chance to explore positive political skills.

**Use probes to help clients move into more beneficial stages of the helping process.** Probes can be used to open up new areas for discussion. They can be used help clients engage in dialogue about any part of the helping process— telling their stories more fully, attacking blind spots, setting goals, formulating action strategies, discussing obstacles to action, and reviewing actions taken. Many clients do not easily move into whatever stage of the helping process might be most useful for them. Probes can help them do so. In the following example, the counselor uses a probe to help a middle-aged couple, Sean and Fiona, who have been complaining about each other, to move on to Stages II and III. Besides complaining, they have talked vaguely about "reinventing" their marriage. They have hinted at doing more things in common.

COUNSELOR: What kinds of things do you like doing together? What are some possibilities?

FIONA: I can think of something, though it might sound stupid to you (she says with a glance toward her husband). We both like doing things for others, you know, caring about other people. Before we were married we talked about spending some time in the Peace Corps together, though it never happened.

SEAN: I wish we had . . . But those days are past.

COUNSELOR: Are they? The Peace Corps may not be an option, but there must be other possibilities. (Neither Fiona nor Sean says anything.) I tell you what. Here are a couple of pieces of paper. Jot down three ways of helping others. Do your own list. Forget what your spouse might be thinking.

The counselor uses probes to get Sean and Fiona to brainstorm possibilities for some kind of service to others. This moves them away from tortuous problem exploration toward opportunity development.

The next client has been talking endlessly about the affair her husband is having. Her husband knows that she knows.

COUNSELOR: You've said you're not going to do anything about it because it might hurt your son. But doing nothing is not the only possible option. Let's just name some of them. Who knows? We might find a gem.

CLIENT: Hmm . . . I'm not sure I know.

COUNSELOR: Well, you know people in the same predicament. You've read novels, seen movies. What are some of the standard things people do? I'm not saying do them. Let's just review them.

CLIENT: Hmm . . . Well, I knew someone in a situation like this who did an outrageous thing. She knew her teenaged daughter was aware of what was going on. So one night at dinner she just said, "Let's all talk about the affair you're having and how to handle it. It's certainly not news to any of us."

COUNSELOR: All right, that's one possible way. Let's hear some more.

This primes the pump. The counselor uses a few more probes to put a number of possibilities on the table. This focus on action brings energy to the session.

In the following example, Jill, the helper, and Justin, the client, have been discussing how Justin is letting his impairment—he has lost a leg in a car accident—stand in the way of his picking up his life again. The session has bogged down a bit.

JILL: Let's try a bit of drama. I'm going to be Justin for a while. You're going to be Jill. As my counselor, ask me some questions that you think might make a difference for me. Me, that is, Justin.

JUSTIN (pausing a long time): I'm not much of an actor, but here goes . . . Why are you taking the coward's way out? Why are you on the verge of giving up? (His eyes tear up.)

Jill gets Justin to formulate the probes. It's her way of asking Justin to move forward and take responsibility for his part of the session. Justin's probes turn out to be challenges, almost accusations, certainly much stronger than anything Jill, at this stage, would have tried. However painful this is for Justin, it's a breakthrough.

**Use probes to challenge clients and help them challenge themselves.** In the last chapter, we saw that sharing empathic highlights can act as mild form of social-influence or challenge. We also saw that effective highlights often act as probes. That is, they can be indirect requests for further information, or ways of steering a client toward a different stage of the helping process. And, as you have probably noticed in the examples used in this chapter, probes can edge much closer to outright challenge. Many probes are not just requests for relevant information. They often place some kind of demand on the client to respond, reflect, review, or reevaluate. Probes can serve as a bridge between communicating understanding to clients and helping them challenge themselves. The following client, having committed himself to standing up to some of his mother's possessive ways, now shows signs of weakening in his resolve.

> HELPER: The other day you talked of "having it out with her"—though that might be too strong a term. But just now you mentioned something about "being reasonable with her." Tell me how these two differ.

> CLIENT (pausing): Well, I think you might be witnessing a case of cold feet . . . She's a very strong woman.

The counselor helps the client revisit his decision to "get tough" with his mother and, if this is what he really wants, what he can do to strengthen his resolve. Using probes as mild forms of challenge is perfectly legitimate, provided you know what you are doing.

## The Relationship Between Sharing Highlights and Using Probes

The trouble with dealing with skills one at a time is that each skill is taken out of context. In the give-and-take of any particular helping session, however, the skills must be intermingled in a natural way. In actual sessions, skilled helpers continually tune in, listen actively, and use a mix of probes and empathy to help clients clarify and come to grips with their concerns, deal with blind spots, set goals, make plans, and get things done. There is no formula for the right mix. That depends on the client, client needs, the problem situation, possible opportunities, and the stage of the helping process.

A word about the relationship between sharing highlights and using probes. Here is a basic guideline: After using a probe to which a client responds, share a highlight that expresses and checks your understanding. Be hesitant to follow one probe with another. The logic of this is straightforward. First, if a probe is effective, it will yield information that needs to be listened to and understood. Second, a shared highlight, if accurate, tends to place a demand on the client to explore further. It puts the ball back in the client's court.

In the following example, the client is a young Chinese American woman whose father died in China and whose mother is now dying in the United States. She has been talking about the traditional obedience of Chinese women and her fears of slipping into a form of passivity in her American life. She talks about her sister, who gives everything to her husband without looking for anything in return. The first counselor sticks to probes.

> COUNSELOR: To what degree is this self-effacing role rooted in your culture?

> CLIENT: Well, being somewhat self-effacing is certainly in my cultural genes. And yet I look around and see many of my North American counterparts adopt a very different style. A style that, frankly, appeals to me. But last year, when I took a trip back to

China with my mother to meet my half-sisters, the moment I landed I wasn't American. I was totally Chinese again.

**COUNSELOR A:**  What did you learn there?

**CLIENT:**  That I am Chinese!

The client says something significant about herself, but instead of responding with understanding, the helper uses another probe. This elicits only a repetition, with some annoyance, of what she had just said. Now, a different approach.

**COUNSELOR B:**  You learned just how deep your cultural roots go.

**CLIENT:**  And if these roots are so deep, what does that mean for me here? I love my Chinese culture. I want to be Chinese and American at the same time. How to do that, well, I haven't figured that out yet. I thought I had, but I haven't.

In this case, an empathic highlight works much more effectively than another probe. Counselor B helps the client move forward.

In the next example, a single middle-aged woman working in a company that has reinvented itself after a downturn in the economy still has a job, but the pay is much less and she is doing work she does not enjoy. She does not have the computer and Internet-related skills needed for the better jobs. She feels both stressed and depressed.

**CLIENT:**  Well, I suppose that I should be grateful for even having a job. But now I work longer hours for less pay. And I'm doing stuff I don't even like. My life is no longer mine.

**HELPER:**  So the extra pressure and stress makes you wonder just how "grateful" you should feel.

**CLIENT:**  Precisely . . . And the future looks pretty bleak.

**HELPER:**  What could you change in the short-term to make things more bearable?

**CLIENT:**  Hmm . . . Well, I know one way. We all keep complaining to one another at work. And this seems to make things even worse. I can get out of that loop. It's a simple way of making life a bit less miserable.

**HELPER:**  So one way is to stop contributing to your own misery by staying away from the complaining chorus . . . What else might you start doing?

**CLIENT:**  Well, there's no use sitting around hoping that what has happened is going to be reversed. I've been really jolted out of my complacency. I assumed, with the economy humming again, I'd find things easy. The economy may be humming, but jobs—good jobs—are still scarce. But I'm still young enough to acquire some more skills. And I do have some skills that I haven't needed to use before. I'm a good communicator and I've got a lot of common sense. I work well with people. There are probably some jobs around here that require those skills.

**HELPER:**  So, given the wake-up call, you think it might be possible to take unused skills and reposition yourself at work.

CLIENT: Repositioning. Hmm, I like that word. It makes a lot of pictures dance through my mind . . . Yes, I need to reposition myself. For instance . . .

This combination of highlights and probing gets things moving. Instead of focusing on the misery of the present situation, the client names a few possibilities for a better future, a Stage-II activity.

You should be careful not to become either an empathic highlight machine, grinding out one highlight after another, or an interrogator, peppering your clients continually with needless probes. All responses to clients, including probes and challenges, are empathic if they are based on a solid understanding of the client's core messages and point of view. All responses that build on, and add to, the client's remarks are implicitly empathic. Since these responses are empathic in effect, they cut down on the need for a steady stream of highlights.

## The Art of Summarizing: Providing Focus and Direction

The communication skills of visibly tuning in, listening, sharing highlights, and probing need to be orchestrated in such a way that they help clients focus their attention on issues that make a difference. The ability to summarize and to help clients summarize the main points of a helping interchange or session is a skill that can be used to provide both focus and challenge.

Brammer (1973) listed a number of goals that can be achieved by judicious use of summarizing—warming up the client, focusing scattered thoughts and feelings, bringing the discussion of a particular theme to a close, and prompting the client to explore a theme more thoroughly. There are certain times when summaries prove particularly useful: at the beginning of a new session, when the session seems to be going nowhere, and when the client needs a new perspective.

**At the beginning of a new session.**  When summaries are used at the beginning of a new session, especially when clients seem uncertain about how to begin, they prevent clients from merely repeating what has already been said before. They put clients under pressure to move on. Consider this example: Liz, a social worker, begins a session with a rather overly-talkative man with a summary of the main points of the previous session. This serves several purposes. First, it shows the client that she had listened carefully to what he had said in the last session and that she had reflected on it after the session. Second, the summary gives the client a jumping-off point for the new session. It gives him an opportunity to add to, or modify, what was said. Finally, it places the responsibility for moving forward on the client. The implied sentiment of the summary is: *Now where do you want to go with this?* Summaries put the ball in the clients' court and give them an opportunity to exercise initiative.

**During a session that is going nowhere.**  A summary can be used to give focus to a session that seems to be going nowhere. One of the main reasons sessions go nowhere is that helpers allow clients to keep discussing the same things over and over again instead of helping them either go more deeply into their stories, focus on possibilities and goals, or discuss strategies that will help them get what they need and want. For instance, Marcia is a coach, consultant, and counselor who is working with the staff of a shelter for the homeless. One of the staff members is showing signs of burnout. In a second meeting with Marcia, she keeps going over the same ground, talking endlessly about stressful incidents that have taken place over the last few months. At one point Marcia provides a summary.

MARCIA: Let's see if I can pull together what you've been saying. The work here, by its very nature, is stressful. You've mentioned a whole string of "incidents," such as being hit by someone you were trying to help or the heated arguments with some of your coworkers. But I believe you've intimated that these are the kinds of things that happen in these places. Shelters are prone to them. They are part of the furniture. They're not going to stop. But they can be very punishing. At times you wish you weren't here. But if you're going to stay, and if these kinds of incidents are not going to stop, maybe some questions might be in order, such as "How do I cope with these kinds of incidents? How do I do my work and get some ongoing satisfaction from it? What changes can we make around here that might lessen the number of these incidents?"

The purpose of the summary is to help the client move beyond "poor me" and find ways of coping with this kind of work. The challenge, in places like shelters, is creating a supportive work environment, developing a sense of organizational and personal purpose, promoting the kind of teamwork that fits the institution's mission, and fostering a culture of coping strategies.

**When the client needs a new perspective.** Often when scattered elements are brought together, the client sees the bigger picture more clearly. In the following example, a man who has been reluctant to go to a counselor with his wife has, in a solo session with the counselor, agreed to a couple of sessions "to please her." In the session, he talks a great deal of his behavior at home, but in a rather disjointed way.

COUNSELOR: I'd like to pull a few things together. You've encouraged your wife in her career, especially when things are difficult for her at work. You also encourage her to spend time with her friends as a way of enjoying herself and letting off steam. You also make sure that you spend time with the kids. In fact, time with them is important for you.

CLIENT: Yeah. That's right.

COUNSELOR: Also, if I have heard you correctly, you currently take care of the household finances. You are usually the one who accepts or rejects social invitations, because your schedule is tighter than hers. And now you're about to ask her to move because you can get a better job in Boston.

CLIENT: When you put it all together like that, it sounds as if I'm running her life . . . She never tells me I'm running her life.

COUNSELOR: Maybe we could talk a little about this when the three of us get together.

CLIENT: Hmm . . . Well, I'd . . . hmm . . . (laughs). I'd better think about all of this before the next session.

The summary provides the client with a mild jolt. He realizes that he needs to face up to the "I am making many decisions for her, and some of them are big" theme implied in the summary.

Pulling the strands of a client's story together and presenting it as a whole can help him or her get the big picture. In the following example, the client is a 52-year-old man who has been talking about a number of problems in living. He has come for help because he has been "down in the dumps" and can't seem to shake it.

HELPER: Let's take a look at what we've seen so far. You're down—not just a normal slump; this time it's hanging on. You worry about your health, but you check out all right physically, so this seems to be more a symptom than a cause of your slump. There are some unresolved issues in your life. One that you seem to be stressing a lot is the fact that your recent change in jobs has meant that you don't see much of your old friends anymore. Since you're single, this leaves you, currently, with a rather bleak social life. Another issue—one you find painful and embarrassing—is your struggle to stay young. You don't like facing the fact that you're getting older. A third issue is the way you—to use your own word—"overinvest" yourself in work, so much so that when you finish a long-term project, suddenly your life is empty.

CLIENT (pauses): It's painful to hear it all that baldly, but that about sums it up. I've suspected I've got some screwed-up values, but I haven't wanted to stop long enough to take a look at it. Maybe the time has come. I'm hurting enough.

HELPER: One way of doing this is by taking a look at what a better future would look like.

CLIENT: That sounds interesting, even hopeful. How would we do that?

The counselor's summary hits home—somewhat painfully—and the client draws his own conclusion. Care should be taken not to overwhelm clients with the contents of the summary. Nor should summaries be used to *build a case* against a client. Helping is not a judicial procedure. Perhaps the foregoing summary would have been more effective if the helper had also summarized some of the client's strengths. That would have provided a more positive context.

**Getting the client to provide the summary.** Summaries can be used when clients don't seem to know where to go next, either in the helping session itself or in an action program out in their real worlds. In cases like this, helpers can, of course, use probes to help them move on. Summaries, however, have a way of keeping the ball in the client's court. Moreover, rather than the helper always providing the summary, it is often better to ask the client to pull together the major points. This helps the client own the helping process, pull together the salient points, and move on. Since this is not meant to be a way of testing clients, the counselor should provide clients with whatever help they need to stitch the summary together. In the following example, the client, who has lost her job and her boyfriend because of her alcohol-induced outbreaks of anger, has been talking about not being able to "stick to the program." The counselor asks her to summarize what she's been doing and the obstacles she has been running into. With the help of the counselor, she stumbles through a summary. At the end of it she says, "I guess it's clear to both of us that I haven't been doing a very good job sticking to the program. On paper my plan looked like a snap. But it seems that I don't live on paper." The counselor and client take a couple of steps back, reviewing goals, plans, obstacles, and execution.

## The Shadow Side of Communication Skills

Up to this point, we have been dealing with basic communication skills. Chapters 6 and 7 consider advanced communication skills and processes that deal with challenging. But first a few words about the shadow side of communication skills. Challenge has its own substantial shadow side which will be discussed at the end of each of the next two chapters.

**Communication skills as necessary but not sufficient.**  Some helpers tend to overidentify the helping process with the communication skills, that is, with the tools that serve it. Being good at communication skills is not the same as being good at helping. Moreover, an overemphasis on communication skills can turn helping into a great deal of talk with very little action—and with few outcomes that made a difference in clients' lives.

Communication skills are essential, of course, but they still must serve both the process and the outcomes of helping. These skills certainly help establish a good relationship with clients. And a good relationship is the basis for the kind of social-emotional reeducation that has been outlined earlier. But you can be good at communication, good at relationship building, even good at social-emotional reeducation and still shortchange your clients, because they need more than that. Some who overestimate the value of communication skills tend to see a skill such as sharing highlights as some kind of "magic bullet." Others overestimate the value of information gathering. This is not a broad indictment of the profession. Rather, it is a caution for beginners.

**The helping relationship versus helping technologies.**  On the other hand, some practitioners underestimate the need for solid communication skills. There is a subtle assumption that the "technology" of their approach, such as treatment manuals, suffices. They listen and respond through their theories and constructs rather than through their humanity. They become technologists instead of helpers. They are like some medical doctors who become more and more proficient in the use of medical technology and less and less in touch with the humanity of their patients. Some years ago I spent ten days in a hospital (an eternity in these days of managed care). The staff were magnificent in addressing my medical needs. But the psychological needs that sprang from my anxiety about my illness were not addressed at all. Unfortunately, my anxieties were often expressed through physical symptoms. Then, those symptoms were treated medically. I asked, "When you have conferences during which patients are discussed, do you say, 'Well, we've thoroughly reviewed his medical status and needs. Now let's turn our attention to what he's going through. What can we do to help him through this experience?'" One resident said, "No, we don't have time." Don't get me wrong. These were dedicated, generous people who had my interests at heart. But they ignored many of my needs. We have a long way to go.

**Developing proficiency in communication skills.** Understanding communication skills and how they fit into the helping process is one thing; becoming proficient in their use is another. Some trainees think that these soft skills should be learned easily and fail to put in the kind of hard work and practice that makes them fluent in them. Doing the exercises in communication skills manuals and practicing these skills in training groups can help, but that isn't enough to make these skills second nature. Techniques such as tuning in, listening, processing, sharing highlights, and probing that are trotted out, as it were, for helping encounters are likely to have a hollow ring to them. These skills must become part of your everyday communication style.

After providing some initial training in communication skills, I tell students, "Now, go out into your real lives and get good at these skills. I can't do that for you." In the beginning, it may be difficult to practice all these skills in everyday life, not because they are so difficult, but because they are relatively rare in conversations. Take sharing highlights. Listen to the conversations around you. If you were to use an unobtrusive counter, pressing a button every time you heard someone share an empathic highlight, you might go days without pressing the button. But you can make sharing highlights a reality in your everyday life. And those who interact with you will notice the difference. They probably will not call it empathy or sharing highlights. Rather, they will says such things as "She really listens to me" or "He takes me seriously."

On the other hand, you will hear many probes in everyday conversations. People are much more comfortable asking questions than providing understanding. However, many of these probes tend to be aimless. Worse, many are disguised criticisms. "Why on earth did you do THAT?" Learning how to integrate purposeful probes with highlights demands practice in everyday life. Life is your lab. Every conversation is an opportunity.

These skills have a place in all the human transactions of life, including business transactions. When businesses are asked what competencies they want to see in those applying for jobs, especially managerial jobs, communication and relationship-building skills are inevitably at, or near, the top of the list. I once ran a training program on these skills for a CPA firm. Although the director of training believed in their value in the business world, many of the account executives did not. They resisted the whole process. I got a call one day from one of them. He had been one of the more notable resisters. He said, "I owe you this call." "Really?" I replied with an edge of doubt in my voice. "Really," he said. He went on to tell me how he had recently called on a potential client, a company that was dissatisfied with its current audit firm and was looking for a new one. During the session the executive [who had given me a hard time] said to himself, "Since we don't have the slightest chance of getting this account, why don't I amuse myself by trying these communication skills?" In his phone call to me, he went on to say, "This morning I got a call from that client. He gave us the account, but in doing so he said, 'You're not getting the account because you were the low bidder. You were not. You're getting the account because we thought that you were the only one that really understood our needs.' So, almost literally, I owe you this call." I forgot to ask him for a share of the fee.

This road to proficiency applies to the basic skills we have being reviewing in the past few chapters and to the advanced skills of challenging outlined and illustrated in the next two chapters.

## EVALUATION QUESTIONS FOR THE USE OF PROBES

**When using probes, how effectively do you do the following?**

- Keep in mind the goals of probing.
- Help clients engage as fully as possible in the therapeutic dialogue.
- Help nonassertive or reluctant clients tell their stories and engage in other behaviors related to managing their problems and developing their opportunities.
- Help clients identify experiences, behaviors, and feelings that give focus to their stories.
- Help clients open up new areas for discussion.
- Help clients explore and clarify points of view, decisions, and proposals.
- Help clients be as concrete and specific as possible.
- Help clients remain focused on relevant and important issues.
- Help clients move on to a further stage of the helping process.
- Use probes to provide mild challenges to clients to examine the way they think, behave, and act, both within helping sessions and in their daily lives.
- Make sure that probing is done in the spirit of empathy.
- Use a mix of statements, open-ended questions, prompts, and requests, not questions alone.
- Follow up a successful probe with an empathic highlight rather than another probe.
- Use whatever judicious mixture of highlights and probing is needed to help clients clarify problems, identify blind spots, develop new scenarios, search for action strategies, formulate plans, and review outcomes of action.

**EXERCISES ON PERSONALIZING**

1. In your dealings with friends, relatives, and colleagues, what kind of questions tend to put you off? Jot down some examples. Why do they annoy you? How frequently do you ask others questions? How often do you follow a question with another question? How often do you follow a question with some indication that you understand the response given to your question?

2. Go back to one of the personal issues you discussed in Chapter 1. What kinds of questions would help you explore the problem situation or unused opportunity? Jot down some examples. What kinds of questions would help you develop more clarity? What kinds of questions would help you move to problem-managing action? Give some examples.

3. Give a brief summary of where you stand with respect to some problem situation you are currently dealing with. Review your summary. What kind of focus does your summary give you? What are the action implications of the summary?

4. In a conversation with a relative, friend, or colleague, take the opportunity to provide the other person with a summary of the points that he or she has been making. Determine how your summary affects the conversation.

# CHAPTER 6

# Helping Clients Challenge Themselves

Challenge: The Basic Concept
  The Goals of Challenging
  The Targets of Challenging
Blind Spots
  Degrees of Awareness
  From Blind Spots to New Perspectives
Linking Challenge to Action
The Shadow Side of Challenging
Exercises on Personalizing

# Challenge: The Basic Concept

Effective helpers must go beyond understanding clients (tuning in, listening, processing, sharing empathic highlights) and helping them clarify their concerns (probing, summarizing). Good counselors also help their clients test reality (challenging). Since helping at its best is a constructive social-influence process, some form of challenge is central to helping. This does not signal a movement from the pro-client stance associated with the communication skills reviewed so far to some kind of tough or client-as-enemy stance in this and the following chapter. The reality is simpler. All helping is some kind of mixture of support and challenge. While challenge without support is harsh and unjustified, support without challenge can end up being empty and counterproductive. None of what follows is an indictment of clients. It is part of the human condition.

Martin (1994) put it well when he suggested that the helping dialogue may add the most value when it is perceived by clients as relevant, helpful, interested, supportive, and *"somehow inconsistent (discordant) with their current theories of themselves and their circumstances"* [italics added] (pp. 53–54). The same point is made by Trevino (1996, p. 203) in the context of cross-cultural counseling.

> Certain patterns of congruency and discrepancy . . . between client and counselor facilitate change. There is a significant body of research suggesting that congruency between counselor and client enhances the therapeutic relationship, whereas discrepancy between the two facilitates change. In a review of the literature on this topic, Claiborn (1982, p. 446) concluded that the presentation of discrepant points of view contributes to positive outcomes by changing "the way the client construes problems and considers solutions."

Challenge adds that discordant note. Notice that the term "challenge" rather then the harder-edged "confrontation" is used here. Most people see both confronting and being confronted as unpleasant experiences. But, at least in principle, they more readily buy the softer but still edgy option of challenge. It has a constructive ring to it. That said, there is also room, as we shall see later, for some form of confrontation.

## The Goals of Challenging

The overall goal of challenging is to help clients do some reality-testing and invest what they learn from this in their futures. If the focus is on the ways clients are *mired down,* the goal goes something like this:

> *Invite clients to challenge themselves to change ways of thinking, expressing emotions, and acting that keep them mired in problem situations and prevent them from identifying and developing opportunities.*

A parallel goal in the spirit of positive psychology is more upbeat. It deals with new perspectives and translating these new perspectives into new ways of acting. It goes something like this:

> *Become partners with your clients in helping them challenge themselves to find possibilities in their problems, to discover unused resources, both internal and*

*external, to invest these resources in the problems and opportunities of their lives, to spell out possibilities for a better future, to find ways of making that future a reality, and to commit themselves to the actions needed to make it all happen.*

This might sound too idealistic, but it contains the spirit of the counselor's role as *catalyst for a better future.*

## The Targets of Challenging

Counselors should help clients challenge whatever stands in the way of understanding and managing problem situations or identifying and developing life-enhancing opportunities. Some of the main targets of challenge are:

   problematic mind-sets

   self-limiting internal behavior

   ways of managing and expressing feelings and emotions

   dysfunctional external behavior

   clients' imperfect understanding of the world in which they live

   discrepancies

   unused strengths and resources

   the predictable dishonesties of everyday life

   inadequate participation in the helping process

Here are a few examples in each of these categories.

**Challenging problematic mind-sets.** Mind-sets here refer to more or less permanent states of mind. They include such things as assumptions, attitudes, beliefs, values, bias, convictions, inclinations, norms, outlook, points of view, unexamined perceptions of self/others/the world, preconceptions, and prejudices. Mind-sets, whether productive or problematic, tend to drive external behavior. Or at least leak out into external behavior. The principle: Invite clients to transform outmoded, self-limiting mind-sets and perspectives into self-liberating and self-enhancing new perspectives that are pregnant with action. Here are examples dealing with two kinds of mind-sets—prejudices and self-limiting assumptions.

*Prejudices.* Candace is having a great deal of trouble with a colleague at work who happens to be Jewish. As she grew up, she picked up the idea that "Jews are treacherous businessmen." The counselor invites her to rethink this prejudicial stereotype. Her colleague may or may not be treacherous, but it's not because he's Jewish. Separating the individual from the prejudicial stereotype helps Candace think more clearly. But it is now clear that Candace has two problems—her troubled relationship with her colleague at work, and her prejudice. She has discovered that her problem with her colleague is not his Jewishness, but she may also need to discover that often in troubled relationships, both parties contribute to the mess. She realizes that she has to go back to the drawing board and examine her behavior and the degree to which it has been affected by her prejudice.

***Self-limiting beliefs and assumptions.*** Albert Ellis (Ellis, 2004; Ellis & MacLaren, 2004) has developed a rational-emotional-behavioral approach (REBT) to helping. He claims that one of the most useful interventions helpers can make is to challenge clients' irrational beliefs. Since clients in some way *talk themselves into* these dysfunctional beliefs, they engage in a form of self-talk. Some of the common beliefs that Ellis contends get in the way of effective living are these:

- ***Being liked and loved.*** I must always be loved and approved by the significant people in my life.

- ***Being competent.*** I must always, in all situations, demonstrate competence, and I must be both talented and competent in some important area of life.

- ***Having one's own way.*** I must have my way, and my plans must always work out.

- ***Being hurt.*** People who do anything wrong, especially those who harm me, are evil and should be blamed and punished.

- ***Being danger-free.*** If anything or any situation is dangerous in any way, I must be anxious and upset about it. I should not have to face dangerous situations.

- ***Being problemless.*** Things should not go wrong in life and if, by chance, they do, there should be quick and easy solutions.

- ***Being a victim.*** Other people and outside forces are responsible for any misery I experience. No one should ever take advantage of me.

- ***Avoiding.*** It is easier to avoid facing life's difficulties than to develop self-discipline; making demands of myself should not be necessary.

- ***Tyranny of the past.*** What I did in the past, and especially what happened to me in the past, determines how I act and feel today.

- ***Passivity.*** I can be happy by avoiding, by being passive, by being uncommitted, and by just enjoying myself.

I am sure that you could add to the list. Ellis suggests that if any of these beliefs are violated in a person's life, he or she tends to see the experience as terrible, awful, even catastrophic. "People pick on me. I hate it. It shouldn't happen. Isn't it awful!" Such "catastrophizing," Ellis says, gets clients nowhere. It is unfortunate to be picked on, but it's not the end of the world. Moreover, clients can often do something about the issues over which they catastrophize.

Take Allison. She is talking with a counselor about how miserable she has been since she suffered a financial reversal in the stock market. She works from the assumption that she should be problemless: "These things shouldn't happen to me!" She also works from the assumption that markets should always go up and there should never be any "corrections." She catastrophizes about her losses in the market: "Look at the state this has left me in!" In reality, while she has suffered some losses, she is still in good financial shape. All of this is an opportunity for her to rethink her approach to wealth and happiness and to come to grips with the latter-day turbulence in financial markets. Learning needs to take the place of catastrophizing.

**Challenging self-limiting internal behavior.**  Some forms of thinking are actually behaviors. They are things we can choose to do or not do. Internally, we daydream, pray, ruminate on things, believe, identify problems, review opportunities, make decisions, formulate plans, make judgments, question motives, approve of self and others, disapprove of self and others, wonder, value, imagine, ponder, think through, create standards, fashion norms, mull over, ignore, forgive, rehearse—we *do* all sorts of things internally. These are behaviors, not just things that happen to us.

The Principle: Invite clients to replace self-limiting and self-defeating with more creative internal behavior. For instance, I can hold a grudge. But I can also forgive. Then I can go on to express my forgiveness externally. The ways in which internal behavior can be self-limiting are legion. We'll consider a few examples.

John daydreams a lot, seeing himself as some kind of hero whom others admire. Thinking about unrealistic success in his social life has taken the place of working for actual success. The new perspective: Daydreaming is not all that bad. It's how you use it. With the help of a counselor, John does not stop daydreaming, but he switches its focus. He daydreams about what a fuller social life might look like. This provides him with some practical strategies for expanding and enriching his interactions with others. He begins to try these out.

Nadia, when given a project, immediately thinks of why it won't work. Then she goes around telling everyone that the project won't work. This annoys her colleagues. With the help of a supervisor she sees how self-limiting her internal behavior is. Internally, her first instinct is to take things apart. She learns that the opposite approach can add value. So she gets into the habit of first trying to see what value the project or program will add to the company and what she might do to make it more realistic. Only after doing some of this internal work does she engage her colleagues in conversation about the project. She discusses projects in a much more balanced way. In the following example Roberto is dissatisfied with certain aspects of his marriage with Maria.

> Before Roberto and Maria got married, they talked a great deal about the cultural difficulties they might face. He still veered toward the Hispanic culture, whereas she had become quite "Anglo" in her attitudes and behavior. She was especially worried about norms relating to the role of the women. Roberto said he would enjoy being married to someone with a "pioneer" spirit. They thought that they had things worked out. That was then. Now, Maria has broken through a number of cultural taboos. She has put herself through college, gotten a job, developed it into a career, and assumed the role of both mother and co-breadwinner. She makes more money than Roberto. His woes include thinking that he is losing face in the community, feeling belittled by his wife's success, and being forced into an "overly democratic" marriage.

If Roberto is going to manage the conflict between himself and his wife better, he needs to challenge himself to review and make changes in some of the ways he thinks.

For some clients, developing new perspectives and changing their internal behavior can be enormously helpful. It certainly works for Bella, a woman whose husband died two years ago. She is suffering from depression, not incapacitating, but still miserable. At one point she says,

BELLA: "You know, I stopped wearing black a year ago. But . . ." (She pauses.)

THERAPIST: "But you're still wearing black inside?"

In a flash Bella has it. Not magic, but now she has the metaphor she needs. She knows that she could stop wearing black "inside." With just a couple more sessions with the therapist, Bella begins to find the peace she is looking for.

**Challenging ways of managing and expressing feelings and emotions.** Managing our emotions and the ways we express them is part of social-emotional intelligence. Some of our emotions are bottled up, some go on inside, and others are quite visible. Sheila becomes depressed when her boss fails to notice the good work she is doing. She feels taken for granted. But at work she puts on a good face. On the other hand, Ira rants and raves about his "stupid" boss to whoever will listen. Everyone, except his boss, knows where he stands. Self-defeating forms of emotional expression or needlessly bottled up emotions need to be challenged. Arthur flies off the handle whenever anyone suggests that a more orderly life would benefit both him and his friends, relatives, and colleagues at work. Arthur needs help in seeing the world as others see it. Cynthia is reluctant to express her delight and praise her son and daughter for their success at school because she doesn't want to "spoil" them. She needs to learn that praising and spoiling are not necessarily linked.

**Challenging dysfunctional external behavior.** External behavior is the stuff people could see if they were looking. For some clients, their external behavior constitutes trouble. When Achilles is with women at work, he engages in behavior which others, including the courts, see as sexual harassment. He thinks he is just being "friendly." Jake is not an alcoholic, but when he has a couple of drinks, he tends to get mean and argumentative. Some clients engage in behaviors that keep them mired in their problems. Clarence, a self-doubting and overly cautious person, is very deferential around his manager. He does not realize that his manager interprets his deference as a lack of ambition. Clarence's behavior keeps him mired in a job he hates. But the manager says nothing, since she would rather have an "obedient" rather than an "aggressive" employee.

Not doing something is also a form of behavior. Clients often fail to make choices and engage in behaviors that would help them cope with problems or develop opportunities. When Clarence is offered an opportunity to update his skills, he turns it down. He also refuses a promotion, saying to himself, "I don't want to get in over my head." His apparently unambitious external behavior is based on a self-defeating way of thinking about himself. His manager sees him bypass these opportunities, but says nothing to him. She says to herself, "Anyway, I've got a hard-working drone. That's something these days. I'll leave well enough alone."

Ryan is having trouble relating to his college classmates. He is aggressive, hogs conversations, tries to get his own way when events are being planned, and criticizes others freely. One of his friends, after a couple of drinks, gets very angry with Ryan and tells him off. "Self-centered," "arrogant," and "pushy," are the kinds of words she uses. Ryan goes into a funk and finally talks things through with the dorm prefect. Once he is helped to see how self-defeating his behavior is, Ryan works with one of the counselors in the student services center to do something about his interpersonal communication style. They discuss ways of being proactive and assertive rather than aggressive. His "edge" has too much of an edge about it. He takes a course in interpersonal communication and finds plenty of incentives to invest what he learns in his interactions with his classmates, in his part-time job, and at home.

One way of helping clients challenge both internal and external actions is to help them explore the consequences of their behaviors. Let's return to Roberto. He has made some mild attempts at sabotaging his wife's career. He refers to his actions as "delaying tactics."

HELPER:  It might be helpful to see where all of this is leading.

ROBERTO:  What do you mean?

HELPER:  I mean let's review the impact your "delaying tactics" have had on Maria and your marriage. And then let's review where these tactics are most likely ultimately to lead.

ROBERTO:  Well, I can tell you one thing. She's become even more stubborn.

Through their discussion Roberto discovers that his sabotage is working against, rather than for him. He is endangering the marriage.

**Challenging clients' imperfect understanding of the world in which they live.**  Clients' failure to see the world as it really is can keep them mired in problem situations and prevent them from identifying and developing opportunities. For instance, parents fail to notice signs indicating that their teenage son has started to use drugs. If, at times, we are blind to others and their needs—often those closest to us—we are also blind to their attitudes towards us and the impact of their behavior toward us. Take Sandra. She interprets her husband's being less insistent when it comes to having sex as a sign that "he was finally coming to his senses." So she is shocked when she learns, by accident, that he is having an affair. All of us have our areas of ignorance and naivete. Clients' failure to fully understand the environment in which they live does not mean that they are stupid. Rather clients, like all of us, have blind spots.

**Challenging discrepancies.**  Various kinds of discrepancies plague our lives. For instance, we don't always do what we say we're going to do. Just review last year's New Year's Resolutions. Discrepancies keep clients mired down in their problem situations. There are discrepancies between:

- what clients think or feel versus what they say
- what they say versus what they do

- their views of themselves versus the views that others have of them
- what they are versus what they claim they want to be
- their stated goals versus what they actually accomplish
- their expressed values versus their actual behavior

And so on. For instance, a helper might assist the following clients to challenge the discrepancies in their lives.

- Tom sees himself as witty; his friends see him as biting.
- Minerva says that physical fitness is important, but she overeats and underexercises.
- George says he loves his wife and family, but he is seeing another woman and stays away from home a great deal.
- Clarissa, unemployed for several months, wants a job, but she doesn't want to participate in a retraining program.

Let's use the example of Clarissa to illustrate how the discrepancy between talking and acting can be challenged. Clarissa has just told the counselor that she has decided against joining the retraining program.

COUNSELOR: I thought that the retraining program would be just the kind of thing you've been looking for.

CLARISSA: Well . . . I don't know if it's the kind of thing I'd like to do . . . The work would be so different from my last job . . . And it's a long program.

COUNSELOR: So you feel the fit isn't good.

CLARISSA: Yes, that's right.

COUNSELOR: Clarissa, you seemed so enthusiastic when you first talked about the program . . . (gently) What's going on?

CLARISSA (pauses): You know, I've gotten a bit lazy . . . I don't like being out of work, but I've gotten used to it.

The counselor sees a discrepancy between what Clarissa is saying and what she is doing. She is actually letting herself slip into into what could be called a *culture of unemployment.* Now that the discrepancy is out in the open, they can work together on how she wants to shape her future.

**Challenging unused strengths and resources.** Challenge should not focus just on clients' problems, but on the "possible self" that every client is. Part of positive psychology is helping clients get in touch with unexploited opportunities and unused or underused resources. Some of these resources are client-based—for instance, talents and abilities not being used—and some are external—for example, failure to identify and use social support in managing problems or developing opportunities. Effective helpers are instinctively asking themselves this question as they partner with their clients—how can I help this client unleash and marshal both internal and external resources? We all have resources we fail to use; they need to be mined.

Strengths are buried even in dysfunctional behavior. For instance, Driscoll (1984) has pointed out that helpers can show clients that even their irrationalities can be a source of strength. Instead of forcing clients to see how stupidly they are thinking and acting, helpers can challenge them to find the logic embedded in even seemingly dysfunctional ideas and behaviors. Then clients can use that logic as a resource to manage problem situations instead of perpetuating them. A psychiatrist friend of mine helped a client see the beauty, as it were, of a very carefully constructed self-defense system. The client, through a series of mental gymnastics and external behaviors, was cocooning himself from real life. My friend helped the client see how inventive he had been and how powerful the system that he had created was. He went on to help him redirect that power into more life-enhancing channels.

**Challenging the predictable dishonesties of everyday life.** The "predictable dishonesties of life" refers to the distortions, evasions, games, tricks, excuse making, and smoke screens that keep clients mired in their problem situations. All of us have ways of defending ourselves from ourselves, from others, and from the world. We all have our little dishonesties. But they are two-edged swords. While lies, whether white or not, may help me cope with difficulties, especially unexpected difficulties, in my interactions with others, they come with a price tag, especially if they become a preferred coping strategy. Blaming others for my misfortunes helps me save face, but it disrupts interpersonal relationships and prevents me from developing a healthy sense of self-responsibility. The purpose of helping clients challenge themselves with respect to the dishonesties of everyday life, whether they take place in the helping sessions or are more widespread patterns of behavior, is not to strip clients of their defenses, which in some cases could be dangerous, but to help them cope with their inner and outer worlds more creatively.

*Distortions.* Some clients would rather not see the world as it is—it is too painful or demanding—and therefore distort it in various ways. The distortions are self-serving. For instance:

- At work, Arnie is afraid of his supervisor and therefore sees her as aloof, whereas in reality she is a caring person. He is working out of past fears rather than current realities.

- Edna sees her counselor in some kind of divine role and therefore makes unwarranted demands on him.

- Nancy sees her getting her own way with her friends as an indication of whether they really like her or not.

Let's take a little closer look at Nancy, who is married to Milan. There are some bumps in their marriage.

Nancy and Milan come from different cultures. They fought a great deal in the early years of their marriage, but then things settled down. Now, squabbles have broken out about the best way to bring up their children. Milan is not convinced that counseling is a good idea, so the counselor is talking to Nancy alone. She has forbidden her 12-year-old son to bicycle to school because she doesn't want "his picture to end up on a milk carton." Milan thought that she was being extremely overprotective. One day he stalks out of the house, yelling back at her, "Why don't you just keep him locked in his room!"

Nancy and her counselor have a session not long after the above incident. Nancy is defending her approach to her son.

**NANCY:** Milan's just too permissive. Now that Jan is entering his teenage years, he needs more guidance, not less. Let's face it, the world we live in is dangerous.

**COUNSELOR:** So from your point of view, this is not the time for letting your guard down . . . Of course, I'm also making the assumption that Milan is not indifferent to Jan's welfare.

**NANCY:** Of course not! Good grief, he cares as much as I do. We just disagree on how to do it. "Safe, not sorry" is my philosophy.

Hopefully, this gets rid of an implied distortion: "I'm interested in my son's welfare, but his father isn't." They continue their dialogue.

**COUNSELOR:** Let's widen the discussion a bit. What other issues do you and Milan disagree on?

**NANCY:** Well, we used to disagree a lot. But we've put that behind us, it would seem. He leaves a lot of the home decisions to me.

**COUNSELOR:** I'm not sure whether you both decided that you should make the decisions at home or if it just happened that way.

**NANCY** (slowly): I suppose it just happened that way . . . I don't really know.

**COUNSELOR:** I'm curious because he seems to be annoyed that you're the one making the decisions about how to bring up your son . . .

**NANCY** (pausing): Like he wants to reassert himself. Take over again.

Another distortion. Perhaps Nancy feels the counselor is getting too close to a sensitive issue that she thought was resolved long ago.

**COUNSELOR:** You got a bit annoyed when I asked whether Milan was as committed to the kids as you . . . Since he cares as much as you do about his son, I'm wondering what the disagreement is really about.

**NANCY:** Like he's drawing a line in the sand, taking a stand on this one? Or what?

**COUNSELOR** (caringly): I don't want to guess what's going through Milan's mind . . . Maybe we could we try once more to get him to come with you.

The counselor has a hunch that the problem is as much about power and getting one's own way as it is about bringing up children. Nancy does seem to have trouble with her own "little dishonesties."

***Games, tricks, and smoke screens.*** If clients are comfortable with their delusions and profit by them, they will obviously try to keep them. If they are rewarded for playing games, inside the counseling sessions or outside, they will continue a game approach to life (see Berne, 1964). Consider some examples:

- Kennard plays the "Yes, but . . ." game. He gets his therapist to recommend some things he might do to control his anger. He then points out why each recommendation will not work. When the therapist calls this game, Kennard says, "Well, I didn't think you guys were supposed to tell clients what to do." A more savvy helper might have sniffed out Kennard's tendency to play games much earlier.

- Dora makes herself appear helpless and needy when she is with her friends, but when they come to her aid, she is angry with them for treating her like a child. When she tries this in an early session, her counselor invites her to examine this "helpless and needy" routine.

The number of games we can play to avoid the work involved in squarely facing the tasks of life is endless. Clients who are fearful of changing will attempt to lay down smoke screens to hide from the helper the ways in which they fail to face up to life. Such clients use communication in order not to communicate. Therefore, helpers do well if they establish an atmosphere that discourages clients from playing games. An attitude of *nonsense is challenged here* should pervade the helping sessions.

***Excuses.*** Snyder, Higgins, and Stucky (1983) examined excuse-making behavior in depth. Excuse making, of course, is universal, part of the fabric of everyday life (see also Halleck, 1988; Higginson, 1999; Snyder & Higgins, 1988; Yun, 1998). Like games and distortions, it has its positive uses in life. Even if it were possible, there is no real reason for setting up a world without myths. On the other hand, avoidance behavior and excuse making contribute a great deal to avoiding the problems of life. Clients routinely provide excuses for why they did something "bad," why they didn't do something "good," and why they can't do something they need to do. Roberto tells the helper that he has engaged in benign attempts to sabotage his wife's career "for her own good" because she would "get hurt" in the Anglo world. The counselor helps him explore the alternative hypothesis that he is not ready for the changes in style that his wife's career and behavior were demanding from him.

This is only skimming the surface of the games, evasions, tricks, distortions, excuses, rationalizations, and subterfuges resorted to by clients (together with the rest of the population). Skilled helpers are caring and empathic, but they do not let themselves be conned. That helps no one.

**Challenging inadequate participation in the helping process.** Ideally, clients become full partners with their counselors in the helping process. They are co-conspirators in the search for a better life. But, often enough, this does not happen. There are two basic reasons for this. First, clients may lack the skills, especially the communication skills, needed to participate fully. If this is the case, you can use your communication skills to help them overcome obstacles to participation. The second reason is that clients are often reluctant to participate fully in the helping process because they are reluctant to change. Reluctance refers to the ambiguity all of us feel when we know that managing our lives better is going to exact a price. Clients, like the rest of us, are not sure whether they want to pay that price. Incentives for not changing often drive out or stand in the way of incentives for changing.

In helping sessions, clients exercise reluctance in many, often covert, ways. They may talk about only safe or low-priority issues, seem unsure of what they want, benignly sabotage the helping process by being overly cooperative, set unrealistic goals and then use their unreality as an excuse for not moving forward, not work very hard at changing their behavior, and may be slow to take responsibility for themselves. They may blame others or the social settings and systems of their lives for their troubles, and play games with helpers. Clients come armored against change to a greater or lesser degree. The reasons for reluctance are many. They are built into the human condition. They affect all of us. Clients have anxieties about the intensity of the helping encounter, the trustworthiness of the counselor, the compatibility of the relationship, the disorganization that delving into problem situations can cause, the pain that can come from revealing their inner lives and their secrets, and change in general. Some clients think that change is impossible, so why try? Others are dissatisfied with the pace of change. For some, counseling is too fast, for others too slow. Some clients are looking for short-term relief. Some come with the idea that counseling is magic and are put off when change proves to be hard work. Future rewards are not compelling.

In a counseling group, I once dealt with a man in his 60s who complained about constant anxiety. He told the story of how he had been treated brutally by his father until he finally ran away from home. He kept saying implicitly: "No one who grows up with scars like mine can be expected to take charge of his life and live responsibly." He had been using his mistreatment as a youth as an excuse to act irresponsibly at work (he had an extremely poor job record), in his relationship with himself (he drank excessively), and in his marriage (he had been uncooperative and unfaithful and yet expected his wife to support him). The idea that he could change, that he could take responsibility for himself even at his age, frightened him, and he wanted to run away from the group. But since his anxiety was so painful, he stayed. And he changed.

In practice, the targets of challenging reviewed above are often mixed together. Consider the case of Minerva, a woman who is seeing a counselor because of her impoverished social life. She believes that the world is filled with dishonest people (a self-limiting mind-set). Whenever she meets someone new, she views that person's behavior through this lens and thinks that he or she is guilty until proved innocent (internal behavior in need of reform). Since she is always on guard, she comes across as cold and indifferent (inadequate management of feelings and emotions). Therefore, when she meets someone new, she is defensive and often questions that person's intentions and actions (external behavior needing change). She also realizes that when she meets someone new, she expects some kind of initial trust on the part of the other person even though she doesn't give it herself (a discrepancy). She fails to see that even her closest friends are becoming uncomfortable around her (failure to see the her world as it really is). In the counseling sessions Minerva remains guarded and is slow to share what she really thinks (inadequate participation).

# Blind Spots

Up to this point the discussion has focused on *what* needs to be challenged, the content, as it were. However, it is also necessary to consider the client's degree of awareness about self-limiting thinking, emotional expression, and behavior. Clients, like the rest of us, don't seem to realize—or realize fully—how they are doing themselves in. That is, they have *blind spots.* Blind spots are part of the human condition. We all have them.

## Degrees of Awareness

Some blind spots appear to be unintentional, while others tend to be self-inflicted. Either way, they stand in the way of change. Blind spots come in a variety of flavors.

**Simple unawareness.** There are things that clients are simply not aware of. Becoming aware of them helps them know themselves better, and both cope with problems and develop opportunities. Serge was surprised when a fellow member of a self-help group called him "talented." He was brought up in a family that prized modesty and "humility." He never thought about himself as talented, creative, resourceful and had little idea how this lack of awareness had narrowed his life. His mind-set stood in the way of change. Katya, on the other hand, was always upbeat. Since she always tried to look at the positive side of things, she "exuded sunshine," as one of her friends put it. What she did not realize was that sometimes her exuberance was inappropriate because it stood in the way of facing problems squarely. One of her colleagues at work thought that Katya's upbeat nature was great but needed to be tempered by a dose of reality.

Mia, one of Katya's friends, had a different problem. She thought that her emotional outbursts were part of her punchy style. She did not realize that sometimes her colleagues at work wanted to strangle her. Casper didn't know that he had an acerbic communication style. If asked, he would describe himself as "assertive." The problem is that his "assertiveness" made people shy away from him. No wonder he was dissatisfied with his social life. Simple unawareness is itself an elastic term. There are degrees. At one end is simple ignorance – "He doesn't have a clue." At the other end the ignorance is not so simple—"Eva probably has some idea that this is not the right thing to do; she's not totally in the dark."

**Self-deception.** Where does simple lack of awareness end and self-deception start? There are things clients would rather not know because, if they knew them, they would be challenged to change their behavior in some way. So they'd rather stay in the dark. Goleman, who now writes extensively about "emotional intelligence" (1995, 1998), early on wrote a book called *Vital Lies, Simple Truths* (1985) on the psychology of self-deception and its pervasiveness in human life. Self-deception and the kind of social-emotional maturity outlined in his book *Emotional Intelligence* are incompatible. Yet, as Eduardo Giannetti (1997) points out in *Lies We Live By: The Art of Self-Deception,* it is ubiquitous. "How," he marvels, "do we carry out such feats as believing in what we don't believe in, lying to ourselves and believing the lie . . .?" (p. viii). Stan thought that he could get away with flirting with other women even though he was engaged. It's just natural for a man," was his excuse. His fiancee saw his behavior as insulting. When she called off the wedding, Stan discovered the seriousness of his self-deception.

**Choosing to stay in the dark.** Choosing to stay in the dark is a common human experience. It is as if someone were to say, "I could find out, but I don't want to, at least not yet." Lots of people, when they have physical symptoms such as pain in their guts, avoid thinking about it. Finding out whether the pain indicates something serious could be uncomfortable. Finding out that the pain is life-threatening might be terrifying. Clients, like the rest of us, often enough choose to stay in the dark. Ilia, recently released from jail, knows that she should have a clear understanding of the conditions of her probation, but chooses not to. When asked about the conditions, she says, "I don't know. No one really explained them to me." She knows that if she gets caught violating any of the conditions she could go back to jail. But she puts that out of her mind. When clients are being vague or evasive with their helpers, they may well not want to know. And it is likely that they don't want you to know, either.

**Knowing, but not caring.** Clients sometimes know that their thinking, forms of emotional expression, and acting are getting them into trouble or keeping them there, but they don't seem to care. Or they know what they should do to get out of trouble, but they don't lift a finger. We can use the term blind spot, at least in an extended sense, to describe this kind of behavior because clients don't seem to fully understand or appreciate the degree to which they are choosing their own misery. They don't appreciate the implications and consequences of not caring. For instance, Tanel says he knows nagging his wife to get a job even though there are two young children at home annoys his wife, but he keeps on anyway. "I can't help it." This creates a great deal of tension, but he continues to focus on his wife's reluctance to get a job rather than the negative consequences of his nagging.

So as you can see, the term blind spot, as used here, is somewhat elastic. We are unaware, we deceive ourselves, we don't want to know, we ignore, we don't care; or, we know, but not fully—that is, we do not fully understand the implications or the consequences of what we know. But it's a good term. It has great face validity. People know what you mean as soon as you use the term.

Helping clients deal with blind spots is one of the most important things you can do as a helper. For instance, if Lester has a prejudice and doesn't advert

## Questions to Uncover Blind Spots

These are the kinds of questions you can help clients ask themselves to develop new perspectives, change internal behavior, and change external behavior:

- What problems am I avoiding?
- What opportunities am I ignoring?
- What's really going on?
- What am I overlooking?
- What do I refuse to see?
- What don't I want to do?
- What unverified assumptions am I making?
- What am I failing to factor in?
- How am I being dishonest with myself?
- What's underneath the rocks?
- If others were honest with me, what would they tell me?

to it, he has a blind spot. If he has a prejudice, knows it (though he probably does not refer to his attitude as a prejudice), and fosters it, then he's engaging in a potentially dysfunctional bit of behavior. If he has a prejudice, fosters it, and lets it spill over into the way he deals with people, Lester has the full dysfunctional package. Contrast this with Bernice. Initially, she is unaware that she is prejudiced; then she becomes aware of her prejudice, tries to get rid of it as part of her internal mental furniture, refuses to act on it, and even learns something about herself and the world as she does all this. She deals with her prejudice creatively. She has turned a problem into an opportunity. Helping clients deal with dysfunctional blind spots can prevent damage, limit damage already done, and turn problems into opportunities. Box 6.1 outlines the kinds of questions you can help clients ask themselves in order to surface blind spots and develop new perspectives.

## From Blind Spots to New Perspectives

Challenge focuses on the kind of understanding that leads to constructive action, together with the constructive action itself. We do our clients a disservice if all that we do is help them identify and explore self-limiting blind spots. The positive-psychology part of challenging is helping clients transform blind spots into new perspectives and helping them translate these new perspectives into more constructive patterns of both internal and external behavior.

There are many upbeat names for this process of transforming blind spots into new perspectives. They include: seeing things more clearly, getting the picture, getting insights, developing new perspectives, spelling out implications, transforming perceptions, developing new frames of reference, looking for meaning, shifting perceptions, seeing the bigger picture, developing different angles, seeing things in context, context breaking, rethinking, getting a more objective view, interpreting, overcoming blind spots, second-level learning, double-loop learning (Argyris, 1999), thinking creatively, re-conceptualizing, discovery, having an "ah-ha" experience, developing a new outlook, questioning assumptions, getting rid of distortions, relabeling, framebreaking, frame-bending, reframing, and making connections. You get the idea. All of these imply some kind of cognitive restructuring which is needed to identify and manage both problems and opportunities. Developing new perspectives, while painful at times, ultimately tends to be prized by clients.

In the following example, the client, Leslie, a fairly religious 83-year-old woman, is a resident of a nursing home. She is talking to one of the aides.

> **LESLIE:** I've become so lazy and self-centered. I can sit around for hours and just reminisce . . . letting myself think of all the good things of the past—you know, the old country and all that. Sometimes a whole morning can go by.
>
> **AIDE:** I'm not sure what's so self-indulgent about that.
>
> **LESLIE:** Well, it's in the past and all about myself. . . I don't know if it's right.
>
> **AIDE:** The way you talk about it, the reminiscing sounds almost like a kind of meditation for you.
>
> **LESLIE:** You mean like a prayer?
>
> **AIDE:** Well, yes. . . like a prayer.

The client sees reminiscing as laziness. The nurse helps her develop a new, more positive perspective. Reminiscing as a kind of meditation on life or even a prayer—Leslie is comfortable with that. With that new slant, Leslie feels that she can "indulge" herself.

Effective helpers assume that clients have the resources to see themselves and the world in which they live in a less distorted way, and to act on what they see. Another way of putting it is that skilled counselors help clients move from what the Alcoholics Anonymous movement calls "stinkin' thinkin'" to healthy thinking. And from "stinkin'" emoting to constructive emotional expression. And from dysfunctional actions to healthy behavior. Consider Carla. Facing menopause, she is lumbered with the outmoded view of menopause as a "deficiency disease." A counselor, without minimizing Carla's discomfort and stress, helps her see menopause as a natural developmental stage of life. Although it indicates the ending of one phase, it also opens up new life-stage possibilities. Looking forward to those possibilities, rather than looking back at what she's lost, helps Carla a great deal.

It's important to note that identifying blind spots is not always the same as developing a new perspective. When Sandra, mentioned earlier, discovered that her husband was having an affair, she discovered a fact. She was blind to how serious her husband was about achieving a more satisfactory sexual life and deceived herself by telling herself that he was finally "coming to his senses." A new perspective for her was a step further: "I now realize more fully that all aspects of marriage are a two-way street and that we have to come up with an arrangement about sexual behavior that is satisfactory to both of us." This might sound pedestrian, but it is a new perspective for her which, if acted on, could help save the marriage.

## Linking Challenge to Action

More and more theoreticians and practitioners are stressing the need to link insight to problem-managing and opportunity-developing action. Wachtel (1989, p.18) put it succinctly: "There is good reason to think that the really crucial insights are those closely linked to new actions in daily life and, moreover, that insights are as much a product of new experience as their cause." Although new perspectives are important, they are not magic. Overstressing insight and self-understanding—that is, new perspectives—can actually stand in the way of action instead of paving the way for it. Unfortunately, the search for insight can too easily become a goal in itself. Much more attention has been devoted to developing insight than to linking insight to action. I do not mean to imply that achieving useful insights into oneself and one's world is not hard work and often painful. But the pain needs to be turned into gain. Constructive behavioral change leading to valued outcomes is the goal. Consider this case.

> Ned, an inveterate micromanager, comes to realize that, despite a company leadership program that most of his direct reports have taken, he still keeps making all the decisions in his team. This slows down the work and makes the department less efficient than it might be. In a leadership training program he gets feedback from a survey filled in by his boss, his direct reports, his peers, and himself. Once he gets over the shock of reading the ratings, he sits down with a coach and talks about what constructive changes he needs to make in his managerial style—more delegating, getting and acting on suggestions from team members, and providing constructive feedback for them. They work together on drawing up an implementation plan.

Effective counselors, as they help their clients develop insights into themselves and their thinking, the ways they express or suppress emotion, and their behavior, maintain a constructive action-centered approach. Clients should be able to say, "I now see much more clearly how I am putting myself in jeopardy, but *I can do something about it,*" and then actually move to action. For instance, Joanna, a woman in her late fifties, has her annual physical exam. The doctor discovers that she has a higher-than-average risk for stroke. She communicates this information to Joanna.

JOANNA: What puts me at risk?

DOCTOR: A few things. The biggest factor is your smoking. Next is your cholesterol level.

JOANNA: So, if I stop smoking and get my diet under control . . .

DOCTOR: It won't make you live forever, but it will make you live better.

JOANNA: I don't want to live forever. And I don't want to be too prudent. That would take all the fun out of life . . . But now that I am a person of a certain age, as they say, I have something to think about. I think that I can rise to the challenge of stopping smoking. And I bet that I can come up with some delicious cholesterol-lowering meals. By the way, what about exercise? Isn't that part of the package?

Joanna has been ignoring the ways in which she is putting her health at risk. That her risk level is higher than average is the basic part of the new perspective. Now that she is approaching her sixties, this realization moves to center stage. Joanna herself adds the "I can do something about it" part.

**An invitation to stop, start, continue.** Challenge is basically simple. It deals with clients thinking, their ways of managing their emotions, and their behavior. It is about "stop, start, and keep going." Since self-challenge is the ideal, challenge on the part of helpers at its best is an *invitation* to clients to:

- *Stop.* Challenge is an invitation to clients to identify and *stop* engaging in self-defeating thinking, self-indulgent emotions, and unconstructive behavior—that is, the kinds of things that keep them mired down in their problem situations or that keep them from identifying and developing opportunities. For instance, Sakae, an Asian American, thinks that he is genetically better than European Americans. He never says this outright, but he communicates this attitude to others often in a variety of not-too-subtle ways. This mars both work and social relationships. And yet he wonders why he feels isolated and depressed. There is a catch. Simply trying to stop an established behavior can be notoriously difficult.

- *Start.* More positively, challenge is an invitation to clients to identify and *start* engaging in self-enhancing kinds of thinking, emotional expression. For instance, Grace—young, intelligent, savvy, and aggressive—was fired from three different jobs, good jobs, for "insubordination." A counselor asked her why she, with all that talent permeated by an engaging, punchy style, was working for other people. Were there not other ways she could channel her energy? Grace, taking a cue from this counseling session, decided to start her own business. It quickly blossomed. Substitution of a productive behavior for an unproductive one, especially when the substituted behavior eliminates the problem behavior, is a form of starting that is often a more powerful intervention than just trying to stop a behavior. For instance, programs that teach a person how to shop for and prepare nutritious, balanced meals run a much better chance of being sustained, especially if they are paired with sensible exercise, than one that merely calls for the elimination of snacks and junk food. Anyway, it's hard eating a bag of potato chips while jogging.

- *Keep going or increase.* Finally, challenge is an invitation to identify and *continue*—and even *increase*—activities that manage problems or develop opportunities, especially when clients don't know what they are doing well or give signs of giving up. For instance, Melissa, a middle-aged woman in an arduous rehabilitation program, feels depressed because she feels she isn't making any progress. The counselor encourages her to keep to her exercises. He suggests that at this stage she needs to define success as simply engaging in the exercises even though she sees no dramatic improvement. "There, I've done them. I've succeeded and I feel good about myself."

Let's return to Roberto, seen earlier nagging his wife. Here are some possibilities in terms of starting, stopping, and continuing internal behaviors. He needs to:

- *stop* telling himself that she is the one with the problem, stop seeing her as the offending party when conflicts arise, and stop telling himself there is no hope for the relationship. He needs to become a co-owner of both the problems and the opportunities of the marriage.

- *start* thinking of his wife as an equal in the relationship, start understanding her point of view, and start imagining what an improved relationship with her might look like. He needs a new communication style with his wife.

- *continue* to take stock of the ways he contributes to their difficulties and increase the number of times he tells himself to let her live her life as fully as he wants to live his own. He needs to identify what he does best in the marriage—he is not a dud as a communicator —and reinforce these behaviors.

That is, Roberto has to get his head straight, mobilize his resources, and turn his reconstructed thinking into action. Of course, given that any relationship is a two-way street and that both parties need to change, Maria has to develop some new perspectives and change some of the ways she thinks and acts.

Roberto also needs to start, continue, increase, and stop certain external behaviors if he is going to do his part in developing a better relationship with his wife. Here are some possibilities. He can:

- *stop* criticizing her in front of others, stop creating crises at home and assigning the blame for them to her, and stop making fun of her business friends.

- *start* activities that will help him develop his own career, start sharing his feelings with her instead of just expressing them in negative ways, start engaging in mutual decision making, and start taking more initiative in household chores and child care.

- *continue* visiting her parents with her and increasing the number of times he goes to business-related functions with her.

With the counselor's help, Roberto begins to challenge himself to develop and implement a set of possibilities that would help him keep his relationship with Maria on an even keel while they work through the issue of Maria's career.

Is challenge enough to stimulate problem-managing action? For some, yes. A few well-placed challenges might be all that some clients need to move to constructive action. Once challenged, they are off to the races. But others may well have the resources to manage their lives better, but not the will. They know what they need to do but are not doing it. A few nudges in the right direction help them overcome their inertia. I have had many one-session encounters that included a bit of listening and thoughtful processing, some sharing of highlights, and a brief challenge that produced a new perspective that sent the client off on some useful course of action. On the other hand, if helping clients challenge themselves to develop new perspectives leads to one profound insight after another but no action and behavioral change, then we are whistling in the wind.

## The Shadow Side of Challenging

Challenge, because of its very nature, has a strong shadow side. The very fact that blind spots are involved puts it squarely in the shadows. We often dislike being challenged, even when it is done well. If helpers are very effective in challenge, they still might well sense some reluctance in their clients' responses. If they are poor at it, they are more likely to experience resistance. Inviting clients to challenge themselves and helping them do it to themselves is your best bet. In this chapter we look at the shadow side in terms of the client's attitudes and behavior. In the next, we consider the shadow side of challenging in terms of helper attitudes and behavior.

Even when challenge is a response to a client's plea to be helped to live more effectively, it can precipitate some degree of disorganization in the client. Different writers refer to this experience under different names: "crisis," "disorganization," "a sense of inadequacy," "disequilibrium," and the like. Counseling-precipitated crises can be beneficial for the client. Whether they are or not depends, to a great extent, on the skill of the helper. But even when a direct challenge or an invitation to self-challenge is accurate and delivered caringly, some clients still dodge and weave. Cognitive-dissonance theory (Festinger, 1957; Draycott & Dabbs, 1998) gives us some insight into the dynamics of this. Since dissonance (discomfort, crisis, disequilibrium) is an uncomfortable state, the client will often try to get rid of it. Outlined below are five of the more typical ways clients experiencing dissonance attempt to rid themselves of their discomfort. Let's examine them briefly as they apply to being challenged.

**Discredit challengers.** The client might confront the helper whose challenges are getting too close for comfort. Some attempt is made to point out that the helper is no better than anyone else. In the following example, the client has been discussing her marital problems and has been invited by the helper to take a look at her own behavior rather than giving example after example of what her spouse does wrong.

> CLIENT: It's easy for you to sit there and suggest that I be more responsible in my marriage. You've never had to experience the misery in which I live. You've never experienced his brutality. You probably have one of those nice middle-class marriages.

Counterattack is a common strategy for coping with challenge. You may even have to field some sarcastic barbs: "Thanks for coming down out of your ivory tower to tell me how to shape up." Effective counselors are not surprised by clients' counter challenges. Nor do they rise to the bait. Respect and empathy always pervade their attempts to help clients face up to difficult issues.

**Persuade challengers to change their views.** In this approach, clients reason with their helpers. They urge them to see what they have said as misinterpretations and to revise their views. In the following example, the client has been talking about the way she blows up at her husband whenever he makes "some stupid mistake." The counselor has invited her to explore the consequences of this pattern of behavior.

> CLIENT: I'm not so sure that my anger at home isn't called for. I think that it's a way in which I'm asserting my own identity. If I were to lie down and let others do what they want, I would become a doormat at home. And, as you have helped me see, assertiveness should be part of my style. I think you see me as a fairly reasonable person. I don't get angry here with you because there is no reason to.

Sometimes a client like this will lead an unwary counselor into an argument about who's really right. In this case, the counselor might help the client explore the difference between assertiveness and aggression. Getting angry may not be the problem but, rather, how she channels her anger.

**Devalue the issue.** This is a form of rationalization. A client who is being invited to challenge himself about his sarcasm points out that he is rarely sarcastic, that "poking fun at others" is just that, good-natured fun, that everyone does it, and that it is a very minor part of his life not worth spending time on. The client has a right to devalue a topic if it really isn't important. The counselor has to be sensitive enough to discover which issues are important and which are not. How would you handle the devaluing this client engages in?

**Seek support elsewhere for the views being challenged.** Some clients, once challenged, go out and gather testimonials supporting their views. In extreme cases clients leave one counselor and go to another because they feel they aren't being understood. They try to find helpers who will agree with them. More commonly, the client remains with the same counselor and offers evidence from others, contesting the helper's point of view.

> CLIENT: I asked my wife about my so-called sarcasm. She said she doesn't mind it at all. And she thinks that my friends see it as humor and as a part of my style.

This is an indirect way of telling the counselor she is wrong. The counselor might well be wrong, but if the client's sarcasm is really dysfunctional in his interpersonal life, the counselor should find some way of pressing the issue.

**Cooperate in the helping session, but then do nothing about it outside.** The client can agree with the counselor as a way of dismissing an issue.

However, the purpose of challenging is not to get the client's agreement but to develop new perspectives that lead to constructive action. Consider this client.

> CLIENT: To tell you the truth, I am pretty lazy and manipulative. And I'm not even very clever about it.

In truth, this client is lazy and manipulative. She has given little evidence so far that she will do anything about it. If it were up to you break through this screen of "honesty," what would you do?

None of this, of course, makes clients evil—or at least no more "evil" than the rest of us. Challenge is a form of "tough love." Supporting clients is not the same as coddling them. And effective challenge is not a form of bullying.

## EVALUATION QUESTIONS ON THE PROCESS OF CHALLENGING

- What are the two goals of challenging and how well do you understand them?
- In what ways is the spirit of challenging a continuation of the spirit of the basic communication skills?
- Defend this statement: Challenge is not a way of showing disrespect for clients or putting them down.
- Which of the targets of challenging outlined in this chapter do you see in your own life?
- What prejudices have you run across in your dealings with others?
- What self-limiting assumptions might stand in the way of someone's being a good helper?
- What unused strengths and resources do you find in your own life?
- Name some of the predictable dishonesties of everyday life you have run across in your dealing with others.
- Describe a game you see people playing in their interactions with others. To what extent are you game-free in your interactions with others?
- What blind spots have you discovered in your own life?
- Describe a new perspective that has helped you move forward in life.
- Name something you should stop doing to improve your life. What is something you might substitute for the behavior you would like to eliminate?
- Name something that you should start doing.
- Name something you should continue to do, maybe even more intensively.
- Describe what you think might be an effective response to one of the five shadow-side ways clients sometimes use to blunt helpers' challenges.

**EXERCISES ON PERSONALIZING**

1. Review a current problem situation you are facing. What kinds of challenges would help you deal more effectively with the problem situation? In what ways does your thinking or your management of feelings and emotions or your behavior need to be challenged?

2. Consider a problem situation or unused opportunity you faced in the past. What kinds of challenges would have helped you to deal with the problem more effectively? Thinking? Emotional management? Behavior?

3. Name some realizations or new perspectives that have made a difference in your life. For instance, one day Maggie woke up to the fact that she did not have to spend the rest of her life with an abusive partner.

4. Name times when you failed to think things through. What caused your blind spots? What affect did they have on your life?

# CHAPTER 7

# Challenging Skills and the Wisdom Needed to Use Them Well

Specific Challenging Skills

    Sharing Advanced Highlights: Capturing the Message Behind the Message

    Information Sharing

    Helper Self-Disclosure and Its Judicious Use

    Immediacy: Discussing the Helper-Client Relationship

    Suggestions and Recommendations

    Confrontation

    Encouragement

From Smart to Wise: Guidelines for Effective Challenging

Linking Challenge to Action

Challenge and the Shadow Side of Helpers

Exercises on Personalizing

# Specific Challenging Skills

There are any number of ways in which helpers can challenge clients to develop new perspectives, change their internal behavior, and change their external behavior. The following are discussed and illustrated in this chapter: (1) sharing the message behind the message, (2) sharing information, (3) helper self-disclosure, (4) immediacy—that is, discussing the helper-client relationship, (5) suggestions and recommendations, (6) confrontation, and (7) encouragement. It has already been noted that both probing and summarizing—and even sharing accurate highlights—can challenge clients to rethink their attitudes, emotionality, and behavior.

## 1. Sharing Advanced Highlights: Capturing the Message Behind the Message

One way of challenging clients is to share with the client your understanding of the message behind the message. For instance, Gordon gets angry when he talks about his interactions with his ex-wife, but as he talks, the helper hears not just anger but also hurt. It may be that Gordon can, with relative ease, talk about and express his anger but is reluctant to talk about his feelings of hurt. When you share basic empathic highlights—provided, of course, that you are accurate—clients recognize themselves almost immediately: "Yes, that's what I meant." However, since sharing more covert messages (let's call them advanced highlights) digs a bit deeper, clients might not immediately recognize themselves in the helper's response. Or they might experience a bit of disequilibrium. That's what makes sharing advanced highlights a form of challenge. For instance, the helper says something like this to Gordon: "It's pretty obvious that you really get steamed when she acts like that . . . But I thought I sensed, mixed in with the anger, a bit of hurt." At that, Gordon looks down and pauses. He finally says, "She can still get to me. She certainly can." This appreciably broadens or deepens the discussion of the problem situation.

Here are some questions helpers can ask themselves to probe a bit deeper as they listen to clients.

- What is this person only half saying?
- What is this person hinting at?
- What is this person saying in a confused way?
- What covert message is behind the explicit message?

Note that advanced empathic listening and processing focuses on what the client is actually saying or at least expressing, however tentatively or confusedly. It is not an interpretation of what the client is saying. Sharing advanced highlights is not an attempt to "psych the client out."

In the hands of skilled helpers, capturing and sharing the message behind the message focuses not just on the problematic dimensions of clients' behavior, but also on unused opportunities and resources. Effective helpers listen for the resources that are buried deeply in clients and often have been forgotten by them. Consider the following example.

The client, a soldier who has been thinking seriously about making the army his career, has been talking to a chaplain about his failing to be promoted. He has seen service in Bosnia, Kosovo, and Iraq, and has performed very well. As he talks, it becomes fairly evident that part of the problem is that he is so quiet and unassuming that it is easy for his superiors to ignore him.

> **SOLDIER:** I don't know what's going on. I work hard, but I keep getting passed over when promotion time comes along. I think I work as hard as anyone else, and I work efficiently, but all of my efforts seem to go down the drain. I'm not as flashy as some others, but I'm just as substantial.

> **CHAPLAIN A:** You feel it's quite unfair to do the kind of work that merits a promotion and still not get it.

> **SOLDIER:** Yeah . . . I suppose there's nothing I can do but wait it out. (A long silence ensues.)

Chaplain A tries to understand the client from the client's frame of reference. He deals with the client's feelings and the experience underlying those feelings. In responding, he shares a basic empathic highlight. But the client merely retreats more into himself. Here's a different approach.

> **CHAPLAIN B:** It's depressing to put out so much effort and still get passed by . . . Tell me more about this "not as flashy bit." What in your style might make it easy for others not to notice you, even when you're doing a good job?

> **SOLDIER:** You mean I'm so unassuming that I could get lost in the shuffle? Or maybe it's the guys who make more noise, the squeaky wheels, who get noticed . . . I guess I've never really thought of selling myself. It's not my style.

From the context, from the discussion of the problem situation, from the client's manner and tone of voice, Chaplain B picks up a theme that the client states in passing in the phrase "not as flashy." They go on to discuss how he might "market himself" in a way that is consistent with his values. Advanced highlights can take a number of forms. Here are some of them.

**Help clients make the implied explicit.** The most basic form of an advanced highlight involves helping clients give fuller expression to what they are implying rather than saying directly. In the following example, the client has been discussing ways of getting back in touch with his wife after a recent divorce, but when he speaks about doing so, he expresses very little enthusiasm.

> **CLIENT** (somewhat hesitatingly): I could wait to hear from her. But I suppose there's nothing wrong with calling her up and asking her how she's getting along.

> **COUNSELOR A:** It seems that there's nothing wrong with taking the initiative to contact her. After all, you'd like to find out if she's doing okay.

> **CLIENT** (somewhat drearily): Yeah, I suppose I could.

Counselor A's response might have been fine at an earlier stage of the helping process, but it misses the mark here, and the client does not move on.

COUNSELOR B: You've been talking about getting in touch with her, but, unless I'm mistaken, I don't hear a great deal of enthusiasm in your voice.

CLIENT: To be honest, I don't really want to talk to her. But I feel guilty—guilty about the divorce, guilty about her going out on her own. Frankly, all I'm doing is taking care of her all over again. And that's one of the reasons we got divorced. I had a need to take care of her and she let me do it. That was the story of our marriage. I don't want to do that anymore.

COUNSELOR B: What would a better way of going about all this be?

CLIENT: I need to get on with my life and let her get on with hers. Neither of us is helpless. (His voice brightens.) For instance, I've been thinking of quitting my job and starting a business with a friend of mine—helping small businesses use the Internet to improve their businesses, you know, Web sites, marketing, all of that.

Counselor B bases her response not only on the client's immediately preceding remark, but also on the entire context of his story. Her response hits the mark and the client moves forward. As with basic highlights, there is no such thing as a good advanced highlight in itself. Rather, does the response help the client clarify the issue more fully, so that he or she might begin to see the need to act differently?

**Help clients identify themes in their stories.** When clients tell their stories, certain themes emerge. Thematic material might refer to feelings (such as themes of hurt, of depression, of anxiety), to thoughts (continually ruminating about a past mistake), to behavior (controlling others, avoiding intimacy, blaming others, overwork), to experiences (such as being a victim, being seduced, being punished, being ignored, being picked on), or some combination of these. Once you see a self-defeating theme or pattern emerging from your discussions, you can share your perception and help the client check it out.

In the following example, a counseling-psychology trainee is talking with his supervisor. The trainee has four clients. In the past week he has seen each of them for the third time. This dialogue takes place in the middle of a supervisory session.

SUPERVISOR: You've had a third session with each of four clients this past week. Even though you're at different stages with each because each started in a different place, you have a feeling, if I understand what you've been saying, that you're going around in circles with a couple of them.

TRAINEE: Yes, I'm grinding my wheels. I don't have a sense of movement.

SUPERVISOR: Any thoughts on what's going on?

TRAINEE: Well, they seem willing enough. And I think I've been very good at listening and sharing highlights. It keeps them talking.

SUPERVISOR: But this doesn't seem to be enough to get them moving forward. I tell you what. Let's listen to one of the tapes.

They listen to a segment of one of the sessions. The trainee turns off the recorder.

TRAINEE: Oh, now I see what I'm doing! It's all sharing highlights with a few uh-huhs. And I thought I was being pushy. But this is as far from pushy as you can get.

SUPERVISOR: So what's missing?

TRAINEE: There are very few probes and nothing close to summaries or mild challenges. Certainly some probes would have given much more focus and direction to the session.

SUPERVISOR: Let me role-play the client as well as I can and see how you might redo the session.

They then spend about fifteen minutes in a role-playing session. The trainee mingles probes with highlights, and the result is quite different.

SUPERVISOR: How close did you get to challenging, even mild challenging?

TRAINEE: I didn't get there at all. . . .You know, I think that I see probes as challenges . . . The thread through all of this is *playing it safe*. I think I'm playing it safe because I don't want to damage the client. I'm afraid to push.

The theme that the supervisor helps the trainee surface is a fear of being "pushy," which explains his "playing it safe" behavior.

**Help clients make connections that may be missing.** Clients often tell their stories in terms of experiences, thoughts, behaviors, and emotions in a hit-or-miss way. The counselor's job, then, is to help them make the kinds of connections that provide insights or perspectives that enable them to move forward in the helping process.

- Her counselor helps Cymae see that she is having difficulty developing strategies for her chosen goals because she is only half-heartedly committed to her goals. They revisit the goals she has set for herself.
- A managerial coach helps Finnbar relate the trouble he is having with a strong woman supervisor to sexist attitudes that leak out into his behavior.
- A therapist helps Joanna see the link between her ingratiating style and her inability to influence her colleagues at work.
- A supervisor helps Dieter see that the persistent anxiety he feels when working with clients is related to the perfectionistic standards he has set for himself.

The following client has a full-time job and is finishing his final two courses for his college degree. His father has recently had a stroke and is incapacitated. He talks about being progressively more anxious and tired in recent weeks. He visits his father regularly. He meets frequently with his mother, his two sisters, and his two brothers to discuss how to manage the family crisis. Under stress, fault lines in family relationships appear. He has deadlines for turning in papers for current courses.

JOHN: I don't know why I'm so tired all the time. And edgy. I'm supposed to be the calm one. I wonder if it's something physical. You know, what's happened to Dad and all that. I never even think about my health.

COUNSELOR: A lot has happened in the past few weeks. Work. School. Your dad's stroke. Juggling schedules.

JOHN (interrupting): But that's what I'm good at. Working hard. Juggling schedules. I do that all the time. And I don't get tired and edgy.

COUNSELOR: Add in your dad's illness . . .

JOHN: You know, I could handle that, too. If I were the only one, you know, just me and Mom, I bet I could do it.

COUNSELOR: All right, so besides your dad's illness, what's so new?

JOHN (slowly): Well, I hate to say it. It's the squabbling. We usually get on pretty well. We all like getting together. But the meetings about Dad, they can be awful. I keep thinking about them at work. And the other evening when I was trying to write a paper for school, I was still ticked off at my older sister.

COUNSELOR: So the family stuff is getting at you no matter what you're doing.

JOHN: I'm just not used to all that. I thought we'd rally together. You know, get support from one another. Sometimes it's just the opposite.

John handles the normal stress of everyday life quite well. But the "family stuff" is acting like a multiplier. They go on to discuss what the family dynamics are like and what John can do to cope with them.

**Share educated hunches based on empathic understanding.** As you listen to clients, thoughtfully process what they say, and put it all into context, hunches about the message behind the message or the story behind the story will naturally begin to form. You can share the hunches that you feel might add value. The more mature and socially competent you become and the more experience you have helping others, the more educated your hunches become. Here are some examples.

Hunches can help clients see the bigger picture. In this example, the counselor is talking with a client who is having trouble because of his perfectionism. He also mentions problems with his brother-in-law, whom his wife enjoys having over. He and his brother-in-law argue and sometimes the arguments have an edge to them. At one point the client describes him as "a guy who can never get anything right." Later the counselor says, "We started out by talking about perfectionism in terms of the inordinate demands you place on yourself. I wonder whether it could be spreading a bit. You should be perfect. But so should everyone else." They go on to discuss the ways his perfectionism is interfering with his social life.

Hunches can help clients see more clearly what they are indirectly expressing or merely implying. In this next example, the counselor is talking to a client who feels that a friend has let her down. "I think I might also be hearing you say that you are more than disappointed—perhaps even betrayed." Since the client has been making every effort to avoid her friend, "betrayal" rings truer than "let down." She gets in touch with the depth of her feelings.

Hunches can help clients draw logical conclusions from what they are saying. A manager is having a discussion with one of his teams members who has expressed, in a rather tentative way, some reservations about one of the team's projects. At one point the manager says, "If I stitch together everything that you've said about the project, it sounds as if you are saying that it was ill-advised in the first place and probably should be shut down. I know that might sound drastic and you've never put it in those words. But if that's how you feel, we should discuss it in more detail."

Hunches can help clients open up areas they are only hinting at. In the next case, a school counselor is talking to a senior in high school. "You've brought up sexual matters a number of times, but you haven't pursued them. My guess is that sex is a pretty important area for you, but perhaps pretty touchy, too."

Hunches can help clients see things they may be overlooking. A counselor is talking to a client who probably has only six months to live. The man is unmarried and has never made a will. He has some money, but has expressed indifference to money matters. "I'm financially lazy," is his theme. He adds, "I'm ready to die." Later in the session, the counselor says, "I wonder if your financial laziness has spread a bit. For instance, you live alone and, if I'm not mistaken, you haven't given anyone power of attorney in health matters either. That could mean that how you die will be in the hands of the doctors." This helps the client begin to rethink how he wants to die. They even discuss finances. He may not be a slave to money, but whatever money he has could go to a good cause.

Hunches can help clients take fuller ownership of partially owned experiences, behaviors, feelings, points of view, and decisions. For example, a counselor is talking to a client who is experiencing a lot of pain in a physical rehabilitation program following an automobile accident. She keeps focusing on how difficult the program is. At one point the counselor says, "You sound as if you have already decided to quit. Or I might be overstating the case . . ." This helps the client enormously. She has been thinking of quitting but she has been afraid to discuss it. They go on to discuss her wanting to give up and her dread of giving up. When the counselor finds out that she has never even mentioned the pain to the members of the rehabilitation staff, they discuss strategies for coping with the pain, including direct conversations with the staff about the pain.

Like all responses, hunches should be based on your understanding of your clients. If your clients were to ask you where your hunches come from—"What makes you think that?"—you should be able to identify the experiential and behavioral clues on which they are based. Of course, sharing advanced empathic highlights is not license to draw inferences from clients' history, experiences, or behavior at will. Nor is it license to load a client with interpretations that are more deeply rooted in your favorite psychological theories than in the realities of the client's world. Sharing advanced empathic highlights constructively takes social competence and emotional intelligence.

## 2. Information Sharing

Sometimes clients are unable to explore their problems fully, set goals, and proceed to action because they lack information of one kind or another. Information can help clients at any stage of the helping process. For instance, in Stage I it helps many clients to know that they are not the first to try to cope with a particular problem. In Stage II, information about various aspects of the problem situation or an unused opportunity (for instance, what it's like to start a business, including the average number of hours entrepreneurs spend on the business per week) can help them further clarify possibilities and set goals. In the implementation stage, information on commonly-experienced obstacles can help clients cope and persevere.

The skill or strategy of information sharing is included under challenging skills because it helps clients develop new perspectives on their problems or shows them how to act. This skill includes both giving information and correcting misinformation. In some cases, the information can prove to be quite confirming and supportive. For instance, a parent who feels responsible following the death of a newborn baby may experience some relief through an understanding of the characteristics of the Sudden Infant Death Syndrome. This information does not *solve* the problem, but the parent's new perspective can help him or her handle self-blame.

In some cases, the new perspectives clients gain from information sharing can be both comforting and painful. Consider the following example.

> Adrian was a college student of modest intellectual means. He made it through school because he worked very hard. In his senior year he learned that a number of his friends were going on to graduate school. He, too, applied to a number of graduate programs in psychology. He came to see a counselor in the student services center after being rejected by all the schools to which he had applied. In the interview, it soon became clear to the counselor that Adrian thought that many, perhaps even most, college students went on to graduate school. After all, most of his closest friends had been accepted in one graduate school or another. The counselor shared with him the statistics of what could be called the educational pyramid—the decreasing percentage of students attending school at higher levels. Adrian did not realize that just finishing college made him part of an elite group. Nor was he completely aware of the extremely competitive nature of the graduate programs in psychology to which he had applied. He found much of this relieving but then found himself suddenly faced with what to do now that he was finishing school. Up to this point he had not thought much about it. He felt disconcerted by the sudden need to look at the world of work.

Giving information is especially useful when lack of accurate information either is one of the principal causes of a problem situation or is making an existing problem worse.

In some medical settings, doctors team up with counselors to give clients messages that are hard to hear and to provide them with information needed to make difficult decisions. For instance, Lester, a 67-year-old accountant, has been given a series of diagnostic tests for possible prostate cancer. He finds out that he does have cancer, but now he faces the formidable task of choosing what to do about it. The doctor sits down and talks with him, and lays out the alternatives. Since there are many different options, including doing nothing, the doctor also describes the pluses and minuses of each option. Later, Lester has a discussion with a counselor. The counselor helps Lester cope with the news, process the information, and begin the process of making a decision.

There are some cautions helpers should observe in giving information. When information is challenging, or even shocking, be tactful and help the client handle the disequilibrium that comes with the news. Do not overwhelm the client with information. Make sure that the information you provide is clear and relevant to the client's problem situation. Don't let the client go away with a misunderstanding of the information. Be supportive; help the client process the information. Finally, be sure not to confuse information giving with advice giving. Professional guidance is not to be confused with telling clients what to do. Neither the doctor nor the counselor tells Lester which treatment to choose. But Lester needs help with the burden of choosing.

## 3. Helper Self-Disclosure and Its Judicious Use

A third skill of challenging involves the ability of helpers to constructively share some of their own experiences, behaviors, and feelings with clients (Edwards & Murdoch, 1994; Hendrick, 1988; Knox, Hess, Petersen, & Hill, 1997; Mathews, 1988; Simon, 1988; Sticker & Fisher, 1990; Watkins, 1990; Weiner, 1983). In one sense, counselors cannot help but disclose themselves: "The counselor communicates his or her characteristics to the client in every look, movement, emotional response, and sound, as well as with every word" (Strong & Claiborne, 1982, p. 173). This is the kind of indirect disclosure that goes on all the time. Effective helpers, as they tune in, listen, process, respond, and become adept at tracking and managing the impressions they are making on clients.

Here, however, we are discussing direct self-disclosure. Research into direct helper self-disclosure has led to mixed and even contradictory conclusions. Some researchers have discovered that helper self-disclosure can frighten clients or make them see helpers as less well-adjusted. Or, instead of helping, self-disclosure on the part of the helper might place another burden on clients. Other studies have suggested that helper self-disclosure is appreciated by clients. Some clients see self-disclosing helpers as "down-to-earth" and "honest."

Direct self-disclosure on the part of helpers can serve as a form of modeling. Self-help groups such as Alcoholics Anonymous use such modeling extensively. Some would say that helper self-disclosure is most appropriate in such settings. This helps new members get an idea of what to talk about and find the courage to do so. It is the group's way of saying, "You can talk here without being judged and getting hurt." Even in one-to-one counseling dealing with alcohol and drug addiction, extensive helper self-disclosure is the norm.

Beth is a counselor in a drug rehabilitation program. She herself was a substance abuser for a number of years but, with the help of the agency where she is now a counselor, she is clean and sober. It is clear to all people with addictions in the program that the counselors there were once substance abusers themselves and are not only rehabilitated but also intensely interested in helping others both rid themselves of drugs and develop a kind of lifestyle that helps them stay drug-free. Beth freely shares her experience, both of being a drug user and of her rather agonizing journey to freedom, whenever she thinks that doing so can help a client.

Other things being equal, counselors who have struggled with addictions themselves often make excellent helpers in programs like this. They know from the inside the games clients afflicted with addictions play. Sharing their experience is central to their style of counseling and is accepted by their clients. It helps clients develop both new perspectives and new possibilities for action. Such self-disclosure is challenging. It puts pressure on clients to talk about themselves more openly or in a more focused way.

Helper self-disclosure is challenging for at least two reasons. First, it is a form of intimacy and, for some clients, intimacy is not easy to handle. Therefore, helpers need to know precisely *why* they are divulging information about themselves. Second, the message to the client is, indirectly, a challenging *You can do it, too*, because revelations on the part of helpers, even when they deal with past failures, often center on problem situations they have overcome or opportunities they have seized. However, done well, such disclosures can be very encouraging for clients.

In the following example, the helper, Rick has had a number of sessions with Tim, a client who has had a rather tumultuous adolescence. For instance, he fell into the wrong crowd and got into trouble with the police a few times. His parents were shocked and his relationship with them became very strained. Rick decides to share some of his own experiences.

RICK: In my junior year in high school I was expelled for stealing. I thought that it was the end of the world. My Catholic family took it as the ultimate disgrace. We even moved to a different neighborhood in the city.

TIM: What did it do to you?

Rick briefly tells his story, a story that includes setbacks not unlike Tim's. But Rick, with the help of a very wise and understanding uncle, was able to put the past behind him. He does not overdramatize his story. In fact, his story makes it clear that developmental crises are normal. How we interpret and manage them is the critical issue.

Since current research on helper self-disclose does not give us definitive answers, we need to stick to common sense. Helper self-disclosure is not a science, but an art. Since clients can misinterpret helpers' self-sharing and its intent, great care needs to be taken to engage in self-disclosure judiciously. Below are some guidelines for doing so.

**Make sure that your disclosures are appropriate.**  Sharing yourself is appropriate if it helps clients achieve treatment goals. Don't disclose more than is necessary. Helper self-disclosure that is exhibitionistic is obviously inappropriate. Self-disclosure on the part of helpers should be a natural part of the helping process, not a gambit. Rick's self-disclosure helps Tim get a different view of the "bad" things that happen in the past. Rick's "we" helps Tim see that he is not the only person who has had problems in the past.

**Be careful of your timing.** Timing is critical. Common sense tells us that premature helper self-disclosure can turn clients off. Rick did not share anything about himself immediately. He waited for a few sessions. However, once he saw an natural opening, he thought that sharing some of his own experiences would help.

**Keep your disclosure selective and focused.** Don't distract clients with rambling stories about yourself. In the following example, the helper is talking to a first-year grad student in a clinical-psychology program. The client is discouraged and depressed by the amount of work he has to do. The counselor wants to help him by sharing his own experience of graduate school.

> COUNSELOR: Listening to you brings me right back to my own days in graduate school. I don't think that I was ever busier in my life. I also believe that the most depressing moments of my life took place then. On any number of occasions, I wanted to throw in the towel. I remember once toward the end of my third year when . . .

It may be that selective bits of this counselor's experience in graduate school would be useful in helping the student get a better conceptual and emotional grasp of his problems, but he has wandered off into the kind of reminiscing that meets his needs rather than the client's. In contrast, Rick's disclosure was selective and focused.

**Don't disclose too frequently.** Helper self-disclosure is inappropriate if it is too frequent. If helpers disclose themselves too frequently, clients may see them as phony and suspect that they have hidden motives. If Rick had continued to share his experiences whenever he saw a parallel with Tim's, Tim might have wondered who was helper and who was client.

**Do not burden the client.** Do not burden an already overburdened client. One novice helper thought that he would help make a client who was sharing some sexual problems more comfortable by sharing some of his own sexual experiences. After all, he saw his own sexual development as not too different from the client's. However, the client reacted by saying, "Hey, don't tell me your problems. I'm having a hard enough time dealing with my own." This novice counselor shared too much of himself too soon. He was caught up in his own willingness to disclose rather than its potential usefulness to the client.

**Remain flexible.** Take each client separately. Adapt your disclosures to differences in clients and situations. When asked directly, clients say that they want helpers to disclose themselves (Hendrick, 1988), but this does not mean that every client in every situation wants it or would benefit from it. Even though Rick's disclosure to Tim was natural, it was an explicit decision on Tim's part.

## 4. Immediacy: Discussing the Helper-Client Relationship

Many, if not most, clients who seek help have trouble with interpersonal relationships, either as a primary or a secondary concern. Some of the difficulties clients have in their day-to-day relationships are also reflected in their relationships with their helpers. For instance, if they are compliant outside, they are often compliant in the helping process. If they become aggressive and angry with authority figures outside, they often do the same with helpers. Therefore, the client's interpersonal style can be examined, at least in part, through an examination of his or her relationship with the helper. If counseling takes place in a group, then the opportunity is even greater. The package of skills enabling helpers to explore their relationships with their clients or vice versa, or enabling clients to do the same with fellow group members and the group leader, has been called "immediacy" by Robert Carkhuff (Carkhuff 1969a, 1969b; Carkhuff & Anthony, 1979). Immediacy can be a useful tool for monitoring and managing the working alliance.

**Types of immediacy in helping and principles for using them.**   Two kinds of immediacy are reviewed here. First, immediacy that focuses on the overall relationship—"How are you and I doing?" Second, immediacy that focuses on some particular event in a session—"What's going on between you and me right now?"

*Relationship-focused: Review your ongoing relationship with the client if this adds value to the helping process.*   General relationship immediacy refers to your ability to discuss with a client where you stand in your overall relationship with him or her and vice versa. The focus is not on a particular incident but on the way the relationship itself has developed and how it is helping or standing in the way of progress. In the following example, the helper is a 44-year-old woman working as a counselor for a large company. She is talking to a 36-year-old man she has been seeing once every other week for about two months. One of his principal problems is his relationship with his supervisor, who is also a woman.

> COUNSELOR: We seem to have developed a good relationship here. I feel we respect each other. I have been able to make demands on you, and you have made demands on me. There has been a great deal of give-and-take in our relationship. You've gotten angry with me, and I've gotten impatient with you at times, but we've worked it out. I'm wondering what our relationship has that might be missing in your relationship with your supervisor.

> CLIENT: Well, for one thing, you listen to me, and I don't think she does. On the other hand, I listen pretty carefully to you, but I don't think I listen to her at all, and she probably knows it. I think she's dumb, and I guess I say that to her in a number of ways even without using the words. She knows how I feel.

The review of the relationship helps the client focus more specifically on his relationship with his supervisor.

Here is another example. Norman, a 38-year-old trainer in a counselor-training program, is talking to Weijun, 25, one of the trainees.

> NORMAN: Weijun, I'm a bit bothered about some of the things that are going on between you and me. When you talk to me, I get the feeling that you are being very careful. You talk slowly. And you seem to be choosing your words, sometimes to the point that what you are saying sounds almost prepared. You have never challenged me on anything in the group. When you talk most intimately about yourself, you seem

to avoid looking at me. I find myself giving you less feedback than I give others. I've even found myself putting off talking to you about all this. Perhaps some of this is my own imagining, but I want to check it out with you.

WEIJUN: I've been afraid to talk about all this, so I keep putting it off, too. I'm glad that you've brought it up. A lot of it has to do with how I relate to people in authority, even though you don't come across as an "authority figure." You don't act the way an authority figure is supposed to act.

In this case, cultural differences play a role. For Weijun, an American who was born in China, giving direct feedback to someone in authority is not natural. However, he does go on to talk to Norman about his misgivings. He thinks that Norman's interventions in the training group are too "unorganized" and that he plays favorites. He has not wanted to bring it up because he fears that his position in the program will be jeopardized. But now that Norman has made the overture, he accepts the challenge.

The interaction can, of course, be initiated by the client, though many clients for obvious reasons would hesitate to do so. Who wants to take on the leader?

**Event-focused immediacy: Address relationship issues as they come up.** Here-and-now immediacy refers to your ability to discuss with clients what is happening between the two of you in the here and now of any given transaction. It is not the entire relationship that is being considered, but rather the specific interaction or incident. In the following example, the helper, a 43-year-old woman, is a counselor in a neighborhood human-services center. Agnes, a 49-year-old woman who was recently widowed, has been talking about her loneliness. Agnes seems to have withdrawn quite a bit, and the interaction has bogged down.

COUNSELOR: I'd like to stop a moment and take a look at what's happening right now between you and me.

AGNES: I'm not sure what you mean.

COUNSELOR: Well, our conversation today started out quite lively, and now it seems rather subdued. I've noticed that the muscles in my shoulders have become tense. I sometimes tense up that way when I feel that I might have said something wrong. It could be just me, but I sense that things are a bit strained between us right now.

AGNES (hesitatingly): Well, a little . . .

Agnes goes on to say how she resented one of the helper's remarks early in the session. She thought that the counselor had intimated that she was lazy. Agnes knows that she isn't lazy. They discuss the incident, clear it up, and move on.

Both kinds of immediacy are at work in the following example. Yalom (Landreth, 1984, p. 323) is talking about his own relationship to one of his clients.

I am quite certain, even before I stopped carrying individual cases, I was doing more and more of what I would call confrontation . . . For example, I recall a client with whom I began to realize I felt bored every time he came in. I had a hard time staying awake during the hour, and that was not like me at all. Because it was a persisting feeling, I realized I would have to share it with him . . . So with a good deal of difficulty and some embarrassment, I said to him, "I don't understand it myself, but when you start talking on and on about your problems in what seems to me a flat tone of voice, I find myself getting very bored." This was quite a jolt to him and he looked very unhappy. Then he began to talk about the way he talked and gradually he came to understand one of the reasons for the way he presented himself verbally. He said, "You know, I think the reason I talk in such an uninteresting way is because I don't think I have ever expected anyone to really hear me." . . .We got along much better after that because I could remind him that I heard the same flatness in his voice I used to hear.

**Situations calling for immediacy.** Part of skilled helping—and, more generally, social intelligence—is knowing when to use any given communication skill. Immediacy can be useful in the following situations:

- When a session is directionless and it seems that no progress is being made.
- When there is tension between helper and client.
- When trust seems to be an issue.
- When diversity, some kind of social distance, or widely differing interpersonal styles between client and helper seem to be getting in the way.
- When dependency seems to be interfering with the helping process: "You don't seem willing to explore an issue until I give you permission to do so or urge you to do so. And I seem to have let myself slip into the role of permission giver and urger."

Make sure that you have a specific reason for using either form of immediacy. Talking about your relationship with a client is not an end in itself.

## 5. Suggestions and Recommendations

This section begins with a few imperatives. Don't tell clients what to do. Don't try to take over their lives. Let clients make their own decisions. These imperatives flow from the values of respect and empowerment. Does this mean, however, that suggestions and recommendations are forbidden? Of course not. I mentioned earlier that there is a natural tension between helpers' desire to have their clients manage their lives better and respecting their freedom. If helpers build strong, respectful relationships with their clients, then stronger interventions can make sense. In this context, suggestions and recommendations can stimulate clients to move to problem-managing action. Helpers move from counseling mode to guidance role. Research has shown that clients will generally go along with recommendations from helpers if those recommendations are clearly related to the problem situation, challenge clients' strengths, and are not too difficult. Effective helpers can provide suggestions, recommendations, and even directives without robbing clients of their autonomy or their integrity.

Here is a classic example of how powerful such an intervention can be, from Cummings's (1979, 2000) work with people with addictions. Substance abusers came to him because they were hurting in many ways. He used every communication skill available to listen to and understand their plight.

> During the first half of the first session the therapist must listen very intently. Then, somewhere in mid-session, using all the rigorous training, therapeutic acumen, and the third, fourth, fifth, and sixth ears, the therapist discerns some unresolved wish, some long-gone dream that is still residing deep in that human being, and then the therapist pulls it out and ignites the client with a desire to somehow look at that dream again. This is not easy, because if the right nerve is not touched, the therapist loses the client. (1979, p. 1123)

So Cummings shared both basic and advanced highlights to let clients know that he understood their plight, their longings, and also their games. They came knowing how to play every game in the book. But Cummings knew all the games, too. Toward the end of the first session he told them they could have a second session—which they invariably wanted—only when they were "clean." The time of the second session depended on the withdrawal period for the kind of substance they were abusing. They screamed, shouted "foul," tried to play games, but he remained adamant. The directive "Get clean, then return" was part of the therapeutic process. And most did return. Clean.

Suggestions, advice, and directives need not always be taken literally. They can act as stimuli to get clients to come up with their own package. One client said something like this to her helper: "You told me to let my teenage son have his say instead of constantly interrupting and arguing with him. What I did was make a contract with him. I told him that I would listen carefully to what he said and even summarize it and give it back to him. But he had to do the same for me. That has produced some useful monologues. And we avoid our usual shouting matches. My hope is to find a way to turn it into dialogue."

In daily human interactions, people feel free to give one another advice. It goes on all the time. But helpers must proceed with caution. Suggestions, advice, and directives are not for novices. It takes a great deal of experience with clients and a great deal of savvy to know when they might work.

## 6. Confrontation

What about clients who keep dragging their feet? Some clients who don't want to change or don't want to pay the price of changing simply terminate the helping relationship. However, those who stay stretch across a continuum from mildly to extremely reluctant and resistant. Or they may be collaborative on some issues, but reluctant when it comes to others. For instance, Hester is quite willing to work on career development but very reluctant to work on improving relationships, even though relationship building is an important part of the career package. "That's my private world," she says of her relationships.

If *inviting* clients to challenge themselves is at one end of the continuum, what's at the other? Where does respecting clients' right to be themselves stop and placing demands on them to live more fully begin? Different helpers answer these questions differently. Therefore, helpers differ, both theoretically and personally, in their willingness to confront. "Traumatic confrontation" (one wonders about the choice of name) is a cognitive behavior modification technique (Lowenstein, 1993) that involves challenging youths to face up to and change dysfunctional behavior. For example, a 12-year-old boy who had become involved in criminal activity after the disappearance of his father was confronted about his behavior. At first he denied everything, but then decided to face up to the situation.

When all is said and done, there is a place in helping for interventions strong enough to merit the term confrontation. Confrontation, as intimated earlier, means challenging clients to develop new perspectives and to change both internal and external behavior even when they show reluctance and resistance to doing so. When helpers confront, they "make the case" for more effective living. Confrontation does not involve "do this or else" ultimatums. More often it is a way of making sure that clients understand what it means not to change—that is, making sure they understand the consequences of persisting in dysfunctional patterns of behavior or of refusing to adopt new behaviors. Welch (1998) suggested that helpers confront the defense mechanism, not the client (something like "loving the sinner but hating the sin"). In this sense, confrontation is actually another way of caring for the client.

Both advice giving and confrontation require high levels of social intelligence and social competence on the part of the helper. In most cases, they should be used sparingly. And when helpers do judge that they might well serve the interests of their clients, they should be guided by the values outlined earlier. That is, confrontation should be empathic and respectful, empower the client, and lead to action. Care should be taken to see the difference between a confrontation that serves the client's needs and confrontation which is mere venting on the part of the helper.

## 7. Encouragement

This section ends on a positive-psychology note. You may not have noticed, but in the last two chapters some form of the word "encouragement" has been used only a couple of times. If the whole purpose of challenging is to help the client move forward, and if encouragement (sugar) works as well as challenge (vinegar), then why don't we hear more about encouragement? The sugar-vinegar analogy is not exactly right, because many clients find challenge both refreshing and stimulating. Challenge certainly does not preclude encouragement. Rather, encouragement itself is a mild form of challenge. Furthermore, encouragement is a form of support and research shows that support is often one of the main ingredients in successful therapy (Beutler, 2000, p. 1004).

Miller and Rollnick (1991; Rollnick & Miller, 1995) introduced an approach to helping called "motivational interviewing." A simple Internet search on "motivational interviewing" reveals an extensive literature, including theory, research, and case studies. Their original work focused on helping clients deal with addictive behavior, but their methodology over the years has been adapted to a much wider range of human problems. Much of the literature highlights the main elements of a problem-management approach. The values of respect, empathy, self-empowerment, and self-healing are emphasized. The spirit of encouragement rather than confrontation pervades the approach. Typically, clients (for instance, pregnant women who are smokers or users of alcohol) receive personal feedback on their problem area, such as how smoking has been affecting their lungs and dangers it has for the child. The literature offers discussions of personal responsibility and ways of encouraging clients to find the motives, incentives, and levers of change that make sense for themselves. Intrinsic motives—that is, motives that clients have internalized for themselves ("I want to be free"), rather than extrinsic motives ("I'll get in trouble if I don't change") are emphasized. This literature also describes ways of helping clients identify obstacles to change and ways of overcoming them.

Common sense suggests that realistic encouragement be included among the set of helping skills. Like most of the skills we have been discussing, encouragement can be used at any stage of the helping process. Clients can be encouraged to identify and talk about their problems and unused opportunities, to review possibilities for a better future, to set goals, to engage in actions that will help them achieve their goals, and to overcome the inevitable obstacles that stand in their way. Effective encouragement is not patronizing. It is not the same as sympathy, nor does it rob the client of autonomy. It respects the client's self-healing abilities. It is a fully-human nudge in the right direction.

## From Smart to Wise: Guidelines for Effective Challenging

Your challenges might be on the mark (smart) and still be ineffective or even hurtful. All challenges should be permeated by the spirit of the client-helper relationship values discussed in Chapter 2; that is, they should be based on understanding, caring (not power games or put-downs); they should be genuine (not tricks or games), and designed to increase the self-responsibility of the client (not ways in which helpers exercise control). They should also serve the stages and tasks of the helping process. Empathy should permeate every kind of challenge. Clearly, challenging well is not a skill that comes automatically. It needs to be learned and then practiced with care. The following principles constitute some basic guidelines for making challenging not just accurate, but wise.

**Keep the goals of challenging in mind.**  Challenge must be integrated into the entire helping process. Keep in mind that the goal is to help clients develop the kinds of alternative perspectives, internal behaviors, and external actions that constitute constructive change. Are the insights relevant to real problems and opportunities rather than merely being dramatic in themselves? To what degree do the new perspectives developed lead to problem-managing and opportunity-developing action?

**Encourage self-challenge.**  Invite clients to challenge themselves, and give them ample opportunity to do so. You can provide clients with probes and structures that help them engage in self-challenge. In the following excerpt, the counselor is talking to a man who has discussed at length his son's ingratitude. There has been something cathartic about his complaints, but it is time to move on.

COUNSELOR:  People often have blind spots in their relationships with others, especially in close relationships. Picture your son sitting with some counselor. He is talking about his relationship with you. What's he saying?

CLIENT:  Well, I don't know . . . I guess I don't think about that very much . . . Hmm . . . He'd probably say . . . well, that he loves me . . . (pauses). And then he might say that since his mother died, I have never really let him be himself. I've done too much to influence the direction of his life rather than let him fashion it the way he wanted . . . Hmm. He'd say that he loves me but he has always resented my "interference."

COUNSELOR:  So, both love for you and resentment for all that control.

CLIENT:  And he'd be right. I'm still doing it. Not with him. He won't let me. But with lots of others. Especially in my business.

The counselor provides a structure that enables the client to challenge himself with respect to his son and then apply what he learns to other settings of life. Would that all clients could respond so easily! Alternatively, the counselor might have asked this client to list three things he thinks he does right and three things he thinks he should reconsider in his relationship with his son. The point is to be inventive with the probes and structures you provide clients to help them challenge themselves.

**Earn the right to challenge.**  Long ago, Berenson and Mitchell (1974) claimed that some helpers don't have the right to challenge others because they are not doing a good job keeping their own houses in order. They made a point. Here are some of the factors that earn you the right to challenge:

- *Develop a working relationship.*  Challenge only if you have spent time and effort building a relationship with your client. If your rapport is poor or you have allowed your relationship with the client to stagnate, then challenge yourself to deal with the relationship more creatively.

- *Make sure you understand the client.*  Empathy drives everything. Effective challenge flows from accurate understanding. Only when you see the world through the client's eyes can you begin to see what he or she is failing to see.

- *Be open to challenge yourself.*  Hesitate to challenge unless you are open to being challenged. If you are defensive in the counseling relationship, or in your relationship with supervisors, or in your everyday life, your challenges might ring hollow. Model the kind of nondefensive attitudes and behavior that you would like to see in your clients.

- *Work on your own life.*  How important is constructive change in your own life? Berenson and Mitchell claimed that only people who are striving to live fully according to their value system have the right to challenge others, for only such persons are potential sources of human nourishment for others. You may disagree with them, but what they say should get all helpers thinking.

In summary, ask yourself, "To what degree am I the kind of person from whom clients would be willing to accept challenges?"

**Be tentative but not apologetic in the way you challenge clients.** Tentative challenges are generally viewed more positively than strong, direct challenges. The principle is this: Challenge clients in such a way that they are more likely to respond rather than react. The same challenge can be delivered in such a way as to invite the cooperation or arouse the resistance of the client. Deliver challenges tentatively, as hunches that are open to review and discussion rather than as accusations. Challenging is certainly not an opportunity to browbeat clients or put them in their place.

On the other hand, challenges that are delivered with too many qualifications—either verbally or through the helper's tone of voice—sound apologetic and can be easily dismissed by clients. I was once working in a career-development center. As I listened to one of the clients, it soon became evident that one reason he was getting nowhere was that he engaged in so much self-pity. When I shared this observation with him, I overqualified it. These are not my exact words, but they must have sounded something like this:

> HELPER: Has it ever, at least in some small way, struck you that one possible reason for not getting ahead, at least as much as you would like, could be that at times you tend to engage in a little bit of self-pity?

I still remember his response. He paused, looked me in the eye, and said, "A little bit of self-pity?" When he paused again, I said to myself, "I've been too harsh!" He continued, "I *wallow* in self-pity." We moved on to explore what he might do to move beyond self-pity to constructive change.

**Challenge unused strengths more than weaknesses.** Berenson and Mitchell (1974) found that successful helpers tend to challenge clients' strengths rather than their weaknesses. What we talk about sets the tone for helping. Individuals who focus on their failures find it difficult to change their behavior. Clients who dwell too much on their shortcomings tend to belittle their achievements, to withhold rewards from themselves when they do achieve, and to live with anxiety. All of this tends to undermine performance.

Challenging strengths is a positive-psychology approach. It means pointing out to clients the assets and resources they have but fail to use. In the following example, the helper is having a one-to-one session with a woman who is a member of a self-help group in a rape crisis center. She is very good at helping others but is always down on herself.

> COUNSELOR: Ann, in the group sessions, you provide a great deal of support for the other women. You have an amazing ability to spot a person in trouble and provide an encouraging word. And when one of the women wants to give up, you are the first to challenge her, gently and forcibly at the same time. But when Ann is dealing with Ann . . .

> ANN: I know where you're headed . . . I know I'm a better giver than receiver. I'm much better at caring than being cared about. I'm not sure why that is . . . Or that it

even matters. I'm sure this is not lost on the other members of the group . . . You know, I've been this way for a long time. I think I've got some bad habits when it comes to dealing with myself. I'm so fearful of being self-indulgent.

The counselor helps Ann place a demand on herself to use her rather substantial resources in her dealings with herself. Since she isn't self-indulgent, it's time to take a look at her resistance to being cared about.

Even adverse life experiences can be a source of strength. For instance, McMillen, Zuravin, and Rideout (1995) studied adult perceptions of benefit from child sexual abuse. Almost half the adults reported some kind of benefit, including increased knowledge of child sexual abuse, protecting other children from abuse, learning how to protect themselves from others, and developing a strong personality—without, of course, discounting the horror of child abuse. Counselors, therefore, can help clients *mine* benefits from adverse experiences, putting to practical use the age-old dictum that "good things can come from evil things." People are more resilient than we make them out to be.

**Build on the client's successes.** Effective helpers do not urge clients to place too many demands on themselves all at once. Rather, they help clients place reasonable demands on themselves and in the process help them appreciate and celebrate their successes. In the following example, the client is a boy in a detention center who is rather passive and allows the other boys to push him around. Recently, however, he has made a few half-hearted attempts to stick up for his rights in the counseling group. The counselor is talking to him alone after a group meeting.

COUNSELOR A: You're still not standing up for your own rights the way you need to. You said a couple of things in there, but you still let them push you around.

This counselor emphasizes the negative and browbeats the client. The following counselor takes a different tack.

COUNSELOR B: Here's what I've noticed. In the group meetings you have begun to speak up. You say what you want to say, even though you don't say it really forcefully. And I get the feeling that you feel good about that. You've got some power. Now the challenge is to find ways of using it more effectively.

CLIENT: I didn't think anyone noticed. You think it was a good start?

COUNSELOR B: Certainly . . . But what you think is more important. And it sounds like you're proud of what you did.

The second counselor emphasizes the client's success, however modest, and goes on to provide some encouragement to do even better.

**Be specific in your challenges.** Specific challenges hit the mark. Vague challenges get lost. Clients don't know what to do about them. Statements such as "You need to pull yourself together and get on with it" may satisfy some helper need, such as the ventilation of frustration, but they do little for clients. Specific statements, on the other hand, can hit the mark. In the following example, the client is experiencing a great deal of stress both at home and at work.

HELPER:    You say that you really want to spend more time at home with the kids and you really enjoy it when you do, but you keep taking on new assignments at work, like the Eclipse project, which will add to your travel schedule. Maybe it would be helpful to talk a bit more about work-life balance.

CLIENT:    Boy, there's that phrase! Work-life balance. The company talks a lot about it, but nothing much happens. I'm not sure there's anyone at work who's got the work-life balance right.

HELPER:    You know what they say about career—"If you're not in charge of your career, no one is." It sounds like the same is true with work-life balance.

CLIENT:    I hadn't thought about it like that . . . But I'm afraid you're right. It's right where it belongs, I suppose, on my shoulders. I've been waiting for my family and my company to figure it out for me.

Some helpers avoid clarity and specificity because they feel that they're being too intrusive. Helping has to be intrusive to make a difference.

**Respect clients' values.**    Challenge clients to clarify their values and to make reasonable choices based on them. Be wary of using challenging, even indirectly, to force clients to accept your values. This violates the empowerment value discussed in Chapter 2. In the following example the client is a 21-year-old woman who has curtailed both her social life, her education, and her career to take care of her elderly mother who is suffering from incipient Alzheimer's.

CLIENT:    I admit that juggling work, home, and school is a real challenge. I keep feeling that I'm not doing justice to any of them.

COUNSELOR A:    You have every right to have a life of your own. Why not get your mother into a nursing facility? You can still visit her regularly. Then get on with life. That's probably what she wants anyway.

Challenging clients to clarify their values is, of course, legitimate. But this counselor does little to help the client clarify her values. She makes suggestions without finding out why the client is doing what she's doing. Counselor B takes a different approach.

COUNSELOR B:    You're trying to juggle four very important areas of yours life—caring for your mother, school, work, and social life. That's a tough assignment. It might help to explore what's driving you in all this. Maybe we could take a look at the values that drive your behavior. Then you could ask yourself about your priorities.

CLIENT:    I've never thought about values as things that drive what I do. I thought we had values and just, well, did them. So let's take a look.

This counselor challenges her gently to find out what she really wants. Helpers can assist clients to explore the consequences of the values they hold, but that is not the same as questioning them.

**Deal honestly, caringly, and creatively with client defensiveness.**    Do not be surprised when clients react strongly to being challenged even when you're trying to help them respond rather than react. If they react negatively, help them work through their reluctance and resistance. Help them share and work through their emotions. If they seem to clam up, try to find out what's going on inside. In the following example, the helper has just delivered a brief summary of the main points of a problem he and his client have been discussing and has gently pointed out the self-destructive nature of some of his client's behaviors.

HELPER: I'm not sure how all this sounds to you.

CLIENT: I thought you were on my side. Now you sound like all the others. And I'm paying you to talk like this to me!

Even though the helper was tentative in his challenge, the client still reacts defensively. Here are two different approaches (A and B) to the client's defensiveness.

A: All I've done is summarize what you have been saying about yourself. And you know you're doing yourself in. Let's look at each point we've been discussing and see if this isn't the case.

This helper takes a defensive, judicial approach. She's about to assemble the evidence. This could well lead to an argument rather than further dialogue. Helper B backs off a bit.

B: So, I'm sounding harsh and unfair to you . . . Kind of dumping on you . . . Let's back up.

This helper backs off without saying that her summary was wrong. She is giving the client some space. It may be that the client needs time to think about what the helper has said. In the B response, the helper tries to get into a constructive dialogue with the angered client.

The principles outlined above are, of course, guidelines, not absolute prescriptions. In the long run, use your common sense. The more flexible you are, the more likely you are to add value to your clients' search for solutions.

## Linking Challenge to Action

More and more theoreticians and practitioners are stressing the need to link insight to problem-managing and opportunity-developing action. Wachtel (1989, p. 18) put it succinctly: "There is good reason to think that the really crucial insights are those closely linked to new actions in daily life and, moreover, that insights are as much a product of new experience as their cause."

Is challenge enough to stimulate problem-managing action? For some, yes. A few well-placed challenges might be all that some clients need to move to constructive action. Once challenged, they are off to the races. Others may well have the resources to manage their lives better, but not the will. They know what they need to do but are not doing it. A few nudges in the right direction help them overcome their inertia. I have had many one-session encounters that included a bit of listening and thoughtful processing, some sharing of highlights, and a brief challenge that produced a new perspective that sent the client off on some useful course of action. On the other hand, if helping clients challenge themselves to develop new perspectives leads to one profound insight after another but no action and behavioral change, then once more we are whistling·in the wind. Helping becomes in intellectual game rather than a serious attempt to partner with clients in their search for a better life.

## Challenge and the Shadow Side of Helpers

Helpers have many characteristics and engage in many behaviors that add value to clients' efforts to manage their lives better. But, helper behavior in counseling sessions can also be part of the shadow side of helping.

**The "MUM effect."**  Initially, some counselor trainees are quite reluctant to help clients challenge themselves. They become victims of what has been called the "MUM effect," the tendency to withhold bad news even when it is in the other's interest to hear it (Rosen & Messer, 1970, 1971; Messer & Rosen, 1972; Messer, Rosen, & Bachelor, 1972; Messer, Rosen, & Messer, 1971). In ancient times, the person who bore bad news to the king was sometimes killed. That obviously led to a certain reluctance on the part of messengers to bring such news. Bad news—and, by extension, the kind of "bad news" that is involved in any kind of invitation to self-challenge—arouses negative feelings in the challenger, no matter how he or she thinks the receiver will react. If you are comfortable with the supportive dimensions of the helping process but uncomfortable with helping as a social-influence process, you could fall victim to the MUM effect and become less effective than you might otherwise be.

**Excuses for not challenging.**  Reluctance to challenge is not a bad starting position. In my estimation, it is a far better approach than being too eager to challenge. However, all helping, even the most client-centered, involves social influence. It is important for you to understand your reluctance (or eagerness) to challenge—that is, to challenge yourself on the issue of challenging and on the very notion of helping as a social-influence process. When trainees examine how they feel about challenging others, here are some of the things they discover:

- I am just not used to challenging others. My interpersonal style has had a lot of the live-and-let-live in it. I have misgivings about intruding into other people's lives.

- If I challenge others, then I open myself to being challenged. I may be hurt, or I may find out things about myself that I would rather not know.

- I might find out that I like challenging others and that the floodgates will open and my negative feelings about others will flow out. I have some fears that deep down I am a very angry person.

- I am afraid that I will hurt others, damage them in some way or other. I have seen others hurt by heavy-handed confrontations.

- I am afraid that I will delve too deeply into others and find that they have problems that I cannot help them handle. The helping process will get out of hand.

- If I challenge others, they will no longer like me. I want my clients to like me.

Vestiges of this kind of thinking can persist long after trainees move out into the field as helpers. People in all sorts of people-oriented occupations, including human-service workers and managers, are bedeviled by the MUM effect and come up with their own set of excuses for not giving feedback. Of course, being willing to challenge responsibly is one thing; having the skills and wisdom to do so is another. One reason managers fail to give feedback is that they do not have the communication, coaching, and counseling skills needed to do it well.

**Helpers' blind spots.**    There is an interesting literature on the humanity and flaws of helpers (Kottler, 2000) that can be of enormous help to both beginners—since prevention is infinitely better than cure—and to old-timers—since you *can* teach old dogs new tricks. Kottler has provided trainees and novices an upbeat view of what passion and commitment in the helping professions should look like.

Since helpers are as human as their clients, they too can have blind spots that detract from their ability to help. There is some evidence that adopting a resource-collaborator role is difficult for some helpers. For instance, in one study (Atkinson and his associates, 1991) counselors were almost unanimous in their preference for a "feeling" approach to counseling, whereas the majority of male clients preferred either a "thinking" or an "acting" orientation. This is not collaboration.

One of the critical responsibilities of supervisors is to help counselors identify their blind spots and learn from them. Once out of training, skilled helpers use different forums or methodologies to continue this process, especially with difficult cases. They take counsel with themselves, asking, "What am I missing here?" They take counsel with colleagues. Without becoming self-obsessive, they scrutinize and challenge themselves and the role they play in the helping relationship. Or, more simply, throughout their careers they continue to learn about themselves, their clients, and their profession.

## Evaluation Questions: The Use of Specific Challenge Skills

**How effectively have I been developing the communication skills listed below that serve the process of challenging?**

- **Sharing advanced empathic highlights.** Sharing hunches with clients about their experiences, behaviors, and feelings to help them move beyond blind spots and develop needed new perspectives.

- **Information sharing.** Giving clients needed information, or helping them search for it, to help them see problem situations in a new light and to provide a basis for action.
- **Helper self-disclosure.** Sharing your own experience with clients as a way of modeling nondefensive self-disclosure and helping them move beyond blind spots.
- **Immediacy.** Discussing aspects of your relationship with your clients to improve the working alliance.
- **Suggestions and recommendations.** Pointing out ways in which clients can more effectively manage problems and develop opportunities, or engage more productively in the stages of the helping process.
- **Confrontation.** Using a solid relationship with the clients to challenge them more forcefully when they show signs of reluctance.
- **Encouragement.** Encouraging clients, at any stage of the helping process, to find within themselves the desire to move forward.

## The Wisdom of Challenging
## How well do I do each of the following as I try to help my clients?

- Invite clients to challenge themselves
- Earn the right to challenge
- Be tactful and tentative in challenging without being insipid or apologetic
- Be specific; develop challenges that hit the mark
- Challenge clients' strengths rather than their weaknesses
- Don't ask clients to do too much too quickly
- Invite clients to clarify and act on their own values

## The Shadow Side of Challenging
## How well do I do the following?

- Identify the games my clients attempt to play with me without becoming cynical in the process
- Become comfortable with the social-influence dimension of the helping role and with the kind of "intrusiveness" that goes with helping
- Incorporate challenge into my counseling style without becoming a confrontation specialist
- Develop the assertiveness needed to overcome the MUM effect
- Challenge the excuses I give myself for failing to challenge clients
- Come to grips with my own imperfections and blind spots both as a helper and as a "private citizen"

**EXERCISES ON PERSONALIZING**

1. Review the seven specific challenging skills outlined in this chapter. Choose two or three and then write out a challenge that you think would help a friend manage a problem situation or develop an unused opportunity more effectively.

2. Without naming the friend, share the challenge with someone else. What kind of second opinion did you get? Review the challenge in the light of the principles of effective challenge. What went well? What could be improved?

3. After reviewing what good challenges look like, write out a somewhat detailed challenge you yourself would benefit from at this time. If possible, share your challenge with a friend and get his or her reaction. How well did you apply the principles of effective challenge to your self-challenge?

# The Stages and Tasks of the Helping Model

*T*he communication skills reviewed in Part Two are critical tools. With them, you can help clients engage in all the stages of the helping model. But those communication skills are not the helping process itself. Part Three is a detailed exposition and illustration of the stages of the helping model. Stage I involves helping clients tell their stories in terms of problem situations and unused opportunities. In Stage I, counselors help clients tell their stories, break through blind spots, develop new perspectives, and work on issues that can make a difference in clients' life.

# CHAPTER 8

# Stage I: Helping Clients Tell Their Stories: An Introduction to Stage I

An Introduction to Stage I

Task 1: Help Clients Tell Their Stories: "What Are My Concerns?"

Guidelines for Helping Clients Explore Problem Situations and Unexploited Opportunities

Action Right From the Beginning

Task 2: Help Clients Develop New Perspectives and Reframe Their Stories: "What Am I Overlooking or Avoiding?"

Task 3: Help Clients Achieve Leverage by Working on Issues That Make a Difference: "What Will Help Me Most?"

Guidelines in the Search for Leverage

Leverage and Action

Is Stage I Enough?

The Shadow Side of Leverage

Exercises on Personalizing

## An Introduction to Stage I

Clients come to helpers because they need help in managing their lives more effectively. Stage I illustrates three ways in which counselors can help clients understand themselves, their problem situations, and their unused opportunities with a view to managing them more effectively. There are three interrelated tasks.

**Task 1: Help clients tell their stories.** "What are my concerns?" Guidelines for helping clients explore problem situations and unexploited opportunities. The questions here are, "What's going on? What are my concerns? What are the issues?"

**Task 2: Help clients develop new perspectives and reframe their stories.** "What am I overlooking or avoiding?" Help clients identify the critical elements of the problem situation or unused opportunity. This often means helping clients identify blind spots, develop new perspectives, and reframe the story itself. The questions here are, "What's really going on? What are my real concerns? What am I failing to see? What are the critical elements of the story?" A judicious use of challenging skills is needed here.

**Task 3: Help clients achieve leverage by working on issues that make a difference.** "What will help me most?" Clients often have multiple problems. If that is the case, help clients work on substantive issues. The questions here are, "What are my key concerns? What should I work on? What will make a difference in my life?"

Even though these tasks are described separately here, in actual helping sessions they are intermingled. It is not a question of moving from the first to the second to the third task in any rigid way. Furthermore, these tasks are not restricted to Stage I for the following reasons: First, clients don't tell all of their stories at the beginning of the helping process. Often the full story "leaks out" over time. Second, blind spots can appear at any stage of the helping process. Blind spots affect choosing goals, setting strategies, and implementing programs. Figure 8.1 outlines the tasks of Stage 1.

Stage I of the helping process can be seen as the assessment stage—finding out what's going wrong, what opportunities lie fallow, what resources are not being used. Client-centered assessment means helping clients understand themselves, find out "what's going on" in their lives, see what they have been ignoring, and make sense out the messiness of their lives. Assessment, in this sense, is not something helpers do to clients. Rather, it is a kind of learning in which, ideally, both client and helper participate through their ongoing dialogue. Other forms of assessment, such as psychological testing and applying psychiatric diagnostic categories, can be useful, but they are beyond the scope of this book. The title of each of the tasks includes a question clients can ask themselves.

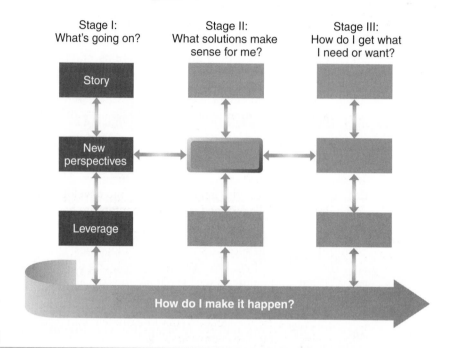

**The Skilled-Helper Model**

Stage I:
What's going on?

Stage II:
What solutions make
sense for me?

Stage III:
How do I get what
I need or want?

Story

New
perspectives

Leverage

How do I make it happen?

FIGURE

**8.1**                           The Helping Model—Stage 1

## Task 1: Help Clients Tell Their Stories: "What Are My Concerns?"

The importance of helping clients tell their stories well should not be underestimated. As Pennebaker (1995b) has noted, "An important. . . feature of therapy is that it allows individuals to translate their experiences into words. The disclosure process itself, then, may be as important as any feedback the client receives from the therapist" (p. 3). Farber, Berano, and Capobianco (2004) outline the bright side of client self-disclosure.

> [M]ost clients feel that therapy is a safe place to disclose, made especially so by the goodness of the therapeutic relationship; that the disclosure process initially generates shame and anticipatory anxiety but ultimately engenders feelings of safety, pride, and authenticity; that keeping secrets inhibits the work of therapy, whereas disclosing produces a sense of relief from physical as well as emotional tension; that disclosures in therapy facilitate subsequent disclosures to one's therapist as well as to family members and friends; and that therapists should actively pursue material that is difficult to disclose (p. 340).

Self-disclosure provides the grist for the mill of problem solving and opportunity development. Here, then, are some of the goals for Stage I:

**Initial Stress Reduction.**  Help clients get things out on the table. This can, and often does, have a cathartic effect that leads to stress reduction. Some clients carry their secrets around for years. Helping them unburden themselves is part of the social-emotional reeducation process alluded to in the previous chapter.

**Clarity.**  Help clients spell out their problem situations and unexploited opportunities with the kind of concrete detail—specific experiences, thoughts, behaviors, and emotions—that enables them to do something about them. Clarity opens the door to more creative options in living. Vague stories lead to vague options and actions.

**Establishing Rapport and Relationship Building.**  Help clients tell their stories in such a way that the helping relationship develops and strengthens. The communication skills outlined in earlier chapters are basic tools for both clarity and relationship building.

**Client Action.**  Right from the beginning, help clients act on what they are learning. Clients do not need so-called grand plans before they can act on their own behalf.

Clients differ radically in their ability to talk about themselves and their problem situations. Reluctance to disclose oneself within counseling sessions is often a window into the client's inability to share himself or herself with others and to be reasonably assertive in the social settings of everyday life. If this is the case, then one of the goals of the entire counseling process is to help clients develop the skills, confidence, and courage they need to share themselves appropriately.

## Guidelines for Helping Clients Explore Problem Situations and Unexploited Opportunities

There are a number of principles that can guide you as you help clients tell their stories.

**Learn to work with all styles of storytelling.**  There are both individual and cultural differences in clients' willingness to talk about themselves. Both affect storytelling. Some clients are highly verbal and quite willing to reveal almost everything about themselves at the first sitting. Take the case of Martina.

Martina, 27, asks a counselor in private practice for an appointment to discuss "a number of issues." Martina is both verbal and willing to talk, and her story comes tumbling out in rich detail. Although the helper uses the skills of attending, listening, sharing highlights, and probing, she does so sparingly. Martina is too eager to tell her story.

Although trained as a nurse, Martina is currently working in her uncle's business because of an offer she "could not turn down." She is doing very well financially, but she feels guilty because service to others has always been a value for her. And, although she likes her current job, she also feels hemmed in by it. A year of study in Europe during college whetted her appetite for "adventure." She feels that she is nowhere near fulfilling the great expectations she has for herself.

She also talks about her problems with her family. Her father is dead. She has never gotten along well with her mother, and now that she has moved out of the house, she feels that she has abandoned her younger brother, who is twelve years younger than she is and whom she loves very much. She is afraid that her mother will "smother" her brother with too much maternal care.

Martina is also concerned about her relationship with a man who is two years younger than she. They have been involved with each other for about three years. He wants to get married, but she feels that she is not ready. She still has "too many things to do" and would like to keep the arrangement they have.

This whole complex story—or at least a synopsis of it—comes tumbling out in a torrent of words. Martina feels free to skip from one topic to another. The way Martina tells her story is part of her enthusiastic style. At one point she stops, smiles, and says, "My, that's quite a bit, isn't it!"

As the helper listens to Martina, he learns a number of things about her. She is young, bright, and verbal and has many resources; she is eager and impatient; some of her problems are probably of her own making; she has some blind spots that could stand in the way of her grappling more creatively with her problems; she has many unexplored options, many unexploited opportunities. That said, the counselor surmises that Martina would probably make her way in life, however erratically, with no counseling at all.

Contrast that example with the following one of a man who comes to a local mental-health center because he feels he can no longer handle his 9-year-old boy.

Nick is referred to the clinic by a doctor in a local clinic because of the trouble he is having with his son. He has been divorced for about two years and is living in a housing project on public assistance. After introductions and preliminary formalities have been taken care of, he just sits there and says nothing; he does not even look up. Since Nick offers almost nothing spontaneously, the counselor uses a relatively large number of probes to help him tell his story. Even when the counselor responds with empathic highlights, Nick volunteers very little. Every once in a while, tears well up in his eyes. When asked about the divorce, he says he does not want to talk about it. "Good riddance" is all he can say about his former wife. Gradually, almost torturously, the story gets pieced together. Nick talks mostly about the trouble his son is getting into, how uncontrollable he seems to be getting, and how helpless he (Nick) feels.

Martina's story is full of possibilities, whereas Nick's is mainly about limitations. In both content and communication style, they are at opposite ends of the scale.

Each client is different and will approach the telling of the story in a different way. Some clients will come voluntarily; others will be sent. Some of the stories you will help clients tell will be long and detailed, others short and stark. Some will be filled with emotion; others will be told coldly, even though they are stories of terror. Some stories will be, at least at first blush, single-issue stories— "I want to get rid of these headaches"—whereas others, like Martina's, will be

## Problem Finding

**Here are some of the kinds of questions counselors can help clients ask themselves to find and specify problem situations. You can probably think of others.**

- What are my concerns?
- What's problematic in my life?
- What issues do I need to face?
- What's troubling me?
- What would those who know me best say of me?
- What's keeping me back from being what I want to be?
- What keeps me from doing what I want to do?
- What do I need to resolve?

multiple-issue stories. Some stories will deal with the inner world of the client almost exclusively—"I hate myself; I'm depressed; I feel lonely"—whereas others will deal with the outer world, for instance, problems with finances, work, or relationships. Still others will deal with a combination of inner and outer concerns.

Some clients will tell the core story immediately, while others will tell a secondary story first to test your reactions. Some clients will make it clear that they trust you just because you are a helper, but you will read mistrust in the eyes of others, sometimes just because you are a helper.

In all these cases, your job is to establish a working relationship with your clients and help them tell their stories as a prelude to helping them manage the problems and take advantage of the opportunities buried in those stories. A story that is brought out into the open is the starting point for possible constructive change. Often the very airing of the story is a solid first step toward a better life.

When clients like Martina pour out their stories, you may let them go on or you may insist on some kind of dialogue. If the client tells the *whole* story in a more or less nonstop fashion, it will be impossible for you to share highlights relating to every core issue the client has brought up. But you can then help the client review the most salient points in some orderly way. Expressions like the following can be used to help the client review the core parts of the story: "You've said quite a bit. Let's see if I've understood some of the main points you've made. First of all . . ." At this point the highlights you share will let the client know that you have been listening intently and that you are concerned about him or her. With clients like Nick, however, it's a different story. Clients who lack the skills needed to tell their stories well, or who are reluctant to do so, constitute a different kind of challenge. Engaging in dialogue with them can be tough work. So you have some idea of the variety of stories you will run into. Box 8.1 provides questions you can help clients ask themselves to identify problem situations.

**Help clients spot unused opportunities.** Note that Martina's story is about both problem situations and opportunities. Many stories are a mixture of both. Early in the history of modern psychology, William James remarked that few people bring to bear more than about 10% of their human potential on the problems and challenges of living. Others since James, though changing the percentages somewhat, have said substantially the same thing, and few have challenged their statements. It is probably not an exaggeration to say that unused human potential constitutes a more serious social problem than emotional disorders, since it is more widespread. If this is the case, then most clients you meet will have unused opportunities that can play a role in helping them manage their problem situation. Pursuing an opportunity can be a way of transcending, rather than *solving*, a problem.

> Bruno, 23, a graduate student in accounting, came from a strong religious background. He said that he was "obsessed" by sexual thoughts. When he gave into them he felt very guilty. He liked the fact his religion put a great deal of emphasis on helping others. He was doing the degree in accounting because he believed that it would help him get a decent job, but he did not find the subject matter intellectually stimulating. Working in a bank and attending school did not give him much time for socializing, but he wasn't a big socializer anyway. Nor did he have time for doing the kind of volunteer work encouraged by his church. He thought he would get to that later. In short, there was a kind of vacuum in his life. Sensing the vacuum as possible opportunity, the counselor asked Bruno how wide his church's focus on helping others extended. Bruno thought for a moment and then replied, "To the whole world." In the discussion that followed, Bruno realized that he didn't know much about the world. Developing a sense of the world was a good idea; it was an opportunity. The counselor doubted that "sexual obsession" was Bruno's main issue. In fact, Bruno set up a program for developing a sense of the world for himself that included reading, an occasional lecture, a course in developing-country finance, and the luxury of a very occasional conference. He made friends with some of the people he met through these activities and eventually met a like-minded young woman whom he began to date. His "obsession" disappeared.

Clients are much more likely to talk about problem situations than about unused opportunities. That's a pity, since clients can manage many problems better by developing unused opportunities instead of dealing directly with their problems. Let's look at some examples.

Lou Anne, a single woman in her mid-20s, was arrested when she broke into a government computer. She was fined and spent some time in jail, a defeated and demoralized hacker. She became friends with Marion, who one day convinced her to go to a religious service. "Just this one time," Lou Anne said. She discovered that the minister was a man with lots of street smarts and who was "not too pious." She liked the service more than she wanted to admit. After a few weeks, she saw this man "for a bit of advice." Once he found out that she was a hacker, he rolled his eyes and said, "Boy, could we use you." She saw him on and off and discussed what she'd do on the outside. Once out, she helped him set up a training program in computer skills for disadvantaged kids. The fact that she had been a hacker appealed to the kids. They listened to her. She supported herself by working as a consultant in the computer security

## Opportunity Finding

**Here are some questions counselors can help clients ask themselves to identify unused opportunities.**

- What are my unused skills/resources?
- What are my natural talents?
- How could I use some of these?
- What opportunities do I let go by?
- What ambitions remain unfulfilled?
- What could I accomplish if I put my mind to it?
- What could I become good at if I tried?
- Which opportunities should I be developing?
- Which role models could I be emulating?

business. This kept her in touch with the latest in hacking techniques, which was of course, still her passion. How fortunate the minister saw talent, however misdirected, rather than depravity. Box 8.2 outlines some of the questions you can help clients ask themselves in order to identify unused opportunities.

**Start where the client starts.**   Clients have different starting points when they launch into their stories. They can start with any stage of the helping process. Join them there. Therefore, "story" is used in its widest sense. It does not mean, narrowly, "This is what happened to me, here's how I reacted, and now this is how I feel." Your job is to stay with your clients no matter where they are, not where you would like them to be. Consider the following:

- Martha starts by saying, "I thought I knew how to handle my son when he reached his teenage years. I knew he might want to try all sorts of crazy things, so I might have to keep the reins pretty tight. And that's what I did. But now things are awful. It's not working. He's out of control." Her starting point is a *failed solution* which has spawned a new problem. Her version of problem management went wrong somewhere.

- Thad says, "I don't know whether I want to be a doctor or a politician—or at least a political scientist. I love both, but I can't do both. I mean, I have to make my mind up this coming year and choose my college courses. I hate being stuck with a decision." Thad's starting point is *choosing a goal.* He has an approach-approach conflict. He wants both goals.

- Kimberley, a human resources executive for a large company, says, "I've found out that our chief executive has been involved in some unethical and, I think, immoral behavior. He's due to retire within the next six months. I don't know whether it's best to bring all this to light or just monitor him till he goes. If I move on him this could blow up into something big and hurt the reputation of the company itself. If I just monitor him till he goes, he gets away with it. I want to do what's best for the company." Her starting point is a dilemma about which *strategy* to use.

- Owen is having problems sticking to his resolve to restrain himself when one of his neighbors on his block "does something stupid." He says, "I know when I speak up" [his euphemism for flying off the handle] "things tend to get worse. I know I should leave it to others who are more tactful than I am. But they don't move quickly enough—or forcefully enough." His starting point is difficulty in *implementing* a course of action to which he has committed himself.

While it is important to start where the client starts, you may have to help any given client back up a bit. Take Owen. He is concerned about implementing a strategy. But Owen probably needs to back up and take a look at his style of dealing with people. What gets him going in the first place?

**Help clients clarify key issues.** To clarify means to discuss problem situations and unused opportunities—including possibilities for the future, goals, strategies for accomplishing goals, plans, implementation issues, and feelings about all of these—as concretely as possible. Vagueness and ambiguity lead nowhere. Clarity means helping clients move from the general to the specific—specific experiences, thoughts, behaviors, intentions, points of view, and decisions.

Consider this case. Janice's husband has been suffering from severe depression for over a year. One day, after Janice suffers a fainting spell, she, too, talks with a counselor. At first, feeling guilty about her husband, she is hesitant to discuss her own concerns. In the beginning she says only that her social life is "a bit restricted by my husband's illness." With the help of empathic highlights and probing on the part of the helper, her story emerges. "A bit restricted" turns, bit by bit, into the following—much fuller—story. This is a summary; Janice did not say this all at once.

> John has some sort of "general fatigue" illness that no one has been able to figure out. It's like nothing I've ever seen before. I move from guilt to anger to indifference to hope to despair. I have no social life. Friends avoid us because it is so difficult being with John. I feel I shouldn't leave him, so I stay at home. He's always tired, so we have little interaction. I feel like a prisoner in my own home. Then I think of the burden he's carrying and the roller-coaster emotions start all over again. Sometimes I can't sleep, then I'm as tired as he. He is always saying how hopeless things are and, even though I'm not experiencing what he is, some kind of hopelessness creeps into my bones. I feel that a stronger woman, a more selfless woman, a smarter woman would find ways to deal with all of this. But I end up feeling that I'm not dealing with it at all. From day to day I think I cover most of this up, so that neither John nor the few people who come around see what I'm going through. I'm as alone as he is.

This is the fuller story spelled out in terms of specific experiences, thoughts, behaviors, and feelings. The actions Janice takes—staying at home, covering her feelings up—are part of the problem, not the solution. But now that the story is out in the open, there is some possibility of doing something about it.

In another case, a woman suffering from bulimia who is now under psychiatric care says that she acted "a little erratically at times" with some of her classmates in law school. The counselor, sharing highlights and using probes, helps her tell her story in much greater detail. Like Janice, she does not say all of this at once, but this is the fuller picture of "a little erratic."

> I usually think about myself as plain looking, even though when I take care of myself some say that I don't look that bad. Ever since I was a teenager, I've preferred to go it alone, because it was safer. No fuss, no muss, and especially no rejection. In law school, right from the beginning, I entertained romantic fantasies about some of my classmates who I didn't think would give me a second look. I pretended to have meals with those who attracted me and then I'd have fantasies of having sex with them. Then I'd purge, getting rid of the fat I got from eating and getting rid of the guilt. But all of this didn't just stay in my head. I'd go out of my way to run into my latest imagined partner in school. And then I'd be rude to him to "get back at him" for what he did to me. That was my way of getting rid of him.

She was not really delusional, but gradually her external behavior, with a kind of twisted logic, began to reinforce her internal fantasies. However, once her story became "public"—that is, once she began talking about it openly with her helper—she began to take back control of her life.

**Help clients discuss the context of their concerns.** Sometimes helping clients explore the background or context of the concern they bring up helps clarify things. This is a further application of the people-in-systems approach mentioned earlier (Conyne & Cook, 2004; Hutchinson, 2003). Consider the following case.

> In a management development seminar, Tarik tells his counselor that he is a manager in a consulting firm. The firm is global and he works in one of its offices in Southeast Asia. He says that he is already overworked, but now his new boss wants him to serve on a number of committees that will take away even more of his precious time. He is also having trouble with one of his subordinates, himself a manager, who Tarik says is undermining his authority in the wider team.

So far we have a garden-variety story, one that could be repeated thousands of time throughout the world.

The counselor, however, suspects there is more to this. Since he is Canadian and the client is from a family with roots in the Middle East, he wants to make sure that he is a learner. He knows that there is an overlay of Western culture in these consulting firms, but he wants to deal with his client as a full person. And so, in sharing highlights and by using a few probes, he learns enough about the background of the manager's story to cast a new light on the problem situation. Here is the fuller story that emerges.

> Tarik is not only a manager in the firm, but also a partner. However, he is a newly-minted partner. The structure in these firms is relatively flat, but the culture is quite hierarchical. And so the clients he has been given to work with are, in large part, the

"dogs" of the region. His boss is not only an American, but he has been there for only four months and, Tarik has heard, is going to stay for only one more year. Since he is near retirement, this is his "fling" in Asia. Though a decent man, he is quite distant and offers Tarik little help. So Tarik believes that his real boss is his boss's boss. But he can't approach him because of company and cultural protocol. The subordinate who is giving him trouble is also a partner. In fact, he has been a partner for several years, but has not delivered the goods. This man thought that he should have been made the manager of the unit Tarik is now running. He has been engaging in a bit of sabotage behind Tarik's back.

A search for some background quickly takes the client's story out of the routine category. Of course, you should not be looking for background just for the sake of looking. The right kind and amount of background provides both richness and context.

**Assess the severity of the client's problems.** Clients come to helpers with problems of every degree of severity. Objectively, problems run from the inconsequential to the life-threatening. Subjectively, however, even a relatively inconsequential problem can be experienced as severe by a client. If a client thinks that a problem is critical—even though by objective standards the problem does not seem to be that bad—then for him or her it *is* critical. In such a case, the client's tendency to "catastrophize"—to judge a problem situation to be more severe than it actually is—itself becomes an important part of the problem situation. One of the tasks of the counselor in this case will be to help the client put the problem in perspective or to teach him or her how to distinguish between degrees of problem severity. Howard (1991, p. 194) put it well.

> In the course of telling the story of his or her problem, the client provides the therapist with a rough idea of his or her orientation toward life, his or her plans, goals, ambitions, and some idea of the events and pressures surrounding the particular presenting problem. Over time, the therapist must decide whether this problem represents a minor deviation from an otherwise healthy life story. Is this a normal, developmentally appropriate adjustment issue? Or does the therapist detect signs of more thoroughgoing problems in the client's life story? Will therapy play a minor, supportive role to an individual experiencing a low point in his or her life course? If so, the orientation and major themes of the life will be largely unchanged in the therapy experience. But if the trajectory of the life story is problematic in some fundamental way, then more serious, long-term story repair might be indicated. So, from this perspective, part of the work between client and therapist can be seen as life-story elaboration, adjustment, or repair.

Savvy therapists not only gain an understanding of the severity of a client's problem or the extent of the client's unused resources, but also understand the limits of helping. What has been this client's highest lifetime level of functioning? What, then, are appropriate expectations? What Howard calls life-story adjustment or repair is not the same as attempting to redo the client's personality.

Years ago, Mehrabian and Reed (1969) suggested the following formula as a way of determining the severity of any given problem situation. It is still useful today.

$$\text{Severity} = \text{Distress} \times \text{Uncontrollability} \times \text{Frequency}$$

The multiplication signs in the formula indicate that these factors are not just additive. Even low-level anxiety, if it is uncontrollable or persistent, can constitute a severe problem; that is, it can severely interfere with the quality of a client's life. In some cases, assessing for possible self-harm or harm to others is called for. The literature on suicide details possible self-harm signs that need to be captured. The literature on violence in social relationships can also be reviewed for signs of possible impending aggression. Even a casual reading of accounts of outbreaks of social violence tells as how easily signs can be missed.

Since clients' stories often unfold over time, assessment is ongoing. But this is true even in medical practice.

**Help clients talk productively about the past.**  Some schools of psychology suggest that problem situations are not clear and fully understood until they are understood in the context of their historic roots. Therefore, helpers in these schools spend a great deal of time helping clients uncover the past. Others disagree with that point of view. Glasser (2000, p. 23) puts it this way: "Although many of us have been traumatized in the past, we are not the victims of our past unless we presently choose to be. The solution to our problem is rarely found in explorations of the past unless the focus is on past successes."

Fish (1995) suggested that attempts to discover the hidden root causes of current problem behavior may be unnecessary, misguided, or even counterproductive. Constructive change does not depend on causal connections in the past. There is evidence to support Fish's contention. Long ago Deutsch (1954) noted that it is often almost impossible, even in carefully controlled laboratory situations, to determine whether event B, which follows event A in time, is actually caused by event A. Therefore, trying to connect present complicated patterns of current behavior with complicated events that took place in the past is an exercise in frustration.

Therefore, asking clients to come up with causal connections between current unproductive behavior and past events could be an exercise in futility for a number of reasons. First, causal connections cannot be proved; they remain hypothetical. Second, there is little evidence suggesting that understanding past causes changes present behavior. Third, talking about the past often focuses mostly on what happened to clients (their experiences) rather than on what they did about what happened to them (their thoughts, intentions, decisions, and behaviors) and therefore interferes with the *bias toward action* clients need to manage current problems.

This is not to say that a person's past does not influence current behavior. Nor does it imply that a client's past should not be discussed. Quite often, being stuck in the past is a function of rejection and resentment. But the fact that past experiences may well *influence* current behavior does not mean that they necessarily *determine* present behavior. Kagan (1996) has challenged what may be called the *scarred for life* assumption: "If orphans who spent their first years in a Nazi concentration camp can become productive adults and if young children made homeless by war can learn adaptive strategies after being adopted by nurturing families" (p. 901), there is hope for us all. As you can imagine, this is one of those issues that members of the helping professions argue about endlessly. Therefore, this is not a debate that is to be settled with a few words here. Here are some suggestions for helping clients talk meaningfully and productively about the past.

***Help clients talk about the past to make sense of the present.*** Many clients come expecting to talk about the past or wanting to talk about the past. There are ways of talking about the past that help clients make sense of the present. But making sense of the present needs to remain center stage. Thus, *how* the past is discussed is more important than whether it is discussed. The man in the following example has been discussing how his interpersonal style gets him into trouble. His father, now dead, played a key role in the development of his son's style.

> HELPER: So your father's unproductive interpersonal style is, in some ways, alive and well in you.

> CLIENT: Until we began talking I had no idea about how alive and well it is. For instance, even though I hated his cruelty, it lives on in me in much smaller ways. He beat my brother. But now I just cut him down to size verbally. He told my mother what she could do and couldn't do. I try to get my mother to adopt my "reasonable" proposals "for her own good"—without, of course listening very carefully to her point of view. There's a whole pattern that I haven't noticed. I've inherited more than his genes.

> HELPER: That's quite an inheritance . . . But now what?

> CLIENT: Well, now that I see what's happened, I'd like to change things. A lot of this is ingrained in me, but I don't think it's genetic in any scientific sense. I've developed a lot of bad habits.

It really does not make any difference whether the client's behavior has been "caused" by his father or not. In fact, if, by hooking the present into the past, he feels in some way that his current nasty style is not his fault, then he has a new problem. Helping is about the future. Now that the problem has been named and is out in the open, it is possible to do something about it. It is about *bad habits*, not sociobiological determinism.

***Help clients talk about the past to be reconciled to or liberated from it.*** A potentially dangerous logic can underlie discussions of the past. It goes something like this: "I am what I am today because of my past. But I cannot change my past. So how can I be expected to change today?"

> CLIENT: I was all right until I was about 13. I began to dislike myself as a teenager. I hated all the changes—the awkwardness, the different emotions, having to be as "cool" as my friends. I was so impressionable. I began to think that life actually must get worse and worse instead of better and better. I just got locked into that way of thinking. That's the same mess I'm in today.

That is not liberation talk. The past is still casting its spell. Helpers need to understand that clients may see themselves as prisoners of their past, but then, in the spirit of Kagan's earlier comments, help them move beyond such self-defeating beliefs.

The following case provides a different perspective. It is about the father of a boy who has been sexually abused by a minister of their church. He finds that he can't deal with his son's ordeal without revealing his own abuse by his father. In a tearful session he tells the whole story. In a second interview he has this to say:

> CLIENT: Someone said that good things can come from evil things. What happened to my son was evil. But we'll give him all the support he needs to get through this. Though I had the same thing happen to me, I kept it all in until now. It was all locked up inside. I was so ashamed and my shame became part of me. When I let it all out last week, it was like throwing off a dirty cloak that I'd been wearing for years. Getting it out was so painful, but now I feel so different, so good. I wonder why I had to hold it in for so long.

This is liberation talk. When counselors help or encourage clients to talk about the past, they should have a clear idea of what their objective is. Is it to learn from the past? Is it to be liberated from it? To assume that there is some "silver bullet" in the past that will solve today's problem is probably asking too much of the past.

***Help clients talk about the past in order to prepare for action in the future.*** The well-known historian A. J. Toynbee had this to say about history: "History not used is nothing, for all intellectual life is action, like practical life, and if you don't use the stuff—well, it might as well be dead." As we will soon see, any discussion of problems or opportunities should lead to constructive action, starting with Stage I and going all the way through to implementation. The insights you help clients get from the past should in some way stir them to action. When one client, Christopher, realized how much his father and one of his high-school teachers had done to make him feel inadequate, he made this resolve: "I'm not going to do anything to demean anyone around me. You know, up to now I think I have, but I called it something else—wit. I thought I was being funny when in fact I was actually being mean." Help clients invest the past proactively in the future.

**As clients tell their stories, search for resources, especially unused resources.** Incompetent helpers concentrate on clients' deficits. Skilled helpers, as they listen to and observe clients, do not blind themselves to deficits, but they are quick to spot clients' resources, whether used, unused, or even abused. These resources can become the building blocks for the future. Consider this example: Terry, a young woman in her late teens who has been arrested several times for street prostitution, is an involuntary, or "mandated," client. The charge this time is possession of drugs. She ran away from home when she was 16 and is now living in a large city on the East Coast. Like many other runaways, she was seduced into prostitution by a man she called "a halfway decent pimp." Now she is very street-smart. She is also cynical about herself, her occupation, and the world. She is forced to see a counselor as part of the probation process. As might be expected, Terry is quite hostile during the interview. She has seen other counselors and sees the interview as a game. The counselor already knows a great deal of the story because of the court record. The dialogue is not easy. Some of it goes like this:

TERRY: If you think I'm going to talk with you, you've got another thing coming. What I do is my business.

COUNSELOR: You don't have to talk about what you do outside. We could talk about what we're doing here in this meeting.

TERRY: I'm here because I have to be. You're here either because you're dumb and like to do stupid things or because you couldn't get a better job. You people are no better than the people on the street. You're just more "respectable."

COUNSELOR: So nobody could really be interested in you.

TERRY: I'm not even interested in me!

COUNSELOR: So, if you're not interested in yourself, then no one else could be, including me.

Terry has obvious deficits. She is engaged in a dangerous and self-defeating lifestyle. But as the counselor listens to Terry, he spots many different resources. Terry is a tough, street-smart woman. The very virulence of her cynicism and self-hate, the very strength of her resistance to help, and her almost unchallengeable determination to go it alone are all signs of resources. Many of her resources are currently being used in self-defeating ways. They are misused resources, but they are resources nevertheless.

Helpers need a resource-oriented mind-set in all their interactions with clients. Contrast two different approaches.

CLIENT: I practically never stand up for my rights. If I disagree with what anyone is saying—especially in a group—I keep my mouth shut.

COUNSELOR A: So clamming up is the best policy . . . What happens when you do speak up?

CLIENT (pausing): I suppose that on the rare occasions when I do speak up, the world doesn't fall in on me. Sometimes others do actually listen to me. But I still don't seem to have much impact on anyone.

COUNSELOR A: So speaking up, even when it's safe, doesn't get you much.

CLIENT: No, it doesn't.

Counselor A, sticking to sharing highlights, misses the resource mentioned by the client. Although it is true that the client habitually fails to speak up, he has some impact when he does speak. Others do listen, at least sometimes, and this is a resource. Counselor A emphasizes the deficit. Let's try another counselor.

COUNSELOR B: So when you do speak up, you don't get blasted, you even get a hearing. Tell me, what makes you think you don't exercise much influence when you speak?

CLIENT (pauses): Well, maybe influence isn't the issue. Usually I don't want to get involved. Speaking up gets you involved.

Note that both counselors share a highlight, but they focus on different parts of the client's message. Counselor A emphasizes the deficit; Counselor B notes an asset and follows up with a probe. This produces a significant clarification of the client's problem situation. Now, not wanting to get involved is an issue to be explored.

The search for resources is especially important when the story being told is bleak. I once listened to a man's story that included a number of bone-jarring life defeats—a bitter divorce, false accusations that led to his dismissal from a job, months of unemployment, serious health concerns, months of homelessness, and more. The only emotion the man exhibited during his story was depression. Toward the end of the session, we had this interchange:

HELPER: Just one blow after another, grinding you down.

CLIENT: Grinding me down, and almost doing me in.

HELPER: Tell me a little more about the "almost" part of it.

CLIENT: Well, I'm still alive, still sitting here talking to you.

HELPER: Despite all these blows, you haven't fallen apart. That seems to say something about the fiber in you.

At the word fiber, the client looked up, and there seemed to be a glimmer of something besides depression in his face. I put a line down the center of a newsprint pad. On the left side I listed the life blows the man had experienced. On the right I put "Fiber." Then I said, "Let's see if we can add to the list on the right side." We came up with a list of the man's resources. His "fiber" included his musical talent, his honesty, his concern for and ability to talk to others, and so forth. After about a half hour, he smiled—weakly, but he did smile. He said that it was the first time he could remember smiling in months.

**Help clients see every problem as an opportunity.** Clients don't come with just problems *or* opportunities. They come with a mixture of both. Although there is no justification for romanticizing pain, the flip side of a problem is an opportunity. Here are a few examples.

- Kevin used his diagnosis of AIDS as a starting point for reintegrating himself into his extended family and challenging the members of his family to come to grips with some of their own problems, problems they had been denying for a long time.

- Beatrice used her divorce as an opportunity to develop a new approach to men based on mutuality. Because she was on her own and had to make her own way, she discovered that she had entrepreneurial skills. She started an arts and crafts company.

- Jerome, after an accident, used a long convalescence period to review and reset some of his values and life goals. He began to visit other patients in the rehabilitation center. This gave him deep satisfaction. He began to explore opportunities in the helping professions.

- Sheila used her incarceration for shoplifting—she called it "time out"—as an opportunity to finish her high school degree and get a head start on college.

- An actor suffering from a traumatic disability found new life by becoming a public advocate for those suffering a similar fate.

- A couple mourning the death of their only child started a day-care center in conjunction with other members of their church.

William Miller (1986) talked about one of the worst days of his life. Everything was going wrong at work. Projects were not working out, people were not responding, the work overload was bad and getting worse—nothing but failure all around. Later that day, over a cup of coffee, he took some paper, put the title "Lessons Learned and Relearned" at the top, and wrote down as many entries as he could. Some hours and seven pages later, he had listed 27 lessons. The day turned out to be one of the best of his life. So he began to keep a daily "Lessons Learned" journal. It helped him avoid getting caught up in self-blame and defeatism. Subsequently, on days when things were not working out, he would say to himself, "Ah, this will be a day filled with learnings!"

Sometimes helping a client spot a small opportunity, be it the flip side of a problem or a standalone, provides enough positive-psychology leverage to put him or her on a more constructive tack.

## Action Right From the Beginning

Shakespeare, in the person of Hamlet, talks about important enterprises—what he calls "enterprises of great pith and moment"—losing "the name of action." Helping can too easily lose the name of action. One of the principal reasons clients do not manage the problem situations of their lives effectively is their failure to act intelligently, forcefully, and prudently in their own best interests.

**The importance of being proactive.** Inactivity can be bad for body, mind, and spirit. Consider the following workplace example:

> A counselor at a large manufacturing concern realized that inactivity did not benefit injured workers. If they stayed at home, they tended to sit around, gain weight, lose muscle tone, and suffer from a range of psychological symptoms such as psychosomatic complaints unrelated to their injuries. Taking a people-in-systems approach (Egan & Cowan, 1979), she worked with management, the unions, and doctors to design temporary, physically light jobs for injured workers. In some cases, nurses or physical therapists visited these workers on the job. Counseling sessions helped to get the right worker into the right job. The workers, active again, felt better about themselves. Both they and the company benefitted from the program.

Counselors add value by helping their clients become proactive. Helping too often entails too much talking and too little action.

Here is a brief overview of a case that illustrates almost the magic of action. Marcus's helpers are his sister-in-law and a friend.

> Life was ganging up on Marcus. Only recently "retired" from his job as office manager of a brokerage firm because the managing partner did not think that he would fit into the new e-commerce strategy of the business, he discovered that he had a form of cancer, the course of which was unpredictable. He had started treatment and was

relatively pain free, but he was always tired. He was also beginning to have trouble walking, but the medical specialists did not know why. It may or may not have been related to the cancer. Healthwise his future was uncertain. After losing his job, he took a marketing position with a software company. But soon, health problems forced him to give it up. Around the same time his daughter was expelled from high school for using drugs. She was sullen and uncommunicative and seemed indifferent to her father's plight. Marcus's wife and daughter did not get along well at all and the home atmosphere was tense whenever they were together. Discussions with his wife about their daughter went nowhere. To top things off, one of his two married sons announced that he was getting a divorce from his wife and was going to fight for custody of their three-year-old son.

Given all these concerns, Marcus, though very independent and self-reliant, more or less said to himself, "I've got to take charge of my life or it could fall apart. And I could use some help." He found help in two people—Sarah, an intelligent, savvy, no-nonsense sister-in-law who was a lay minister of their church, and Sam, a friend of many years who was a doctor in family practice.

Sam helped Marcus maintain his self-reliance by helping him find out as much as he could about his illness through the Internet. Without becoming preoccupied with his search, Marcus learned enough about his illness and possible treatments to become a partner with his doctors in choosing therapeutic interventions. He also drew up a living will and gave Sam power of attorney in order to avoid getting caught in the trap of using every possible remedy in a futile attempt to prolong his life at the end. Sam also helped him pursue some intellectual interests—art, literature, theater—that Marcus hadn't had time for because of business.

Here's a man who, with some help from his friends, uses adversity as an opportunity to get more fully involved with life. He redefines life, in part, as good conversations with family and friends. Refusing to become a victim, he stays active. Marcus did die two years later, but he managed to live until he died.

**Using the time between sessions productively.** If helping proves to be short-term for whatever reason, then counselors must help clients use the time between sessions as productively as possible. "How can I leverage what I do within the session to have an impact on what the client does the rest of the week?" Or month, as the case may be. This does not deny that good things are happening within the session. On the contrary, it capitalizes on whatever learning or change takes place there.

Many helpers give their clients homework from time to time as a way of helping them act on what they're learning in the sessions (Kazantzis, 2000; Scheel, Hanson, & Razzhavaikina, 2004). Mahrer and his associates (1994) reviewed the methods helpers use to do this. They came up with 16 methods. Among them are:

- Suggest some homework task and ask the client to carry it out and report back because he or she is now a "new person."

- Wait until the client comes up with a post-session task and then help him or her clarify and focus it.

- Highlight the client's readiness—or seeming willingness—to carry out some task, but leave the final decision to the client.
- Use a contractual agreement to move the client to some appropriate activity.

Counselors provide help in defining the activity and custom fitting it to the client's situation. Some helpers exhort clients to carry out some action. Some even mandate some kind of action as a condition for further sessions—"I'll see you after you've attended your first Alcoholics Anonymous meeting."

For further suggestions on how to incorporate homework into your helping sessions see Broder (2000). He not only offers many techniques, but discusses why some clients resist the idea of homework and suggests ways of addressing such resistance. For instance, Broder, in conjunction with Albert Ellis, the originator of Rational Emotive Behavior Therapy, has developed a series of audiotapes dealing with the many problems clients encounter. Clients use these tapes to further their understanding of what they learned in the therapy session and to engage in activities that will help them resolve their conflicts (see www.therapistassistant.com).

There is no need to call what the client does between sessions "homework." The term *homework* puts some clients off. It is also too "teacher-ish" for some helpers. It sounds like an add-on rather than something that flows organically from what takes place within the helping sessions. But *flows* is a wide term. The principle behind homework is more important than the name, and the principle is clear: Use every stage and every task of the helping process as a stimulus for problem-managing and opportunity-developing action. Use the term *homework* if it works for you. "Assign" it if this works for you and your client. But have a clear picture of why you are assigning any particular task. Don't routinely assign homework for its own sake.

Homework often has a predetermined cast to it. But encouraging clients to act in their own behalf can be a much more spontaneous exercise. Take the case of Mildred.

Mildred, 70, single, was a retired teacher. In fact, she had managed to find ways of teaching even after the mandatory retirement age. But she was ultimately "eased out." Given her savings, her pension, and social security, she was ready for retirement financially. But she had not prepared socially or emotionally. She was soon depressed and described life as "aimless." She wandered around for a few months and finally, at the urging of a friend, saw a counselor, someone about her own age. The counselor knew that Mildred had many internal resources, but they had all seemed to go dormant. In the second session, he said to her, "Mildred, you're allowing yourself to become a wreck. You'd never stand for this kind of behavior from your students. You need to get off your butt and seize life again. Tell me what you're going to do. I mean right after you get out of this session with me."

This was a bit of a shock treatment for Mildred. A revived Mildred looked at him and asked, "Well, what do you do?" They went on to discuss what life can offer after 70. Her first task, she decided, was to get into community in some way. The school had been her community but it was gone. She went back to the next counseling session with some options.

You can use homework assignments or find other, perhaps more organic, ways of helping clients move to action. Your role demands that you be a catalyst for client action.

## EVALUATION QUESTIONS FOR TASK 1

### How effectively am I doing the following?

### Establishing a Working Alliance

- Developing a collaborative working relationship with the client.
- Using the relationship as a vehicle for social-emotional reeducation.
- Not doing for clients what they can do for themselves.

### Helping Clients Tell Their Stories

- Using a mix of tuning in, listening, empathy, probing, and summarizing to help clients tell their stories, share their points, discuss their decisions, and talk through their proposals as concretely as possible.
- Using probes when clients get stuck, wander about, or lack clarity.
- Understanding blocks to client self-disclosure and providing support for clients who have difficulty talking about themselves.
- Helping clients talk productively about the past.

### Building Ongoing Client Assessment into the Helping Process

- Getting an initial feel for the severity of a client's problems and his or her ability to handle them.
- Noting and working with client resources, especially unused resources.
- Understanding clients' problems and opportunities in the larger context of their lives.

### Helping Clients Move to Action

- Helping clients develop an action orientation.
- Helping clients spot early opportunities for changing self-defeating behavior or engaging in opportunity-development behavior.

### Integrating Evaluation into the Helping Process

- Keeping an evaluative eye on the entire process with the goal of adding value through each interaction and making each session better.
- Finding ways of getting clients to participate in and own the evaluation process.

## Task 2: Help Clients Develop New Perspectives and Reframe Their Stories: "What Am I Overlooking or Avoiding?"

Often enough, counselors need to use the challenging skills outlined in Chapters 6 and 7 to help clients follow the principles outlined in Task 1. When clients present themselves to helpers, there is no instant or easy way of reading what is in their hearts. Clients reveal this over the course of helping sessions. And some helpers are much better than others in discovering "what is really going on." As we have seen, clients approach storytelling in quite different ways. Let's consider a couple of shadow-side versions.

Some clients who tell stories that are general, partial, and ambiguous may or may not have ulterior motives. For instance, one subtext in the shadows is, "If I tell my story too clearly and reveal myself, warts and all, then I will be expected to do something about it." The accountability issue lurks in the background. Another issue is the accuracy of the story. At one end are clients who tell their stories as honestly as possible. At the other end are clients who, for whatever reason, fudge. Or lie. Levitt and Rennie (2004) found that clients frequently hold back pertinent information. Clients' most frequent secrets are negative thoughts and feelings about themselves or about their therapy.

Fudging seems to have something to do with self-image. Some clients are not especially concerned about what their helpers think of them. They have no particular need to be seen in a favorable light. Other clients ask themselves, at least subconsciously, "What will the helper think of me?" Some are extremely concerned about what their helpers think of them and will skew their stories to present themselves in the best light. Kelly (2000; 2002; also see Kelly, Kahn, & Coulter, 1996) sees therapy, at least in part, as a self-presentational process. She suggests that clients benefit by perceiving that their therapists have favorable views of them. Therefore, if therapy contributes to clients' positive identity development, then we should expect some fudging. Hiding some of the less desirable aspects of themselves, intentionally or otherwise (see Kelly, 2000b), becomes a means to an end. Hill, Gelso, and Mohr (2000) object to this hypothesis, suggesting that research shows that clients don't hide much from their therapists. Arkin and Hermann suggest (2000) that all of this is really more complicated than the others realize. Well.

Take an example. Relatively few male victims of childhood sexual abuse discuss the residue of that in adulthood (Holmes, Offen, & Waller, 1997). This used to be explained by the now-discounted myths that few males are sexually abused and that abuse has little impact on males. Rather, it seems that it is just harder for males, at least in North American society, to admit both the abuse and its effects. We know that some clients fudge and some clients lie. We can only hypothesize why. But we do know that the vagaries of client self-presentation muddy the waters.

Throw all of this together with their various combinations and permutations, add in all other variables that might affect client storytelling, and you have an infinite number of self-presentation styles. In the end, each client has his or her own style. Add the enormous diversity found in clients and in story content, and it is clear to savvy helpers that each client represents an N = 1 (sole subject) research project. The competent, caring, and savvy helper tackles each project without naivete and without cynicism.

## Guidelines for Challenging the Way Clients Tell Their Stories

Here are some ways to challenge clients to tell their stories productively.

**Invite clients to own their problems and unused opportunities.**  It is all too common for clients to refuse to take responsibility for their problems and unused opportunities. Instead, they have a whole list of outside forces and other people who are to blame. Therefore, clients need to challenge themselves or be challenged to own the problem situation. Here is the experience of one counselor who had responsibility for about 150 young men in a youth prison within the confines of a larger central prison.

> I believe I interviewed each of the inmates. And I did learn from them. What I learned had little resemblance to what I had found when I read their files containing personal histories, including the description of their crimes. What I learned when I talked with them was that they didn't belong there. With almost universal consistency they reported a "reason" for their incarceration that had little to do with their own behavior. An inmate explained with perfect sincerity that he was there because the judge who sentenced him had also previously sentenced his brother, who looked very similar; the moment this inmate walked into the courtroom he knew he would be sentenced. Another explained with equal sincerity that he was in prison because his court-appointed lawyer wasn't really interested in helping him. (Miller, 1984, pp. 67–68)

This is perhaps an extreme form of a common phenomenon. But we don't have to go behind prison walls to find this lack of ownership. Lack of ownership is common.

Take the case of a client who feels that her business partner has been pulling a fast one on her. He's made a deal on his own. She's alarmed, but she hasn't done anything about it so far. Let's say that three different helpers, A, B, and C share a highlight. Consider how they differ.

> **HIGHLIGHT A:**  You feel angry because he unilaterally made the decision to close the deal on his terms.

> **HIGHLIGHT B:**  You're angry because your legitimate interests were ignored.

> **HIGHLIGHT C:**  You're furious because you were ignored, your interests were not taken into consideration, maybe you were even financially victimized, and you let him get away with it.

Challenging clients in terms of ownership means helping clients understand that in some situations, they may have some responsibility for creating, or at least perpetuating, their problem situations. Statement C does precisely that—"You let him get away with it."

Not only problems, but also opportunities need to be seized and owned by clients. As Wheeler and Janis (1980) noted, "Opportunities usually do not knock very loudly, and missing a golden opportunity can be just as unfortunate as missing a red-alert warning" (p. 18). Consider Tess and her brother, Josh.

> Tess and Josh's father died years ago. Their mother died about a year ago. She left a small country cottage to both of them. They have been fighting over its use. Some of the fighting has been quite bitter. They have never been particularly close, but up till now they have just had squabbles, not all-out war. Without admitting it, Tess has been shocked by her own angry and bellicose behavior. But not shocked enough to do anything about it.
>
> Then Josh has a heart attack. Tess knows that this is an opportunity to do something about their relationship. But she keeps putting it off. She even realizes that the longer she puts it off, the harder it will be to do something about it. In a counseling sessions, she says:
>
> TESS: I thought that this was going to be our chance to patch things up, but he hasn't said anything.
>
> COUNSELOR: So, nothing from his camp . . . What about yours?
>
> TESS: I think it might already be too late. We're falling right back into our old patterns.
>
> COUNSELOR: I didn't think that's what you wanted.
>
> TESS (ANGRILY): Of course that's not what I wanted! . . .That's the way it is.
>
> COUNSELOR: Tess, if someone put a gun to your head and said, "Make this work or I'll shoot," what would you do?
>
> TESS (AFTER ALONG PAUSE): You mean it's up to me . . .
>
> COUNSELOR: If you mean that I'm assigning this to you as a task, then no. If you mean that you can still seize the opportunity no matter what Josh does, well . . .
>
> TESS: I think I'm really angry with myself . . . I know deep down it's up to me . . . no matter what Josh does. And I keep putting it off.

Josh, of course, is not in the room. So it's about Tess's ownership of the opportunity. And she senses, now more deeply, that "missing a golden opportunity can be just as unfortunate as missing a red-alert warning."

**Invite clients to state their problems as solvable.** Jay Haley (1976, p. 9) said that if "therapy is to end properly, it must begin properly—by negotiating a solvable problem." Or by exploring a realistic opportunity, someone might add. It is not uncommon for clients to state problems as unsolvable. This justifies a "poor-me" attitude and a failure to act.

> UNSOLVABLE PROBLEM: In sum, my life is miserable now because of my past. My parents were indifferent to me and at times even unjustly hostile. If only they had been more loving, I wouldn't be in this mess. I am the failed product of an unhappy environment.

Of course, clients will not use this rather stilted language, but the message is common enough. The point is that the past cannot be changed. As we have seen, clients can change their attitudes about the past and deal with the present consequences of the past. Therefore, when a client defines the problem exclusively as a result of the past, the problem cannot be solved. "You certainly had it rough in the past and are still suffering from the consequences now" might be the kind of response that such a statement is designed to elicit. The client needs to move beyond such a point of view.

A solvable or manageable problem is one that clients can do something about. Consider a different version of the foregoing unsolvable problem.

> SOLVABLE PROBLEM: Over the years I've been blaming my parents for my misery. I still spend a great deal of time feeling sorry for myself. As a result, I sit around and do nothing. I don't make friends, I don't involve myself in the community, I don't take any constructive steps to get a decent job.

This message is quite different from that of the previous client. The problem is now open to being managed because the client has stated it almost entirely as something she does or fails to do. The client can stop wasting her time blaming her parents, since she cannot change them; she can increase her self-esteem through constructive action and therefore stop feeling sorry for herself; and she can develop the interpersonal skills and courage she needs to enter more creatively into relationships with others.

This does not mean that all problems are solvable by the direct action of the client. A teenager may be miserable because his self-centered parents are constantly squabbling and seem indifferent to him. He certainly can't solve the problem by making them less self-centered, stopping them from fighting, or getting them to care for him more. But he can be helped to find ways to cope with his home situation more effectively by developing fuller social opportunities outside the home. This could mean helping him develop new perspectives on himself and family life and challenging him to act, both internally and externally, in his own behalf.

**Invite clients to move on to the next needed stage of the helping process.** We have touched on this topic already in our discussion of probing. There is no reason to keep going over the same issue with clients. You can help clients challenge themselves to:

- clarify problem situations by describing specific experiences, behaviors, and feelings when they are being vague or evasive;

- talk about issues—problems, opportunities, goals, commitment, strategies, plans, actions—when they are reluctant to do so;

- develop new perspectives on themselves, others, and the world when they prefer to cling to distortions;

- review possibilities and critique them, develop goals, and commit themselves to reasonable agendas when they would rather continue wallowing in their problems;

- search for ways of getting what they want, instead of just talking about what they would prefer;

- spell out specific plans instead of taking a scattered, hit-or-miss approach to change;

### From Blind Spots to New Perspectives

**These are the kinds of things clients will be able to say to themselves when you help them develop new perspectives.**

- Here's a new angle . . .
- Here's something I've not thought of . . .
- Here's something I've overlooked . . .
- To be completely honest . . .
- Here's one way I've been fooling myself. . .
- Here's an important piece of the puzzle . . .
- Here's the real story . . .
- Here's the complete story . . .
- Oh, now I see that . . .

- persevere in the implementation of plans when they are tempted to give up;
- review what is and what is not working in their pursuit of change "out there."

In sum, counselors can help clients challenge themselves to engage more effectively in all the stages and tasks of problem management during the sessions themselves, and in the changes they are pursuing in everyday life. Box 8.3 outlines what goes through clients' mind when they develop a new perspective.

## Task 3: Help Clients Achieve Leverage by Working on Issues That Make a Difference: "What Will Help Me Most?"

Helping is expensive, both financially and psychologically. It should not be undertaken lightly. Therefore, a word is in order about what might be called the *economics* of helping. We use the term *leverage* to introduce the economics of helping. How can we help our clients get the most out of the helping process? Helpers need to ask themselves, "Am I adding value through each of my interactions with this client?" Clients need to be helped to ask themselves, "Am I working on the right things? Am I spending my time well in these sessions and between sessions?" The question here is not "Does helping help?" but "Is helping working in this situation? Is it worth it?"

It's important that helpers determine whether the client is ready to invest in constructive change. And to what degree. Change requires work on the part of clients. If they do not have the incentives to do the work, they might begin and then trail off. If this happens, it's a waste of resources. For instance, Helmut and Gretchen are mildly dissatisfied with their marriage and seem to be looking for a "psychological pill" that will magically make things better. There seem to be few incentives for the work required to reinvent the marriage. On the other hand, Beth is an intelligent but bored empty-nester. Her husband travels a great deal and she has too much time on her hands. She's quite dissatisfied with her current lifestyle. She has plenty of incentives to deal with her malaise and to identify and develop opportunities for a fuller life. Leverage comes from having a good client-helper relationship, working on the right problems and opportunities, choosing the right goals, and pursuing these goals through the right strategies—all ending up in constructive change.

## Guidelines in the Search for Leverage

Clients often need help to get a handle on complex problem situations. A 41-year-old depressed man with a failing marriage, a boring, run-of-the-mill job, deteriorating interpersonal relationships, health concerns, and a drinking problem cannot work on everything at once. Priorities need to be set. The blunt questions for determining priorities go something like this: "Where is the biggest payoff? Where should the limited resources of both client and helper be invested? Where to start?"

> Andrea, a woman in her mid-30s, is referred to a neighborhood mental-health clinic by a social worker. During her first visit, she pours out a story of woe, both historical and current—brutal parents, teenage drug abuse, a poor marriage, unemployment, poverty, and the like. Andrea is so taken up with getting it all out that the helper can do little more than sit and listen.

Where is Andrea to start? How can the time, effort, and money invested in helping provide a reasonable return to her? What are the economics of helping in Andrea's case? The following principles of leverage serve as guidelines for choosing issues to work on. These seven principles overlap; more than one may apply at the same time.

- Determine whether helping is called for.
- If there is a crisis, first help the client manage the crisis.
- Begin with the problem that seems to be causing the most pain for the client.
- Begin with issues the client sees as important.
- Begin with some manageable sub-problem of a larger problem situation.
- Begin with a problem that, if handled, will lead to some kind of general improvement in the client's overall condition.
- Focus on a problem for which the benefits will outweigh the costs.

Underlying all these principles is an attempt to make clients' initial experience of the helping process rewarding so that they will have the incentives they need to continue to work. Examples of the use and abuse of these principles follow.

**1. Determine whether helping is called for.** The literature says relatively little about screening—that is, about deciding whether any given problem situation or opportunity deserves attention. The reasons are obvious. Helpers-to-be are rightly urged to take their clients and their clients' concerns seriously. They are also urged to adopt an optimistic attitude, an attitude of hope, about their clients. Finally, they are schooled to take their profession seriously and are convinced that their services can make a difference in the lives of clients. For those and other reasons, the first impulse of the average counselor is to try to help clients no matter what the problem situation might be.

There is something very laudable in this. It is rewarding to see helpers prize people and express interest in their concerns. It is rewarding to see helpers put aside the almost instinctive tendency to evaluate and judge others and to offer their services to clients just because they are human beings. However, like any other profession, helping can suffer from the *law of the instrument.* A child, given a hammer, soon discovers that almost everything needs hammering. Helpers, once equipped with the models, methods, and skills of the helping process, can see all human problems as needing their attention. In fact, in many cases counseling may be a useful intervention but yet be a luxury expense which cannot be justified.

Under the term *differential therapeutics,* Frances, Clarkin, and Perry (1984) discussed ways of fitting different kinds of treatment to different kinds of clients. They also discussed the conditions under which "no treatment" is the best option. In the no-treatment category, they included clients who have a history of treatment failure or who seem to get worse from treatment, such as:

- criminals trying to avoid or diminish punishment by claiming to be suffering from psychiatric conditions;
- patients with malingering or fictitious illness;
- chronic nonresponders to treatment;
- clients likely to improve on their own;
- healthy clients with minor chronic problems;
- reluctant and resistant clients who try one dodge after another to avoid help.

Although a decision needs to be taken in each case, and although some might dispute several of the categories proposed, the possibility of no treatment deserves serious attention.

The no-treatment or no-further-treatment option can do a number of useful things: interrupt helping sessions that are going nowhere or are actually destructive; keep both client and helper from wasting time, effort, and money; delay help until the client is ready to do the work required for constructive change; provide a breather period that allows clients to consolidate gains from previous treatments; provide clients with an opportunity to discover that they can do without treatment; keep helpers and clients from playing games with themselves and one another; provide motivation for the client to find help in his or her own daily life. However, a decision on the part of helping professionals not to treat or to discontinue treatment that is proving fruitless is countercultural, and therefore difficult to make.

It goes without saying that screening can be done in a heavy-handed way. Statements such as the following are not useful:

- "Your concerns are actually not that serious."
- "You should be able to work that through without help."
- "I don't have time for problems as simple as that."

Whether such sentiments are expressed or implied, they obviously indicate a lack of respect and constitute a caricature of the screening process.

Practitioners in the helping and human service professions are not alone in grappling with the economics of treatment. Doctors face clients day in and day out with problems that run from the life-threatening to the inconsequential. Statistics suggest that more than half of the people who come to doctors have nothing physically wrong with them. Doctors, consequently, have to find ways to screen patients' complaints. I am sure that the best find ways to do so that preserve the dignity of their patients.

Effective helpers, because they are empathic, pick up clues relating to a client's commitment but they don't jump to conclusions. They test the waters in various ways. If clients' problems seem inconsequential, they probe for more substantive issues. If clients seem reluctant, resistant, and unwilling to work, they challenge clients' attitudes and help them work through their resistance. But in both cases they realize that there may come a time, and it may come fairly quickly, to judge that further effort is uncalled for due to lack of results. It is better, however, to help clients make such a decision themselves, or challenge them to do so. In the end, if the helper has to call a halt, his or her way of doing so should reflect basic counseling values.

**2. If there is a crisis, first help the client manage the crisis.**  Although crisis intervention is sometimes viewed as a special form of counseling, it can also be seen as a rapid application of the three stages of the helping process to the most distressing aspects of a crisis situation.

*Principle violated:* Zachary, a student near the end of his second year of a four-year doctoral program in counseling, gets drunk one night and is accused of sexual harassment by a student whom he met at a party. Knowing that he has never sexually harassed anyone, he seeks the counsel of a faculty member whom he trusts. The faculty member asks him many questions about his past, his relationship with women, how he feels about the program, and so on. Zachary becomes more and more agitated and then explodes: "Why are you asking me all these silly questions?" He stalks out and goes to a fellow student's house.

*Principle used:* Seeing his agitation, his friend says, "Good grief, Zach, you look terrible! Come in. What's going on?" He listens to Zachary's account of what has happened, interrupting very little, merely putting in a word here and there to let his friend know he is with him. He sits with Zachary when he falls silent or cries a bit, and then slowly and reassuringly talks Zach down, engaging in an easy dialogue that gives his friend an opportunity gradually to find his composure again. Zach's student friend has a friend in the university's student services department. They call him up, go over, and have a counseling and strategy session on the next steps in dealing with the harassment charges.

The friend's instincts are much better than those of the faculty member. He does what he can to defuse the immediate crisis and helps Zach take the next crisis-management step.

**3. Begin with the problem that seems to be causing the most pain for the client.** Clients often come for help because they are hurting, even though they are not in crisis. Their hurt, then, becomes a point of leverage. Their pain also makes them vulnerable. If it is evident that they are open to influence because of their pain, seize the opportunity, but move cautiously. Their pain may also make them demanding. They can't understand why you cannot help them get rid of it immediately. This kind of impatience may put you off, but it, too, needs to be understood. Such clients are like patients in the emergency room, each seeing himself or herself as needing immediate attention. Their demands for immediate relief may well signal a self-centeredness that is part of their character and therefore part of the broader problem situation. It may be that their pain is, in your eyes, self-inflicted, and ultimately you may have to challenge them on it. But pain, whether self-inflicted or not, is still pain. Part of your respect for clients is your respect for their vulnerability.

> *Principle violated:* Rob, a man in his mid-20s, comes to a counselor in great distress because his wife has just left him. The counselor's first impression is that Rob is an impulsive, self-centered person with whom it would be difficult to live. The counselor immediately challenges him to take a look at his interpersonal style and the ways in which he alienates others. Rob seems to listen, but he does not return.

> *Principle used:* Rob goes to a second counselor, who also sees a number of clues indicating self-centeredness and a lack of maturity and discipline. However, she listens carefully to his story, even though it is one-sided. She explores with him the incident that precipitated his wife's leaving. Instead of adding to his pain by making him come to grips with his selfishness, she focuses on what he wants for the future, especially the immediate future. Of course, Rob thinks that his wife's return is the most important part of a better future. She says, "I assume she would have to be comfortable with returning." He says, "Of course." She asks, "What could you do on your part to help make her more inclined to want to return?" His pain provides the incentive for working with the counselor on how he might need to change, even in the short term. Of course, the counselor realizes that she is hearing only one side of the story, a story that might well be very complicated.

The second helper does not use pain as a club. However guilty he might be, Rob didn't need his nose rubbed in his pain. The helper uses his pain as a point of leverage. What is Rob willing to do to rid himself of his pain and create the future he says he wants?

**4. Begin with issues the client sees as important.** The client's frame of reference is a point of leverage. Given the client's story, you may think that he or she has not chosen the most important issues for initial consideration. However, helping clients work on issues that are important in their eyes sends an important message: "Your interests are important to me."

> *Principle violated:* A woman comes to a counselor complaining about her relationship with her boss. She believes that he is sexist. Male colleagues not as talented as she get the best assignments and a couple of them are promoted. After listening to her story, the counselor has her explore her family background for "context." After listening, he believes that she probably has some leftover developmental issues with her father and an older brother that affect her attitude toward older men. He pursues this line of thinking with her. She is confused and feels put down. When she does not return for a second interview, the counselor says to himself that his hypothesis has been confirmed.

> *Principle used:* The woman seeks out a lawyer who deals with equal-opportunity cases. The lawyer, older and not only smart but wise, listens carefully to her story and probes for missing details. Then he gives her a snapshot of what such cases involve if they go to litigation. Against that background, he helps her explore what she really wants. Is it more respect? More pay? A promotion? Revenge? A different kind of boss? A better use of her talents? A job in a company that does not discriminate? Once she names her preferences, he discusses with her the options for getting what she wants. Litigation is not one of them.

The first helper substituted his own agenda for hers. He turned out to be somewhat sexist himself. The second helper accepted her agenda and helped her broaden it. He saw no leverage in litigation, but he did suggest that her problem situation could be an opportunity to reset her career. He suspected that there were plenty of firms eager to employ people of her caliber.

### 5. Begin with some manageable sub-problem of a larger problem situation.
Large, complicated problem situations often remain vague and unmanageable. Dividing a problem into manageable bits can provide leverage. Most larger problems can be broken down into smaller, more manageable, sub-problems.

> *Principle violated:* Aaron and Ruth, in their mid-50s, have a 25-year-old son living at home who has been diagnosed as schizophrenic. Aaron is a manager in a manufacturing concern that is in economic difficulty. His wife has a history of panic attacks and chronic anxiety. All these problems have placed a great deal of strain on the marriage. They both feel guilty about their son's illness. The son has become quite abusive at home and has been stigmatized by people in the neighborhood for his "odd" behavior. Aaron and Ruth have also been stigmatized for "bringing him up wrong." They have been seeing a counselor who specializes in a *systems* approach to such problems. They are confused by his everything–is–related–to–everything–else approach. They are looking for relief but are exposed to more and more complexity. They finally come to the conclusion that they do not have the internal resources to deal with the enormity of the problem situation and drop out.

> *Principle used:* A couple of weeks pass before they screw up their courage to make contact with a psychiatrist, Fiona, whom a family friend has recommended highly. Although Fiona understands the systemic complexity of the problem situation, she also understands their need for some respite. She first sees the son and prescribes some antipsychotic medication. His odd and abusive behavior is greatly reduced.

Even though Aaron and his wife are not actively religious, she also arranges a meeting with a rabbi from the community known for his ecumenical activism. The rabbi puts them in touch with an ecumenical group of Jews and Christians who are committed to developing a "city that cares" by starting up neighborhood groups. This is a positive psychology approach. Involvement with one of these groups helps diminish their sense of stigma. They still have many concerns, but they now have some relief and better access to both internal and community resources to help them with those concerns.

The psychiatrist helps then target two manageable sub-problems—the son's behavior and the couple's sense of isolation from the community. Some immediate relief puts them in a much better position to tackle their problems longer term. There are, indeed, serious family systems issues here, but theories and methodologies should not take precedence over the client's immediate needs.

**6. Begin with a problem that, if handled, will lead to some kind of general improvement in the client's overall condition.** Some problems, when addressed, yield results beyond what might be expected. This is the *spread effect.*

*Principle violated:* Jeff, a single carpenter in his late 20s, comes to a community mental-health center with a variety of complaints, including insomnia, light drug use, feelings of alienation from his family, a variety of psychosomatic complaints, and temptations toward exhibitionism. He also has an intense fear of dogs, something that occasionally affects his work. The counselor sees this last problem as one that can be managed through the application of a behavior modification methodology. He and Jeff spend a fair amount of time in the desensitization of the phobia. Jeff's fear of dogs abates quite a bit and many of his symptoms diminish. But gradually the old symptoms reemerge. His phobia, though significant, is not related closely enough to his primary concerns to involve any kind of significant spread effect.

*Principle used:* Predictably, Jeff's major problems reemerge. One day he becomes disoriented and bangs a number of cars near his job site with his hammer. He is overheard saying, "I'll get even with you." He is admitted briefly to a general hospital with a psychiatric ward. The immediate crisis is managed quickly and effectively. During his brief stay, he talks with a psychiatric social worker. He feels good about the interaction, and they agree to have a few sessions after he is discharged. In their talks, the focus turns to his isolation. This has a great deal to do with his lack of self-esteem—"Who would want to be my friend?" The social worker believes that helping Jeff to get back into community may well help with other problems. He has managed his problem by staying away from close relationships with both men and women. They discuss ways in which he can begin to socialize. Instead of focusing on the origins of Jeff's feelings of isolation, the social worker, taking an opportunity-development approach, helps Jeff involve himself in mini-experiments in socialization. As Jeff begins to get involved with others, his symptoms begin to abate.

The second helper finds leverage in Jeff's lack of human contact. The mini-experiments in socialization reveal some underlying problems. One of the reasons that others shy away from him is his self-centered and abrasive interpersonal style. When he repeats his questions—"Who would want to be my friend?" the helper responds, "I bet a lot of people would . . . if you were a bit more concerned about them and a bit less abrasive. You're not abrasive with me. We get along fine. What's going on?"

**7. Focus on a problem for which the benefits will outweigh the costs.** This is not an excuse for avoiding difficult problems, but it is a call for balance. If you demand a great deal of work from both yourself and the client, then the expectation is that there will be some kind of reasonable payoff for both of you.

> *Principle violated:* Margaret discovers to her horror that Hector, her husband, is HIV-positive. Tests reveal that she and her recently-born son have not contracted the disease. This helps cushion the shock. But she has difficulty with her relationship with Hector. He claims that he picked up the virus from a "dirty needle." But she didn't even know that he had ever used drugs. The counselor focuses on the need for the "reconstruction of the marital relationship." He tells her that some of this will be painful because it means looking at areas of their lives that they had never reviewed or discussed. But Margaret is looking for some practical help in reorienting herself to her husband and to family life. After three sessions she decides to stop coming.

> *Principle used:* Margaret still searches for help. The doctor who is treating Hector suggests a self-help group for spouses, children, and partners of HIV-positive patients. In the sessions, Margaret learns a great deal about how to relate to someone who is HIV-positive. The meetings are very practical. The fact that Hector is not in the group helps. She begins to understand herself and her needs better. In the security of the group, she explores mistakes she has made in relating to Hector. Although she does not "reconstruct" either her relationship with Hector or her own personality, she does learn how to live more creatively with both herself and him. She realizes that there will probably be further anguish, but she also sees that she is getting better prepared to face that future.

Reconstructing both relationships and personality, even if possible, is a very costly and chancy proposition. The cost-benefit ratio is out of balance. Once more, it may be a question of a helper more committed to his or her theories than to the needs of clients. Margaret gets the help she needs from the group and from sessions she and Hector have with a psychiatric aide.

The leverage mind-set that pervades these principles is second nature in effective helpers.

## Leverage and Action

Helping clients identify and deal with high-leverage issues, if done well, should also help them move toward the little actions that lead to a more formal plan for constructive change.

> One client, a man in his early 30s, discussed all sorts of problems and unused opportunities with a counselor. He skipped from one topic to another, usually settling on the issue that had caused him the most trouble the previous week. The counselor always listened attentively. At the beginning of the third session she said, "I have a hypothesis I'd like to explore with you." She went on to name what she saw as a thread that wove its way through most of the issues he discussed—a reluctance on his part to commit himself fully to anyone or anything. He was thunderstruck, because he instantly recognized the truth. In the next few weeks he began to explore his problems and concerns from this perspective. Time after time he saw himself withdrawing when he should have been committing himself, whether to a project or to a person. Gradually he began to commit himself in little ways. For instance, when an intimate woman friend of his began talking about where their relationship was going, he said, "There is something in me that wants to change the subject or run away. But no. Let's begin talking about the future and what future we might

## Questions on Leverage

**Help clients ask themselves such questions as:**

- What problem or opportunity should I really be working on?
- Which issue, if faced, would make a substantial difference in my life?
- Which problem or opportunity has the greatest payoff value?
- Which issue do I have both the will and the courage to work on?
- Which problem, if managed, will take care of other problems?
- Which opportunity, if developed, will help me deal with critical problems?
- What is the best place for me to start?
- If I need to start slowly, where should I start?
- If I need a boost or a quick win, which problem or opportunity should I work on?

*have together."* At least he committed himself to opening a discussion about the relationship, if not to the relationship itself.

Discussing issues that make a difference can galvanize some clients into using resources that have lain dormant for years. Box 8.4 pulls together some questions you can help clients ask themselves about leverage.

## Evaluation Questions on Leverage

### How well am I doing the following?

- Helping clients focus on issues that have payoff potential for them
- Maintaining a sense of movement and direction in the helping process
- Avoiding unnecessarily extending the problem identification and exploration stage
- Moving to other stages of the helping process as clients' needs dictate
- Encouraging clients to act on what they are learning

### Is Stage I Enough?

Some clients seem to need only the opportunity to tell their stories. That is, in a relatively limited amount of time they meet with a counselor and get help in telling and exploring their problem situation. They tell their story in greater or lesser detail, perhaps clear up some blind spots, develop some new perspectives, and then go off and manage quite well on their own. That is, they set goals for themselves, draw up a plan of action, and get on with it. Here are two ways in which clients may need only the first stage of the helping process.

**A declaration of intent and the mobilization of resources.**  For some clients, the very fact that they approach someone for help may be sufficient to help them begin to pull together the resources needed to manage their problem situations more effectively. For these clients, going to a helper is a declaration, not of helplessness, but of intent: "I'm going to do something about what is bothering me."

> Gerard was a young man from Ireland living illegally in New York. Even though he had a community of fellow countrymen to provide him support, he felt, as he put it, "hunted." He moved from one low-paying job to another, always living with the fear that something terrible was going to happen. One day he talked to a priest about his concerns. This was really the first time he had shared the burden that was weighing him down. The priest listened carefully but gave him no advice. But the talk provided the stimulus Gerard needed. "I know that I've got to do something about my life," he said at the end of their discussion. He quickly decided to return to Ireland. This unleashed pent-up inner resources. Once in Ireland, he took a degree in computing technology, then moved to Britain, and eventually got a job with the English arm of a German computer firm. Just as important, he felt "whole" again.

Merely seeking help can trigger a resource-mobilization process in some clients. Once they begin to mobilize their resources, they begin to manage their lives quite well on their own.

**Coming out from under self-defeating emotions.**  Some clients come to helpers because they are incapacitated, to a greater or lesser degree, by negative feelings and emotions. Often when helpers show such clients respect, listen carefully to them, and understand them in a nonjudgmental way, those self-defeating feelings and emotions subside. That is, the clients benefit from the counseling relationship as a process of social-emotional reeducation and repair. Once that happens, they are able to call on their own inner and environmental resources and begin to manage the problem situation that precipitated the incapacitating feelings and emotions. In short, they move to action. These clients, too, seem to have been able to resolve their problem merely by telling their story.

> One woman, let's call her Katrina, was very depressed after undergoing a hysterectomy. She had only two relatively brief sessions with a counselor. Katrina, helped to sort through the emotions she was feeling, discovered that the predominant emotion was shame. She felt wounded and exposed to herself and guilty about her incompleteness. Once she saw what was going on, she was able to pick up her life once more.

Of course, not all clients move on once given the opportunity to identify and explore their concerns. Many need the kind of help to sort out what they really need and want, and how they can get it.

## The Shadow Side of Leverage

No one knows how much time is wasted in counseling sessions discussing problems that won't make much of a difference in clients' lives. Some helpers, perhaps unknowingly, encourage clients to discuss problems that fit their theories and their approaches to therapy rather than the needs of clients. They fit the client to the *technology*, rather than joining clients in the search for issues that will make a difference in clients' lives. Some counselors misuse the principles of, or fail to spot opportunities for, leverage. They:

- begin with the framework of the client, but never get beyond it;
- deal effectively with the issues the client brings up, but fail to challenge clients to consider significant issues they are avoiding;
- help clients explore and deal with problems, but do not help clients translate them into opportunities, or move on to explore other unused opportunities;
- recognize the client's pain as a source of leverage, but then overfocus on the pain or allow it to mask other dimensions of the client's problems;
- help clients achieve small victories, but fail to help clients build on their successes;
- help clients start with small, manageable problems, but fail to help clients face more demanding problems or to see the larger opportunities embedded in these problems;
- help clients deal with a problem or opportunity in one area of life (for instance, self-discipline in an exercise program), but fail to help them generalize what they learn to other, more difficult areas of living (for instance, self-control in interpersonal relationships).

Client behaviors also contribute to the shadow side of leverage. Some clients, for whatever set of reasons, don't bring up their key concerns. Effective helpers can spot clues that suggest that clients are hiding some concerns, skirting important issues, or even lying. Then, with tact, they can help clients approach these issues. But, in the end, counselors are not mind readers. Clients can avoid talking about their real concerns if they want to. One way of handling this kind of avoidance behavior is to point out its possibility in the initial helping contract shared with the client. The contract could say something like this:

> "Sometimes clients don't bring up things that really bother them. Or, they are uncertain whether they want to tackle a particular problem. They feel that they are not ready. They may even feel caught, and lie about their problems. Since counselors are not mind readers, all this might escape the notice of the counselor. All I can say is that I am willing to help you explore any issue that you believe will make a difference in your life. I am quite aware of human foibles, both my own and those of my clients. I will do my best to deal with sensitive issues in a caring and respectful way."

Choose your own words. One client told me, "I'm glad that you mentioned that no-fault part about lying. In the last couple of sessions I haven't told you the entire truth. But now I'm ready." Then we got down to business.

It should be clear by now that the three tasks outlined in this chapter are interrelated, but are not sequential steps. There are ways of being with clients to help them get a clear, objective, and realistic picture of their concerns.

# EXERCISES ON PERSONALIZING

1. Write out the story of a difficult relationship you have had or are having. Make it as clear and as accurate as you can. Then, review your story in light of the principles of effective story-telling. Use the principles in this chapter to make it a better story.

2. To what degree is the problem situation, as outlined by you, owned by you? To what degree is the story stated as solvable?

3. Now, write out the story of the same difficult relationship as seen through the eyes of the other party. How do the two stories compare? Reframe the story to indicate your ownership (your part) of the problem situation. Restate the story in a way that indicates that it is solvable.

4. What aspect of the relationship, if worked on, might help it improve? What is a good point of entry?

5. If you have several problems, issues, or concerns, which one would you work on first? Why?

# CHAPTER 9

# Stage II: Helping Clients Identify, Choose, and Shape Problem-Managing Goals

**The Preferred Picture: "What Do I Need and Want?"**
Solution–Focused Helping
The Power of Goal Setting
The Three Tasks of Stage II

**Task 1: Help Clients Discover Possibilities for a Better Future: "What Would My Life Look Like If It Were Better?"**
The Psychology of Hope
Possible Selves
Skills for Identifying Possibilities for a Better Future
The Case of Brendan: Dying Better

**Task 2: Help Clients Craft Problem-Managing Goals: "What Solutions Are Best for Me?"**
From Creativity to Innovation
Guidelines for Helping Clients Design and Shape Their Goals

**Task 3: Help Clients Find the Incentives to Commit Themselves to a Better Future: "What Am I Willing to Pay for What I Want?"**
Guidelines for Helping Clients Commit to Goals

**Stage II and Action**

**The Shadow Side of Goal Setting**

**Exercises on Personalizing**

# The Preferred Picture: "What Do I Need and Want?"

In Stage II, the counselor help clients determine what they need and want instead of what they now have. "Now I engage in dangerous sex practices. I'm a sitting duck for disease and for harming others. I need to adopt and continually use safe-sex practices. Maybe it's time for me to take another look at the place of sex in my life." Stage II is about a better future.

## Solution-Focused Helping

O'Hanlon and Weiner-Davis (2003) claim that a trend "away from explanations, problems, and pathology, and toward solutions, competence, and capabilities" is continuing to emerge in the helping professions. An earlier study showed that clients were interested in solutions to their problems and in feeling better, whereas many helpers were concerned about the origin of problems and transforming them through insight (Llewelyn, 1988). Solution-focused approaches to helping (de Jong & Berg, 2002; Murphy, 1997; Pichot, 2003; Sharf, 2001; Sharry, 2003) tackle this disconnect. Intensive discussion of problem situations is often based on a "working through" mentality, whereas action or "solution" approaches are based on the assumption that many problems need to be dealt with, or even transcended, rather than worked through. At any rate, the goal of helping, as stated in Chapter 1, is "problems managed," not just "problems explored and understood" and "opportunities developed," not just "opportunities identified and discussed." Here is a quick overview of what these solution-focused approaches have in common.

**Philosophy.** In relating with clients, focus on resources rather than deficits; on success rather than failure; on credit rather than blame; on solutions rather than problems. Use common sense. Don't let theory get in the way of helping clients.

**Role of helper.** Helpers are consultants, catalysts, guides, facilitators, assistants. It is very helpful to adopt, at least temporarily, the client's world view. This helps lessen reluctance and resistance. Sharing highlights helps demonstrate your understanding of the client's world. Your job is to notice and amplify life-giving forces within the client, and any sign of constructive change in behavior. Become a detective for good things. Listen to problems, but listen even more to the opportunity buried within the problem. Use questions that inspire and encourage clients to give positive examples.

**View of clients.** Clients are people like the rest of us. See them as people with complaints about life, not packages of symptoms. Don't assume that they will arrive ambivalent about change and resistant to therapy. Clients have a reservoir of wisdom, learned and forgotten, but still available. Clients have resources and strengths to resolve complaints. Clients will have their own view of life, just as everyone else. Respect the reality they construct, even though they might have to move beyond it. In a way, clients are experts in their own lives. Help them feel competent to solve their own problems. When helpers see clients as problems to be solved, they impoverish them and take their power away.

**The nature of problems and how we talk about things.** Clients, like the rest of us, become what they talk about. If you always encourage them to talk about problems, they run the risk of becoming *problem people.* Then helping turns into remedying pathology and deficits. Help them see their problems as *complaints.* We all have complaints. In other words, help them *normalize* their problems. They are the ordinary difficulties of life. For instance, overeating is showing too much enthusiasm for the wonderful texture, taste, and comfort of food. Hyperactivity is energy that at times gets the better of us and interferes with rest, relaxation, and relationships. A perfectionist is person who loves quality but who goes too far. Help clients see problems as external to themselves, not things that define and control their lives. Problems are complaints that bother us rather than define us.

**Dealing with past.** There is no escape from past trauma. It did happen. However, if you help clients dwell on it, they will become captives of it. If fact, many will arrive as captives. They need to liberate themselves from the past. That said, clients should get an organized or integrated view of past bad experiences, but they should not explore the origins and causation. Looking for deep, underlying causes for symptoms is a mistake. Focus on a client's ability to survive the problem situation.

**Pursuing insights.** Insight is not necessary for change. Too often one insight follows another, but none of them leads to problem-managing action. Therefore, do not concentrate on insight generation. Insights tend to focus on problems, not solutions. "Now I understand why my brother and I don't get along. We're vying for our mother's attention." Do not let clients keep defining themselves in terms of the problem. Rather, engage them in generating *outcome* scenarios.

**Exploring and exploiting competencies, successes, and "normal times."** Help clients identify ways of thinking, behaving, and interacting that have worked in the past. And, since clients are not continually manifesting problem behavior, help them explore the times when they are free of such behavior. Help them identify what has been working during these misery-free periods and capitalize on it. Have them recall successes from the past, for instance, when they have handled disagreements more creatively. "When your marriage was good, what was it like?" Catch clients being competent and resourceful and help them take a good look at themselves at such times. Notice competencies revealed in a client's story and behavior.

**Dreaming: possibilities for a better future.** The principle is this: the future we anticipate is the future we create. The helper and the client should partner in the systematic search for possibilities and unused potential. The client's imagination needs to be *provoked* to discover new ways of approaching life. Questions should stimulate clients to think as creatively as possible about this better future. "What images capture your hopes for your future?" "What can you do to keep these hopeful images alive?" "If you no longer needed help, how would that show up in your actions?" "If you did the right thing for yourself this week and were filmed, what would the film's highlights show?" "On the assumption that you would like to move forward, what might you do to push the envelope a bit in the coming week?"

**Designing solutions.** There is no one correct way to live one's life. Help clients actively design solutions that will turn their possibilities into opportunity-developing realities. Solutions do not have to be complex. Rather, help clients to look for simple solutions to complex problems. Often, a small change is all that is necessary. Small changes can make a big difference. Solutions don't have to take care of everything. Troubles don't have to be totally solved. Help clients find systemic solutions. A change in one part of the system can produce good results in another part of the system. Look for interventions that break up patterns of self-limiting behavior. Don't hesitate to design solutions that get rid of symptoms. Getting rid of symptoms is not shallow, useless, or dangerous. Part of the solution should be clients' ability to grapple with future problems on their own. Clients should leave therapy with identified tools to do so.

**Delivering.** Implementation is everything. The pace of change will be different for each client. Some need help to ease themselves into solutions gradually. The smallest action is a step forward. On the other hand, rapid change is possible. Don't shortchange clients. Solutions often require that clients develop new ways of relating to their social environment. Where will support come from? Who needs to be engaged to make things work?

Of course, not everyone would buy totally into this package. For instance, some say that it runs the risk of being a don't-worry-just-be-happy approach, that it's too pie-in-the-sky. Others say, "Let's get real." Change comes from dealing with problems. People are used to dealing with problems. They talk about problems. A solution-focused approach might be too new for some. On the other hand, traditional problem-focused approaches to helping also come in for their knocks. Its critics say that it is painfully slow, asks clients to look back at yesterday's failures, looks for the causes of problems, rarely results in a new vision, places blame and therefore promotes defensiveness, and uses deficit-focused language.

However, wedding the positive-psychology approach of solution-focused therapies to a problem-management and opportunity-development approach to problem solving—the approach used in this book—faces down all these criticisms. There is no question of either/or. The interplay of the two provides the most robust system. The problem-management and opportunity-development approach provides the backbone of helping. But the use of its stages are dictated by client need. The solution-focused philosophy gives direction to *how* clients and helpers partner in using these stages to create a better future. Flexibility is the key. The arbiter is common sense and social intelligence. Do what is best for the client.

There is a semantic problem with the word "solution." It means two distinct things. First and foremost it means an end state—outcomes, results, accomplishments, impact. Take Pinta. Her eating was out of control. She was killing herself with food. But now she is eating moderately and has lost a lot of weight. A new approach to eating is in place. This is a solution in the end-state sense. "Solution" also means a strategy a client uses to get to the end state. For instance, Pinta joined a Twelve-Step program for overeaters and faithfully attended the meetings. "My eating is out of control," she said at the first meeting. In the group she learned a variety of ways to get back in control of her eating. Stage II of the helping process deals with solutions in the primary sense—end states, accomplishments, goals, outcomes. Stage III focuses on solutions in the secondary sense—means, actions, strategies.

## The Power of Goal Setting

Goal setting, whether it is called that or not, is part of everyday life. We all do it all the time.

> Why do we formulate goals? Well, if we didn't have goals, we wouldn't do anything. No one cooks a meal, reads a book, or writes a letter without having a reason, or several reasons, for doing so. We want to get something we want through our actions, or we want to prevent or avoid something we don't want. *These desires are beacons for our actions* [italics added]; they tell us which way to go. When formalized into goals, they play an important role in problem solving. (Dorner, 1996, p. 49)

Even not setting goals is a form of goal setting. If we don't name our goals, that does not mean that we don't have any. Instead of overt goals, then, we have a set of covert, or default, goals. They may be enhancing or limiting. We don't like the sagging muscles and flab we see in the mirror. But not deciding to get into better shape is a decision to continue the fitness drift. Since life is filled with goals—chosen goals or goals by default—it makes sense to make them work for us rather than against us.

Goals, at their best, mobilize our resources; they get us moving. They are a critical part of the self-regulation system. If they are the right goals for us, they get us headed in the right direction. According to Locke and Latham (1984), helping clients set goals empowers them in four ways, described below.

**Goals help clients focus their attention and action.** A counselor at a refugee center in London described Simon, a victim of torture in a Middle Eastern country, to her supervisor as aimless and minimally cooperative in exploring the meaning of his brutal experience. Her supervisor suggested that she help Simon explore possibilities for a better future. The counselor started one session by asking, "Simon, if you could have one thing you don't have, what would it be?" Simon came back immediately, "A friend." During the rest of the session, he was totally focused. What was uppermost in his mind was not the torture but the fact that he was so lonely in a foreign country. When he did talk about the torture, it was to express his fear that torture had "disfigured" him, if not physically, then psychologically, thus making him unattractive to others.

**Goals help clients mobilize their energy and effort.** Clients who seem lethargic during the problem-exploration phase often come to life when asked to discuss possibilities for a better future. A patient in a long-term rehabilitation program who had been listless and uncooperative said to her counselor after a visit from her minister, "I've decided that God and God's creation, and not pain, will be the center of my life. This is what I want." That was the beginning of a new commitment to the arduous program. She collaborated more fully in exercises that helped her manage her pain. Clients with goals are less likely to engage in aimless behavior. Goal setting is not just a "head" exercise. Many clients begin engaging in constructive change after setting even broad or rudimentary goals.

**Goals provide incentives for clients to search for strategies to accomplish them.** Setting goals, a Stage-II task, leads naturally into a search for means to accomplish them, a Stage-III task. Lonnie, a woman in her 70s, had been described by her friends as "going downhill fast." She decided, after a heart-problem scare that proved to be a false alarm, that she wanted to live as fully as possible until she died. She searched out ingenious ways of redeveloping her social life, including remodeling her house and taking in two young women from a local college as boarders. Her broad goal was to get involved in life instead of becoming life's victim.

**Clear and specific goals help clients increase persistence.** Not only are clients with clear and specific goals energized to do something, but they also tend to work harder and longer. An AIDS patient who said that he wanted to be reintegrated into his extended family managed, against all odds, to recover from five hospitalizations to achieve what he wanted. He did everything he could to buy the time he needed. Clients with clear and realistic goals don't give up as easily as clients with vague goals or with no goals at all.

Therefore, setting goals, whether formally or informally, provides clients with a sense of direction. People with a sense of direction:

    have a sense of purpose;

    live lives that are going somewhere;

    have self-enhancing patterns of behavior in place;

    focus on results, outcomes, and accomplishments;

don't mistake aimless action for accomplishments;

have a defined, rather than an aimless, lifestyle.

Locke and Latham (1990) pulled together years of research on the motivational value of setting goals. Although this value is incontrovertible, the number of people who disregard problem-managing and opportunity-developing goal setting and its advantages are legion. The challenge for counselors is to help clients do it well.

## The Three Tasks of Stage II

In many ways Stages II and III, together with the Action Arrow, are the most important parts of the helping model because they are about outcomes, accomplishments, and impact. It is here that counselors help clients develop and implement programs for constructive change. The payoff for identifying and clarifying both problem situations and unused opportunities lies in doing something about them. The skills helpers need to help clients do precisely that—engage in constructive change—are reviewed and illustrated in Stages II and III. In these stages, counselors help clients ask and answer the following two commonsense but critical questions. *What do you want?* is the first. *What do you have to do to get what you want?* is the second.

Problems can make clients feel hemmed in and closed off. To a greater or lesser extent they have no future, or the future they have looks troubled. The tasks of Stage II outline three ways in which helpers can partner with their clients in crafting a better future.

**Task 1: Help clients discover possibilities for a better future.**  "What would my life look like if it were better?" "What are some of the things I think I need or want?" Counselors, in helping clients, ask themselves these questions, help them develop a sense of hope. As Gelatt (1989) noted, "The future does not exist and cannot be predicted. It must be imagined and invented" (p. 255).

**Task 2: Help clients craft problem-managing goals.**  "What are the best solutions for me?" "Out of all the possibilities for a better future, which ones make most sense for me?" Here, clients set goals and craft *a viable change agenda* from among the possibilities. Helping them shape this agenda is the central task of helping.

**Task 3: Help clients explore their commitment to goals.**  "What am I willing to pay for what I want?" Help clients discover incentives for commitment to their change agenda. It is a further look at the economics of personal change.

Figure 9.1 highlights these three tasks of the helping process. Even though these three tasks are spelled out separately in this chapter and the next, they are usually intermingled in the client-helper dialogue about a better future.

**The Skilled-Helper Model**

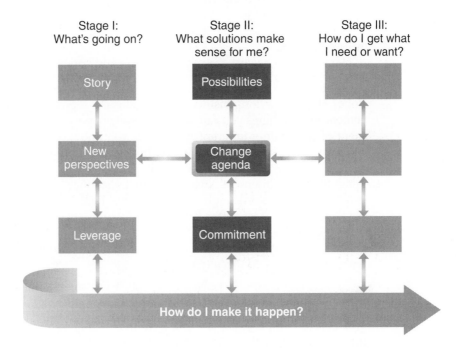

FIGURE

## 9.1
The Helping Model—Stage II

## Task 1: Help Clients Discover Possibilities for a Better Future: "What Would My Life Look Like If It Were Better?"

The goal of Task 1 is to help clients develop a sense of direction by exploring possibilities for a better future. I once was sitting at the counter of a late-night diner when a young man sat down next to me. The conversation drifted to the problems he was having with a friend of his. I listened for a while and then asked, "Well, if your relationship was just what you wanted it to be, what would it look like?" It took him a bit to get started, but eventually he came up with several snapshots of the kind of relationship he could live with. Then he stopped, looked at me, and said, "You must be a professional." I believe he thought that because this was the first time in his life that anyone had ever asked him to describe some possibilities for a better future.

Too often the exploration and clarification of problem situations are followed, almost immediately, by the search for solutions in the secondary sense—actions that will help deal with the problem or develop the opportunity. Some solution-focused approaches suggest starting with a review of possible actions clients could take to manage some problem situation. The upside of such an approach is the bias toward action. The downside is the lack of clarity about desired outcomes. *Actions to achieve what?*

## The Psychology of Hope

Hope as part of human experience is as old as humanity. Who of us has not started sentences with "I hope . . ."? Who of us has not experienced hope or lost hope? Hope has a long history as a religious concept. St. Paul said hope that centers around things you can see is not really hope, thus highlighting the element of uncertainty inherent in hope. If you know that tomorrow you will receive the Oscar, you can no longer hope for it. You know it's a sure thing. Hope plays a key role in both developing and implementing possibilities for a better future. An Internet search reveals that scientific psychology is more interested in hope than one might initially believe. Rick Snyder (2000), who, as we have seen earlier, has written extensively about the positive and negative uses of excuses in everyday life, has become a kind of champion for hope (for an extensive bibliography of all aspects of hope see Snyder's Web site, http://www. psych.ukans.edu/faculty/rsnyder/).

**The nature of hope.**  Snyder starts with the premise that human beings are goal directed. Hope, according to Snyder is the process of thinking about one's goals—Serena is determined that she will give up smoking, drinking, and the "soft" drugs she occasionally uses now that she is pregnant. Hope is also about having the will, desire, or motivation to move toward these goals—Serena is serious about her goal because she has seen the damaged children of mothers on drugs and she is also, at heart, a decent, caring person. Finally, hope involves thinking about the strategies for accomplishing one's goals—Serena knows that two or three of her friends will give her the support she needs and she is willing to join an arduous Twelve-Step program to achieve her goal. If we say that Serena has *high hopes,* we mean that her goal is clear, her sense of agency (or urgency) is high, and that she is realistic in planning the pathways to her goal. Both a sense of agency and some clarity around pathways are required. Hope, of course, has emotional connotations. Serena feels a mixture of positive emotions—elation, determination, satisfaction—knowing that *the will* (agency) and *the way* (pathways) have come together. Success is in sight even though she knows that there will be barriers—for instance, the ongoing lure of tobacco, wine, and soft drugs.

**The benefits of hope.**  Snyder (1995, pp. 357–358) combed the research literature in order to discover the benefits of hope as he defines it. Here is what he found.

> The advantages of elevated hope are many. Higher as compared with lower hope people have a greater number of goals, have more difficult goals, have success at achieving their goals, perceive their goals as challenges, have greater happiness and less distress, have superior coping skills, recover better from physical injury, and report less burnout at work, to name but a few advantages.

Counselors who do not spend a significant part of their time with clients helping them develop possibilities, clarify goals, devise strategies or pathways, and develop the sense of agency needed to bring all of this to fruition are short-changing their clients. Since Stages II and III deal with possibilities, goals, commitment, pathways, and overcoming barriers, they are hope-centered.

## Possible Selves

One of the characters in Gail Godwin's novel *The Finishing School* warns against getting involved with people who have "congealed into their final selves." Clients come to helpers, not necessarily because they have congealed into their final selves—if this is the case, why come at all?—but because they are stuck in their current selves. Counseling is a process of helping clients get unstuck and develop a sense of direction. Consider the case of Ernesto. He was very young but very stuck, for a variety of sociocultural and emotional reasons.

> A counselor first met Ernesto in the emergency room of a large urban hospital. He was throwing up blood into a pan. He was a member of a street gang, and this was the third time he had been beaten up in the last year. He had been so severely beaten this time that it was likely he would suffer permanent physical damage. Ernesto's style of life was doing him in, but it was the only one he knew. He was in need of a new way of living, a new scenario, a new way of participating in city life. This time he was hurting enough to consider the possibility of some kind of change.

Markus and Nurius (1986) used the term *possible selves* to represent "individuals' ideas of what they might become, what they would like to become, and what they are afraid of becoming" (p. 954). The counselor worked with Ernesto, not by helping him explore the complex sociocultural and emotional reasons he was in this fix but, principally, by helping him explore his *possible selves* in order to discover a different purpose in life, a different direction, a different lifestyle. Task 1 of Stage II is about possible selves. The notion of possible selves has captured the imagination of many helpers and of those interested in human development, such as teachers. Do a literature search on the Internet using the term *possible selves research literature,* and you will find very rich and useful citations.

## Skills for Identifying Possibilities for a Better Future

At its best, counseling helps clients move from problem-centered mode to *discovery* mode. Discovery mode involves creativity and divergent thinking. According to Taylor and his associates (1998), however, not just any kind of mental stimulation will do. Rather, creative thinking is helpful to the degree that it "provides a window on the future by enabling people to envision possibilities and develop plans for bringing those possibilities about. In moving oneself from a current situation toward an envisioned future one, the anticipation and management of emotions and the initiation and maintenance of problem-solving activities are fundamental tasks" (p. 429).

**Creativity and helping.** One of the myths about creativity is that some people are creative and others are not. Each of us has some modicum of creativity that can be stimulated. Clients can be more creative than they are—it is a question of finding ways to help them be so. Stages II and III help clients tap into their dormant creativity. A review of the requirements for creativity (see Cole & Sarnoff, 1980; Robertshaw, Mecca, & Rerick, 1978, pp. 118–120) shows, by implication, that people in trouble often fail to use whatever creative resources they might have. The creative person is characterized by:

- Optimism and confidence, whereas clients are often depressed and feel powerless;

- Acceptance of ambiguity and uncertainty, whereas clients may feel tortured by ambiguity and uncertainty and want to escape from them as quickly as possible;

- A wide range of interests, whereas clients may be people with a narrow range of interests or people whose normal interests have been severely narrowed by anxiety and pain;

- Flexibility, whereas clients may have become rigid in their approach to themselves, others, and the social settings of life;

- Tolerance of complexity, whereas clients are often confused and looking for simplicity and simple solutions;

- Verbal fluency, whereas clients are often unable to articulate their problems, much less their goals and ways of accomplishing them;

- Curiosity, whereas clients may not have developed a searching approach to life, or may have been hurt by being too venturesome;

- Drive and persistence, whereas clients may be all too ready to give up;

- Independence, whereas clients may be quite dependent or counterdependent;

- Nonconformity or reasonable risk taking, whereas clients may have a history of being very conservative and conformist, or may get into trouble with others and with society precisely because of their particular brand of nonconformity.

Of course, creative people can have emotional and behavioral problems but they are, generally speaking, creative in spite of and not because of their problems. Or they are creative in the ways they manage their problems.

A review of some of the principal obstacles or barriers to creativity (see Azar, 1995) brings further problems to the surface. Innovation is hindered by:

- Fear—clients are often quite fearful and anxious.

- Fixed habits—clients may have self-defeating habits or patterns of behavior that may be deeply ingrained.

- Dependence on authority—clients may come to helpers looking for the "right answers," or be quite counterdependent (the other side of the dependency coin) and fight efforts to be helped with a variety of games.

- Perfectionism—clients may come to helpers precisely because they are hounded by this problem and can accept only ideal or perfect solutions.

- Social networks—being "different" sets clients apart when they want to belong.

It is easy to say that imagination and creativity are most useful in Stages II and III, but it is another thing to help clients stimulate their own, perhaps dormant creative potential.

**Divergent thinking.** Many people habitually take a convergent-thinking approach to problem solving—that is, they look for the *one right answer.* Such thinking has its uses, of course. However, many of the problem situations of life are too complex to be handled by convergent thinking. Such thinking limits the ways in which people use their own, and environmental, resources. Divergent thinking, on the other hand—thinking outside the box—assumes that there is always more than one answer. Consider the following case.

> Quentin wanted to be a doctor, so he enrolled in the pre-med program at school. He did well, but not well enough to get into medical school. When he received the last notice of refusal, he said to himself, "Well, that's it for me and the world of medicine. Now what will I do?" When he graduated, he took a job in his brother-in-law's business. He became a manager and did fairly well financially, but he never experienced much career satisfaction. He was glad that his marriage was good and his home life rewarding, because he derived little satisfaction from his work.

Not much divergent thinking went into handling this problem situation. No one asked Quentin what he really wanted. For Quentin, becoming a doctor was the "one right career." He didn't give serious thought to any other career related to the field of medicine, even though there are dozens of interesting and challenging jobs in the field of health care.

The case of Caroline, who also wanted to become a doctor but failed to get into medical school, is quite different from Quentin's.

> Caroline thought to herself, "Medicine still interests me; I'd like to do something in the health field." With the help of a medical career counselor, she reviewed the possibilities. Even though she was in pre-med, she had never realized that there were so many careers in the field of medicine. She decided to take whatever courses and practicum experiences she needed to become a nurse. Then, while working in a clinic in the hills of Appalachia—an invaluable experience for her—she managed to get an M.A. in family-practice nursing by attending a nearby state university part-time. She chose this specialty because she thought it would enable her to be closely associated with providing a broad range of services to patients; in addition, she would be in a position of increased responsibility for the delivery of these services.
>
> When Caroline graduated, she entered private practice with a doctor as a nurse practitioner in a small Midwestern town. Since the doctor divided his time among three small clinics, Caroline had a great deal of responsibility in the clinic where she practiced. She also taught a course in family-practice nursing at a nearby state school and conducted workshops in holistic approaches to preventive medical self-care. Still not satisfied, she began and finished a doctoral program in practical nursing. She taught at a state university and continued her practice. Needless to say, her persistence paid off with an extremely high degree of career satisfaction.

A successful professional career in health care remained Caroline's aim throughout. A great deal of divergent thinking and creativity went into the elaboration of that initial aim into specific goals, and into coming up with the courses of action to accomplish them. But for every success story, there are many more failures. Quentin's case is probably the norm, not Caroline's. For many, divergent thinking is either uncomfortable or too much work.

**Brainstorming: A tool for divergent thinking.**  One excellent way of helping clients think divergently and more creatively is brainstorming. Brainstorming is a simple idea-stimulation technique for exploring possibilities. Brainstorming in Stages II and III is a tool for helping clients develop both possibilities for a better future, and ways of accomplishing goals. There are certain rules that help make this technique work: suspend judgment, produce as many ideas as possible, use one idea as a takeoff point for others, get rid of normal constraints to thinking, and produce even more ideas by clarifying items on the list. Here, then, are the rules:

*Suspend your own judgment, and help clients suspend theirs.* When brainstorming, do not let clients criticize the ideas they are generating and, of course, do not criticize them yourself. There is some evidence that this rule is especially effective when the problem situation has been clarified and defined and goals have not yet been set. In the following example, a woman whose children are grown and married is looking for ways of putting meaning into her life.

> CLIENT:  One possibility is that I could become a volunteer, but the very word makes me sound a bit pathetic.

> HELPER:  Add it to the list. Remember, we'll discuss and critique them later.

Having clients suspend judgment is one way of handling the tendency on the part of some to play a "Yes, but" game with themselves. That is, they come up with a good idea and then immediately show why it isn't really a good idea, as in the preceding example. By the same token, avoid saying such things as "I like that idea," "This one is useful," "I'm not sure about that idea," or "How would that work?" Premature approval and criticism cut down on creativity. A marriage counselor was helping a couple brainstorm possibilities for a better future. When Nina said, "We will stop bringing up past hurts," Tip, her husband, replied, "That's your major weapon when we fight. You'll never be able to give that up." The helper said, "Add it to the list. We'll look at the realism of these possibilities later on."

*Encourage clients to come up with as many possibilities as possible.* The principle is that quantity ultimately breeds quality. Some of the best ideas come along later in the brainstorming process. Cutting the process short can be self-defeating. In the following example, a man in a sex-addiction program has been brainstorming activities that might replace his preoccupation with sex.

> CLIENT:  Maybe that's enough. We can start putting it all together.

> HELPER:  It doesn't sound like you were running out of ideas.

> CLIENT:  I'm not. It's actually fun. It's almost liberating.

> HELPER:  Well, let's keep on having fun for a while.

> CLIENT (pausing):  Ha! I could become a monk.

Later on, the counselor, focusing on this "possibility," asked, "What would a modern-day monk who's not even a Catholic look like?" This helped the client explore the concept of sexual responsibility from a completely different perspective and to rethink the place of religion and service to others in his life. And so, within reason, the more ideas the better. Helping clients identify many possibilities for a better future increases the quality of the possibilities that are eventually chosen and turned into goals. In the end, however, do not invoke this rule for its own sake. Possibility generation is not an end in itself. Use your clinical judgment, your social intelligence, to determine when enough is enough. If a client wants to stop, often it's best to stop.

**Help clients use one idea to stimulate others.** This is called *piggybacking*. Without criticizing the client's productivity, encourage him or her to both develop strategies already generated, and to combine different ideas to form new possibilities. In the following example, a client suffering from chronic pain is trying to come up with possibilities for a better future.

CLIENT: Well, if there is no way to get rid of all the pain, then I picture myself living a full life without pain at its center.

HELPER: Expand that a bit for me.

CLIENT: The papers are filled with stories of people who have been living with pain for years. When they're interviewed, they always look miserable. They're like me. But every once in a while there is a story about someone who has learned how to live creatively with pain. Very often, they are involved in some sort of cause which takes up their energies. They don't have time to be preoccupied with pain.

When one client with multiple sclerosis brought of this possibility, "I'll have a friend or two with whom I can share my frustrations as they build up," the helper asked, "What would that look like?" The client replied, "Not just a complaining session or just a poor-me thing. It would be a normal part of a give-and-take relationship. We'd be sharing both joys and pain of our lives like other people do."

**Help clients let themselves go and develop some "wild" possibilities.** When clients seem to be "drying up," or when the possibilities being generated are quite pedestrian, you might say, "Okay, now draw a line under the items on your list and write the word wild under the line. Now let's see if you can come up with some really wild possibilities." Later, it is easier to cut suggested possibilities down to size than to expand them. The wildest possibilities often have within them at least a kernel of an idea that will work. In the following example, an older single man who is lonely is exploring possibilities for a better future.

CLIENT: I can't think of anything else. And what I've come up with isn't very exciting.

HELPER: How about getting a bit wild? You know, some crazy possibilities.

CLIENT: Well, let me think. . . . I'd start a commune and would be living in it . . . or near it. And . . .

Clients often need permission to let themselves go, even in harmless ways. They repress good ideas because they might sound foolish. Helpers need to create an atmosphere where such apparently foolish ideas will be not only accepted, but also encouraged. Help clients come up with conservative possibilities, liberal possibilities, radical possibilities, and even outrageous possibilities.

It's not always necessary to use brainstorming explicitly. You, as helper, can keep these rules in mind and then, by sharing highlights and using probes, you can get clients to brainstorm even though they don't know that's what they're doing. A brainstorming mentality is useful throughout the helping process.

**Future-oriented probes.**   One way of helping clients invent the future is to ask them, or get them to ask themselves, future-oriented questions related to their current poorly-managed problems or undeveloped opportunities. The following questions are examples of different ways to help clients find answers to the questions "What do you want?" "What do you need?" These questions focus on outcomes—that is, on what will be in place after the clients act.

- *What would this problem situation look like if you were managing it better?* Ken, a college student who has been a loner, has been talking about his general dissatisfaction with his life. A couple of the things he said: "I'd be having fewer anxiety attacks. And I'd be spending more time with people rather than by myself."

- *What changes in your present lifestyle would make sense?* Cindy, who described herself as a "bored homemaker," replied, "I would not be drinking as much. I'd be getting more exercise. I would not sit around and watch the soaps all day. I'd have something meaningful to do."

- *What would you be doing differently with the people in your life?* Lonnie, a graduate student at a university near his parents' home, realized that he had not yet developed the kind of autonomy suited to his age. He mentioned these possibilities: "I would not be letting my mother make my decisions for me. I'd be sharing an apartment with one or two friends."

- *What patterns of behavior would be in place that are not currently in place?* Bridget, a depressed resident in a nursing home, had this suggestion: "I'd be engaging in more of the activities offered here in the nursing home." Rick, who is suffering from lymphoma, says: "Instead of seeing myself as a victim, I'd be on the Web finding out every last thing I can about this disease and how to deal with it. I know there are new treatment options. And I'd also be getting a second or a third opinion. You know, I'd be managing my lymphoma instead of just suffering from it."

- *What current patterns of behavior would be eliminated?* Bridget, the nursing home resident, adds these to her list: "I would not be putting myself down for incontinence I cannot control. I would not be complaining all the time. It gets me and everyone else down!"

- *What would you have that you don't have now?* Sissy, a single woman who has lived in a housing project for 11 years, said, "I'd have a place to live that's not rat-infested. I'd have some friends. I wouldn't be so miserable all the time." Drew, a man tortured by perfectionism, muses, "I'd be wearing sloppy clothes, at least at times, and like it. More than that, I'd have a more realistic sense of the world and my place in it. The world is messy, it's chaotic much of the time. I'd find the beauty in the chaos."

- *What accomplishments would be in place that are not in place now?* Ryan, a divorced man in his mid-30s, said, "I'd have my degree in practical nursing. I'd be doing some part-time teaching. I'd be close to someone that I'd like to marry."

- *What would this opportunity look like if you developed it?* Enid, a woman with a great deal of talent who has been given one modest promotion in her company and who feels like a second-class citizen, had this to say: "In two years I'll be an officer of this company or have a very good job in another firm."

It is a mistake to suppose that clients will automatically gush with answers. Ask the kinds of questions just listed, or encourage them to ask themselves the questions, but then help them answer them. Create the therapeutic dialogue around possibilities for a better future. Many clients don't know how to use their innate creativity. Thinking divergently is not part of their mental lifestyle.

**Exemplars and models as a source of possibilities.** Some clients can see future possibilities better when they see them embodied in others. You can help clients brainstorm possibilities for a better future by helping them identify exemplars or models. By models I don't mean superstars or people who do things perfectly. That would be self-defeating. In the next example, a marriage counselor is talking with a middle-aged, childless couple. They are bored with their marriage. When he asks them, "What would your marriage look like if it looked a little better?" he can see that they are stuck.

COUNSELOR: Maybe the question would be easier to answer if you reviewed some of your married relatives, friends, or acquaintances.

WIFE: None of them have super marriages. (Husband nods in agreement.)

COUNSELOR: No, I don't mean super marriages. I'm looking for things you could put in your marriage that would make it a little better.

WIFE: Well, Fred and Lisa are not like us. They don't always have to be doing everything together.

HUSBAND: Who says we have to be doing everything together? I thought that was your idea.

WIFE: Well, we always are together. If we weren't always together, we wouldn't be in each other's hair all the time.

COUNSELOR: All right, who else do you know who are doing things in their marriage that appeal to you. Anyone.

HUSBAND: You know Ron and Carol do some volunteer work together. Ron was saying that it gets them out of themselves. I bet they have better conversations because of it.

COUNSELOR: Now we're cooking . . . What else? What couple do you find the most interesting?

Even though it was a somewhat torturous process, these two people were able to come up with a range of possibilities for a better marriage. The counselor had the couple write ideas down so they wouldn't lose them. At this point, the purpose was not to get the clients to commit themselves to these possibilities, but to identify them.

In the following case, the client finds herself making discoveries by observing people she had not identified as models at all.

> Fran, a somewhat withdrawn college junior, realizes that when it comes to interpersonal competence, she is not ready for the business world she intends to enter when she graduates. She wants to do something about her interpersonal style and a few nagging personal problems. She sees a counselor in the Office of Student Services. After a couple of discussions with him, she joins a "lifestyle" group on campus that includes some training in interpersonal skills. Although she expands her horizons a bit from what the members of the group say about their experiences, behaviors, and feelings, she tells her counselor that she learns even more by watching her fellow group members in action. She sees behaviors that she would like to incorporate in her own style. A number of times she says to herself while in the group, "Ah, there's something I never thought of." Without becoming a slavish imitator, she begins to incorporate some of the patterns she sees in others into her own style.

Models or exemplars can help clients more specifically name what they want. Models can be found anywhere: among the client's relatives, friends, and associates, in books, on television, in history, in movies. Counselors can help clients identify models, choose those dimensions of others that are relevant, and translate what they see into realistic possibilities for themselves. Box 9.1 outlines the kinds of questions you can help clients ask themselves to discover possibilities for a better future.

## The Case of Brendan: Dying Better

Brendan, a heavy drinker, had extensive and irreversible liver damage and it was clear that he was getting sicker. But he wanted to "get some things done" before he died. Brendan's action orientation helped a great deal. Over the course of a few months, a counselor helped him to name some of the things he wanted before he died or on his journey toward death. Brendan came up with the following possibilities:

- "I'd like to have some talks with someone who has a religious orientation, like a minister. I want to discuss some of the bigger issues of life and death."

## Questions for Exploring Possibilities

BOX
**9.1**

**Help clients ask themselves these kinds of questions:**

- What are my most critical needs and wants?
- What are some possibilities for a better future?
- What outcomes or accomplishments would take care of my most pressing problems?
- What would my life look like if I were to develop a couple of key opportunities?
- What should my life look like a year from now?
- What should I put in place that is currently not in place?
- What are some wild possibilities for making my life better?

- "I don't want to die hopeless. I want to die with a sense of meaning."
- "I want to belong. You know, to some kind of community, people who know what I'm going through, but who are not sentimental about it. People not disgusted with me because of the way I've done myself in."
- "I'd like to get rid of some of my financial worries."
- "I'd like a couple of close friends with whom I could share the ups and downs of daily life. With no apologies."
- "As long as possible, I'd like to be doing some kind of productive work, whether paid or not. I've been a flake. I want to contribute even if just in an ordinary way."
- "I need a decent place to live, maybe with others."
- "I need decent medical attention. I'd like a doctor who has some compassion. One who could challenge me to live until I die."
- "I need to manage these bouts of anxiety and depression better."
- "I want to be get back with my family again. I want to hug my dad. I want him to hug me."
- "I'd like to make peace with one or two of my closest friends. They more or less dropped me when I got sick. But at heart, they're good guys."
- "I want to die in my home town."

Of course, Brendan didn't name all these possibilities at once. Through understanding and probes, the counselor helped him name what he needed and wanted and then helped him stitch together a set of goals from these possibilities (Stage II) and ways of accomplishing them (Stage III).

**EVALUATION QUESTIONS FOR TASK 1**

- To what degree am I an imaginative person?
- In what ways can I apply the concept of possible selves to myself?
- What problems do I experience as I try to help clients use their imaginations?
- Against the background of problem situations and unused opportunities, how well do I help clients focus on what they want?
- To what degree do I prize divergent thinking and creativity in myself and others?
- How effectively do I use empathic highlights, a variety of probes, and challenge to help clients brainstorm what they want?
- Besides direct questions and other probes, what kinds of strategies do I use to help clients brainstorm what they want?
- How effectively do I help clients identify models and exemplars that can help them clarify what they want?
- How well do I help clients act on what they are learning?

## Task 2: Help Clients Craft Problem-Managing Goals: "What Solutions Are Best for Me?"

Once possibilities for a better future have been developed and reviewed, clients need to make some choices; that is, they need to choose one or more of those possibilities and turn them into a program for constructive change.

### From Creativity to Innovation

Task 1 is, in many ways, about **creativity**—getting rid of boundaries, thinking beyond one's limited horizon, moving outside the box, developing a range of possibilities. Task 2 is about **innovation** —that is, turning possibilities into a practical program for change. If implemented, a goal constitutes the *solution* with a big *S* for the client's problem or unused opportunity. Consider the following case.

Bea, an African American woman, was arrested when she went on a rampage in a bank and broke several windows. She had exploded with anger because she felt that she had been denied a loan mainly because she was black and a single mother. In discussing the incident with her minister, she comes to see that she has become very prone to anger. Almost anything can get her going. She also realizes that venting her anger leads to a range of negative consequences. But she is constantly "steamed up" about "the system." To complicate the picture, she tends to take her anger out on those around her, including her friends and her two children.

The minister helps her look at four possible ways of dealing with her anger—venting it, repressing it, channeling it, or simply giving up and ignoring the things she gets angry at, including the injustices around her. Giving up is not in her makeup. Merely venting her anger seems to do little but make her more angry. Repressing her anger, she reasons, is just another way of giving up, and that is demeaning. And she's not very good at repressing anyway. The "channeling" option needs to be explored. In the end, Bea takes a constructive approach to channeling her frustrations. She joins a political action group involved in community organizing. She learns that she can channel her anger without giving up her values or her intensity. She also discovers that she is good at influencing others and getting things done. She begins to feel better about herself. The "system" doesn't seem to be such a fortress anymore.

Since goals can be highly motivational, helping clients set realistic goals is one of the most important tasks in the helping process.

## Guidelines for Helping Clients Design and Shape Their Goals

Practical goals do not usually leap out fully formed. Goals, once chosen, need to be shaped or crafted. Here is a set of so-called shaping tools you can use in your dialogue with clients. Goals need to be:

stated as *outcomes,* rather than activities;

*specific* enough to be verifiable and to drive action;

*substantive* and challenging;

both *venturesome* and *prudent;*

*realistic* in regard to resources needed to accomplish them;

*sustainable* over a reasonable time period;

*flexible* without being wishy-washy;

*congruent* with the client's *values,* rather than products of other people's values;

set in a reasonable *time frame.*

Just how this package of goal characteristics will look in practice will differ from client to client. There is no one formula. Be careful not to get lost in the details of these characteristics. Effective helpers will keep them in the back of their minds and, in a second-nature manner, turn them into helpful "sculpting" probes whenever needed. Now a word about each.

**Help clients state what they need and want as outcomes or accomplishments.** The goal of counseling, as emphasized again and again, is about outcomes and impact—that is, solutions with a big *S.* "I want to start doing some exercise" is an activity rather than an outcome. "Within six months I will be running three miles in less than thirty minutes at least four times a week" is an outcome. It is a pattern of behavior that will be in place by a certain time. If a client says, "My goal is to get some training in interpersonal communication skills," then she is stating her goal as a set of activities—a solution with a small *s*—rather than as an accomplishment. But if she says that she wants to become a better listener as a wife and mother, then she is stating her goal as an accomplishment, even though "better listener" needs further clarification. Goals stated as outcomes provide direction for clients.

You can help clients describe what they need and want by using this *past-participle* approach—drinking stopped, number of marital fights decreased, phobia controlled, anger habitually channeled. Stating goals as outcomes or accomplishments is not just a question of language. Helping clients state goals as accomplishments rather than activities helps them avoid directionless and imprudent action. If a woman with breast cancer says that she thinks she should join a self-help group, she should be helped to see what she wants to get out of such a group. Joining a group and participating in it are activities. She wants support. Goals, at their best, are expressions of what clients need and want. Clients who know what they want are more likely to work, not just harder, but also smarter. Consider the case of Chester, a former Marine suffering from post-traumatic stress disorder.

> Chester was involved in the Kosovo peacekeeping effort. During a patrol, he and three of his buddies shot and killed four civilians who were out to kill their neighbors. Afterward, he began acting in strange ways, wandering around at times in a daze. He was sent home and given a medical discharge. Although he seemed to recover, he lived an aimless life. He went to college but dropped out during the first semester. He became rather reclusive but never really engaged in odd behavior. Rather he was sinking into the landscape. He moved in and out of a number of low-paying jobs. He also became less careful about his person. He said to a counselor, "You know, I used to be very careful about the way I dressed. Kind of proud of myself in the marine tradition. Don't get me wrong; I'm not a bum and don't smell or anything, but I'm not myself." The whole direction of Chester's life was wrong; he was headed for serious trouble. He was bothered by thoughts about the war and had taken to sleeping whenever he felt like it, day or night, "just to make it all stop."

Ed, Chester's counselor, had a good relationship with Chester. He helped Chester tell his story and challenged some of his self-defeating thinking. He went on to help Chester focus on what he wanted from life. They moved back and forth between Stage I and Stage II, between a discussing his problems and searching for possibilities for a better future. Eventually, Chester began talking about his real needs and wants—that is, what he needed to accomplish to "get back to his old self." Here is an excerpt from their dialogue.

> CHESTER:  I've got to stop hiding in my hole. I'm going to get out and see people more. I'm going to stop feeling so damn sorry for myself. Who wants to be with a nothing!
>
> COUNSELOR:  What will Chester's life look like a year or two from now?
>
> CHESTER:  One thing for sure. He will be seeing women again. He might not be married, but he will probably have a special girlfriend. And she will see him as an ordinary guy.

Here Chester talks about changes as patterns of behavior that will be in place. He is starting to paint a picture of what he wants to be. He is designing some goals. The counselor's probe reinforces this outcome approach.

**Help clients move from broad aims to clear and specific goals.** Specific, rather than general, goals tend to drive behavior. Therefore, broad goals need to be translated into more specific goals and tailored to the needs and abilities of each client. Skilled helpers use probes to help clients move from the general to the specific.

Chester said that he wanted to become "more disciplined." His counselor helped him make that more specific.

> COUNSELOR: What areas do you want to focus on?
>
> CHESTER: Well, if I'm going to put more order in my life, I need to look at the times I sleep. I've been going to bed whenever I feel like it and getting up whenever I feel like it. It was the only way I could get rid of those thoughts and the anxiety. But I'm not nearly as anxious as I used to be. Things are calming down.
>
> COUNSELOR: So "more disciplined" means a more regular sleep schedule because there's no particular reason now for not having one.
>
> CHESTER: Yeah, sleeping whenever I want is just a bad habit. And I can't get things done if I'm asleep.

Chester goes on to translate "more disciplined" into other problem-managing needs and wants related to school, work, and care of his person. Greater discipline, once translated into specific patterns of behavior, will have a decidedly positive impact on his life.

Counselors often add value by helping clients move from good intentions to broad but more specific aims and, finally, to specific goals. Here is an example of a client who moves from *good intentions,* to a *broad aim,* then on to a *specific goal.* Good intentions, while "good," should never be mistaken for concrete goals. In the following example, the client, Dermot, has been discussing his relationship with his wife and children. The counselor has been helping him see that his "commitment to work" is really a lack of work/life balance that is not appreciated by his family. Dermot is open to challenge and a fast learner.

> DERMOT: Boy, this session has been an eye-opener for me. I've really been blind. My wife and kids don't see my investment—rather, my overinvestment—in work as something I'm doing for *them.* I've been fooling myself, telling myself that I'm working hard to get them the good things in life. In fact, I'm spending most of my time at work because I like it. My work is mainly for me. It's time for me to realign some of my priorities.

The last statement is a good intention, an indication on Dermot's part that he wants to do something about a problem, now that he sees it more clearly. It may be that Dermot will now go out and put a different pattern of behavior in place without further help from the counselor. Or he may benefit from some help in realigning his priorities.

A broad aim is more than a good intention. It has content; that is, it identifies the area in which the client wants to work and makes some general statement about that area.

**DERMOT:** I don't think I'm spending so much time at work in order to run away from family life. But family life is deteriorating because I'm just not around enough. I must spend more time with my wife and kids. Actually, it's not just a case of must. I want to.

Dermot moves from a declaration of intent to a broad aim—spending more time at home. But he still has not created a picture of what that would look like.

To help Dermot move toward greater specificity, the counselor uses such probes as "Tell me what 'spending more time at home' will look like."

**DERMOT:** I'm going to consistently spend three out of four weekends a month at home. During the week I'll work no more than two evenings.

**COUNSELOR:** So you'll be at home a lot more. Tell me what you'll be doing with all this time.

Notice how much more specific Dermot's statement is than "I'm going to spend more time with my family." He sets a goal as a specific pattern of behavior he wants to put in place. But his goal, as stated, deals with quantity, not quality. The counselor's probe is really a challenge. It's not just the amount of time Dermot is going to spend with his family, but also the kinds of things he will be doing. *Quality time*, some call it. A client trying to come to grips with work/life balance once said to me, "My family, especially my kids, don't make the distinction between quantity and quality. For them, quantity is quality. Or there's no quality without a chunk of quantity." This warrants further discussion because, maybe, the family wants a relaxed, rather than an intense, Dermot at home.

Helping clients move from good intentions to more and more specific goals is a shaping process. Consider the example of a couple whose marriage has degenerated into constant bickering, especially about finances.

- *Good intention.* "We want to straighten out our marriage."
- *Broad aim.* "We want to handle our decisions about finances in a much more constructive way."
- *Specific goal.* "We try to solve our problems about family finances by fighting and arguing. We'd like to reduce the number of fights we have and begin making mutual decisions about money. We yell at each other instead of talking things out. We need to set up a month-by-month budget. Otherwise, we'll be arguing about money we don't even have. We'll have a trial budget ready the next time we meet with you."

Having sound household finances is a fine goal. A goal in itself. Reducing unproductive conflict is also a fine goal. In this case, however, installing a sound, fair, and flexible household budget system is also instrumental to establishing peace at home. Declarations of intent, broad goals, and specific goals can all drive constructive behavior, but specific goals have the best chance. Is it possible to get clients to be too specific about their goals? Yes, if they get lost in the planning details, and crafting the goal becomes more important than the goal itself.

If the goal is clear enough, the client will be able to determine progress toward it. For many clients, being able to measure progress is an important incentive. If goals are stated too broadly, the counselor and client can have difficulty in determining both progress and accomplishment. "I want to have a better relationship with my wife" is a very broad goal, difficult to verify. "I want to socialize more, you know, with couples we both enjoy" comes closer, but "socialize more" needs more clarity.

It is not always necessary to count things to determine whether a goal has been reached, though sometimes counting is helpful. Helping is about living more fully, not about counting activities. At a minimum, however, counselor and client need to be able to verify desired outcomes in some way. For instance, a couple might say something like "Our relationship is better, not because we've stop squabbling. In fact, we've discovered that we like to squabble. But life is better because the meanness has gone out of our squabbling. We accept each other more. We listen more carefully, we talk about more personal concerns, we are more relaxed, and we make more mutual decisions about issues that affect us both." This couple does not need a scientific experiment to verify that they have improved their relationship.

**Help clients establish goals with substance, goals that make a difference.** Outcomes and accomplishments are meaningless if they do not have the required impact on the client's life. The goals clients choose should have substance to them—that is, some significant contribution toward managing the original problem situation or developing some opportunity.

> Vitorio ran the family business. His son, Anthony, worked in sales. After spending a few years learning the business and getting an MBA part-time at a local university, Anthony wanted more responsibility and authority. His father never thought that he was "ready." They began arguing quite a bit, and their relationship suffered from it. Finally, a friend of the family persuaded them to spend time with a consultant-counselor who worked with small family businesses. He spent relatively little time listening to their problems. After all, he had seen this same problem over and over again—the reluctance and conservatism of the father, the pushiness of the son.
>
> Vitorio wanted the business to stay on a tried-and-true course. Anthony wanted to be the company's marketer, to move it into new territory. After a number of discussions with the consultant-counselor, they settled on this scenario: They would create a marketing department, headed by Anthony. He could divide his time between sales and marketing as he saw fit, provided that he maintained the current level of sales. Vitorio agreed not to interfere. They would meet once a month with the consultant-counselor to discuss problems and progress. Vitorio insisted that the consultant's fee come from increased sales. After some initial turmoil, the bickering decreased dramatically. Anthony easily found new customers, although they demanded modifications in the product line, which Vitorio reluctantly approved. Both sales and margins increased to the point that they needed another person in sales.

Not all issues in family businesses are handled as easily. In fact, a couple of years later, Anthony left the business and founded his own. But the goal package they worked out—the deal they cut—made quite a difference, both in the father-son relationship and in the business.

Goals often have substance to the degree that they help clients *stretch* themselves. As Locke and Latham (1984, pp. 21, 26) noted, "Extensive research . . . has established that, within reasonable limits, the . . . more challenging the goal, the better the resulting performance . . . People try harder to attain the hard goal. They exert more effort. . . . In short, people become motivated in proportion to the level of challenge with which they are faced. . . . Even goals that cannot be fully reached will lead to high effort levels, provided that partial success can be achieved and is rewarded." Consider the following case.

> A young woman ended up as a quadriplegic because of an auto accident. In the beginning, she was full of self-loathing—"the accident was all my fault; I was just stupid." She was close to despair. Over time, however, with the help of a counselor, she came to see herself not as a victim of her own "stupidity," but as someone who could bring hope to young people with life-changing afflictions. In her spare time, she visited young patients in hospitals and rehabilitation centers, got some to join self-help groups, and generally helped people like herself to manage an impossible situation in a more humane way. One day she said to her counselor, "The best thing I ever did was to stop being a victim and become a fellow traveler with people like myself. The last two years, though bitter at times, have been the best years of my life." She had set her goals quite high—becoming an outgoing helper instead of remaining a self-centered victim—but they proved to be quite realistic.

Of course, when it comes to goals, "challenging" should not mean "impossible." There seems to be a curvilinear relationship between goal difficulty and goal performance. If the goal is too easy, people see it as trivial and ignore it. If the goal is too difficult, they feel justified in giving up. However, this difficulty-performance ratio differs from person to person. What is small for some is big for others.

**Help clients set goals that are prudent.** Although the helping model described in this book encourages a bias toward action on the part of clients, action needs to be both directional and wise. Discussing and setting goals should contribute to both direction and wisdom. The following case begins poorly but ends well.

> Harry was a sophomore in college who was admitted to a state mental hospital because of some bizarre behavior at the university. He was one of the disc jockeys for the university radio station. He came to the notice of college officials one day when he put on an attention-getting performance that included rather lengthy dramatizations of grandiose religious themes. In the hospital, the staff soon discovered that this quite pleasant, likable young man was actually a loner. Everyone who knew him at the university thought that he had many friends, but in fact he did not. The campus was large, and his lack of friends went unnoticed.
>   Harry was soon released from the hospital but returned weekly for therapy. At one point he talked about his relationships with women. Once it became clear to him that his meetings with women were perfunctory and almost always took place in groups—he had imagined that he had a rather full social life with women—Harry launched a full program of getting involved with the opposite sex. His efforts ended in disaster, however, because Harry had some basic sexual and communication problems. He also had serious doubts about his own worth and, therefore, found it difficult to make a gift of himself to others. He ended up in the hospital again.
>   The counselor helped Harry get over his sense of failure by emphasizing what Harry could learn from the "disaster." With the therapist's help, Harry returned to the problem-clarification and new-perspectives part of the helping process and then

established more realistic short-term goals regarding getting back "into community." The direction was the same—establishing a realistic social life—but the goals were now more prudent because they were downsized and more manageable. Harry attended socials at a local church, where a church volunteer provided support and guidance.

Harry's leaping from problem clarification to action without taking time to discuss possibilities and set reasonable goals was part of the problem rather than part of the solution. His lack of success in establishing solid relationships with women actually helped him see his problem with women more clearly. There are two kinds of prudence—playing it safe is one, doing the wise thing is the other. Problem management and opportunity development should be venturesome. They are about making wise choices, rather than playing it safe.

**Help clients formulate realistic goals.** While stretch goals can help clients energize themselves, goals set too high can do more harm than good. Locke and Latham (1984, p. 39) put it succinctly:

> Nothing breeds success like success. Conversely, nothing causes feelings of despair like perpetual failure. A primary purpose of goal setting is to increase the motivation level of the individual. But goal setting can have precisely the opposite effect if it produces a yardstick that constantly makes the individual feel inadequate.

A goal is realistic if the client has access to the *resources* needed to accomplish it, the goal is under the client's *control*, and external circumstances do not prevent its accomplishment.

*Resources: Help clients choose goals for which the resources are available.* It does little good to help clients develop specific, substantive, and verifiable goals if the resources needed for their accomplishment are not available. Consider the case of Rory, who, because of merger and extensive restructuring, has had to take a demotion. He now wants to leave the company and become a consultant.

> *Insufficient resources:* Rory does not have the assertiveness, marketing savvy, industry expertise, or interpersonal style needed to become an effective consultant. Even if he did, he does not have the financial resources needed to tide him over while he develops a business.

> *Sufficient resources:* Challenged by the outplacement counselor, Rory changes his focus. Graphic design is an avocation of his. He is not good enough to take a technical position in the company's design department, but he does apply for a supervisory role in that department. He is good with people, very good at scheduling and planning, and knows enough about graphic design to discuss issues meaningfully with the members of the department.

Rory combines his managerial skills with his interest in graphic design to move in a more realistic direction. The move is challenging, but it can have a substantial impact on his work life. For instance, the opportunity to hone his graphic design skills will open up further career possibilities.

*Control: Help clients choose goals that are under their control.* Sometimes clients defeat their own purposes by setting goals that are not under their control. For instance, people commonly believe that their problems would be solved if only other people would not act the way they do. In most cases, however, we do not have any direct control over the ways others act. Consider the following example.

> Tony, a 16-year-old boy, felt that he was the victim of his parents' inability to relate to each other. Each tried to use him in the struggle, and at times he felt like a Ping-Pong ball. A counselor helped him see that he could probably do little to control his parents' behavior, but that he might be able to do quite a bit to control his *reactions* to his parents' attempts to use him. For instance, when his parents started to fight, he could simply leave instead of trying to help. If either tried to enlist him as an ally, he could say that he had no way of knowing who was right. Tony also worked at creating a good social life outside the home. That helped him weather the tensions he experienced at home.

Tony needed a new way of managing his interactions with his parents to minimize their attempts to use him as a pawn in their own interpersonal game. Goals are not under clients' control if they are blocked by external forces that they cannot influence. "To live in a free country" may be an unrealistic goal for a person living in a totalitarian state because he cannot change internal politics, nor can he change emigration laws in his own country or immigration laws in other countries. "To live as freely as possible in a totalitarian state" might well be an aim that could be translated into realistic goals.

**Help clients set goals that can be sustained.**   Clients need to commit themselves to goals that have staying power. One separated couple said that they wanted to get back together again. They did so only to get divorced again within six months. Their goal of getting back together again was achievable but not sustainable. Perhaps they should have asked themselves, "What do we need to do not only to get back together but also to stay together? What would our marriage have to look like to become, and remain, workable?" In discretionary-change situations, the issue of sustainability needs to be visited early on.

Many Alcoholics Anonymous-like programs work because of their one-day-at-a-time approach. The goal of being, let us say, drug-free has to be sustained only over a single day. The next day is a new era. In a previous example, Vitorio and Anthony's arrangement had enough staying power to produce good results in the short term. It also allowed them to reset their relationship and to improve the business. The goal was not designed to produce a lasting business arrangement because, in the end, Anthony's aspirations were bigger than the family business.

**Help clients choose goals that have some flexibility.**   In many cases, goals have to be adapted to changing realities. Therefore, there might be some trade-offs between goal specificity and goal flexibility in uncertain situations. Napoleon noted this when he said, "He will not go far who knows from the first where he is going." Sometimes making goals too specific or too rigid does not allow clients to take advantage of emerging opportunities.

> Even though he liked the work, and even the company he worked for, Jessie felt like a second-class citizen. He thought that his supervisor gave him most of the dirty work and that there was an undercurrent of prejudice against Hispanics in his department.

Jessie wanted to quit and get another job, one that would pay the same relatively good wages he was now earning. A counselor helped Jessie challenge his choice. Even though the economy was booming, the industry in which Jessie was working was lagging behind. There were few jobs available for workers with Jessie's set of skills. The counselor helped Jessie choose an interim goal that was more flexible and more directly related to coping with his present situation. The interim goal was to use his time preparing himself for a better job outside this industry. In six months to a year he could be better prepared for a more viable career. Jessie began volunteering for special assignments that helped him learn some new skills, and took some crash courses dealing with computers and the Internet. He felt good about what he was learning and more easily ignored the prejudice.

Counseling is a living, organic process. Just as organisms adapt to their changing environments, clients need to adapt their choices to their changing circumstances.

**Help clients choose goals consistent with their values.** Although helping is a process of social influence, it remains ethical only if it respects, within reason, the values of the client. Values are criteria we use to make decisions. Helpers may challenge clients to reexamine their values, but they should not encourage clients to perform actions that are not in keeping with their values.

The son of Vincente and Consuela Garza is in a coma in the hospital after an automobile accident. He needs a life-support system to remain alive. His parents are experiencing a great deal of uncertainty, pain, and anxiety. They have been told that there is practically no chance that their son will ever come out of the coma. One possibility is to terminate the life-support system. While the counselor should not urge them to terminate the life-support system if that is counter to their values, she can help them explore and clarify the values involved. In this case, the counselor suggests that they discuss their decision with their clergyman. In doing so, they learn that the termination of the life-support system would not be against the tenets of their religion. Now they are free to explore other values that relate to their decision.

Some problems involve a client's trying to pursue contradictory goals or values. Chester, the ex-Marine, wants to get an education, but he also wants to make a decent living as soon as possible. The former goal would put him in debt, but failing to get a college education would lessen his chances of securing the kind of job he wants. The counselor helps him identify and use his values to consider some trade-offs. Chester chooses to work part-time and go to school part-time. He chooses a job in an office instead of a job in construction. Even though the latter pays better, it would be much more exhausting and would leave him with little energy for school.

**Help clients establish realistic time frames for accomplishing goals.** Goals that are to be accomplished "sometime or other" probably won't be accomplished at all. Therefore, helping clients put some time frames in their goals can add value. Greenberg (1986) talked about immediate, intermediate, and final outcomes. Here's what they look like when applied to Janette's problem situation. She suffers in a variety of ways because she lets others take advantage of her. She needs to become more assertive in standing up for her own rights.

- *Immediate outcomes* are changes in attitudes and behaviors evident in the helping sessions themselves. For Janette, the helping sessions constitute a safe forum for her to become more assertive. In her dialogues with her counselor, she learns and practices the skills she needs to be more assertive.

- *Intermediate outcomes* are changes in attitudes and behaviors that lead to further change. It takes Janette a while to transfer her assertiveness skills to both the workplace and to her social life. She chooses relatively safe situations to practice being more assertive. For instance, she stands up to her mother more.

- *Final outcomes* refer to the completion of the overall program for constructive change through which problems are managed and opportunities developed. It takes more than two years for Janette to become assertive in a consistent, day-to-day way.

The next example deals with a young man who has been caught shoplifting. Here, too, there are immediate, intermediate, and final outcomes.

> Jensen, a 22-year-old on probation for shoplifting, was seeing a counselor as part of a court-mandated program. An immediate need in his case was overcoming his resistance to his court-appointed counselor and developing a working alliance with her. Because of the counselor's skill and her unapologetic tough but caring attitude, he quickly came to see her as being "on his side." Their relationship became a platform for establishing further goals. An intermediate outcome was attitudinal in nature. Brainwashed by what he saw on television, Jensen thought that America owed him some of its affluence and that personal effort had little to do with it. The counselor helped him see that his attitude of entitlement was unrealistic, and that hard work played a key role in most payoffs. There were two significant final outcomes in Jensen's case. First, he made it through the probation period free of any further shoplifting attempts. Second, he acquired and kept a job that helped him pay his debt to the retailer.

Taussig (1987) talked about the usefulness of setting and executing minigoals early in the helping process. Consider the case of Gaston.

> Gaston, a 16-year-old school dropout and loner, was arrested for arson. Though he lived in the inner city and came from a single-parent household, it was difficult to discover just why he had turned to arson. He had torched a few structures that seemed relatively safe to burn. No one was injured. Was his behavior a cry for help? Social rage expressed in vandalism? Just a way of getting some kicks? The social worker assigned to the case found these questions too speculative to be of much help. Instead of looking for the root causes of Gaston's malaise, she tried to help him set some simple goals that appealed to him and that could be accomplished relatively quickly. One goal was to set up some social support. The counselor helped Gaston join a social club at a local youth center. A second goal was to find a role model. Gaston struck up a friendship with one of the more active members of the center, a dropout who had gotten a high school equivalency degree. He also received some special attention from one of the adult monitors of the center. This was the first time he had experienced the presence of a strong adult male in his life. A third goal was to broaden his view of the world. A group of college students who did volunteer work in both the black and white communities invited Gaston and a couple of the other boys to work with them in a housing facility for the elderly located in a white neighborhood. This was the first time he had been engaged in any kind of work outside the black community. The

## A Checklist for Shaping Goals

1. Is the goal stated in *outcome* or *results* language?
2. Is the goal specific enough to drive behavior? How will I know when I have accomplished it?
3. If I accomplish this goal, will it make a difference? Will it really help manage the problems and opportunities I have identified?
4. Does this goal have "bite" while remaining prudent?
5. Is it doable?
6. Can I sustain this goal over the long haul?
7. Does this goal have some flexibility?
8. Is this goal in keeping with my values?
9. Have I set a realistic time frame for the accomplishment of the goal?

experience helped him push back the walls a bit. He saw white people with real needs. The accomplishment of these minigoals helped Gaston become a bit more realistic about the world around him. He enjoyed the camaraderie of the volunteer group and began experiencing himself in a new, more constructive way.

The suggestion here is not that goal setting is a facile answer to intractable social problems, but that the achievement of sequenced minigoals can go a long way toward making a dent in intractable problems.

There is no such thing as a set time frame for every client. Some goals need to be accomplished now, some soon; others are short-term goals; still others are long-term. Consider the case of a priest who had been unjustly accused of child molestation.

- *A "now" goal:* Some immediate relief from debilitating anxiety attacks and keeping his equilibrium during the investigation and court procedures
- *A "soon" goal:* Obtaining the right kind of legal aid
- *A short-term goal:* Winning the court case
- *A long-term goal:* Reestablishing his credibility in the community and learning how to live with those who would continue to suspect him

There is no particular formula for helping all clients choose the right mix of goals at the right time and in the right sequence. Although helping is based on problem-management principles, it remains an art.

It is not always necessary, then, to make sure that each goal in a client's program for constructive change has all the characteristics outlined in this chapter. These characteristics constitute a checklist counselors can use to help clients design goals sensibly. Box 9.2 outlines some questions that you can help clients ask themselves to design and shape goals.

**EVALUATION QUESTIONS FOR TASK 2**

- To what degree am I helping clients choose specific goals from among a number of possibilities?
- How well do I challenge clients to translate good intentions into broad goals, and broad goals into specific, actionable goals?
- To what extent do I help clients shape their goals so that they have the characteristics outlined in Box 9.2?
- How effectively do I help clients establish goals that take into consideration both their needs and wants?
- To what degree do I help clients become aware of goals that are naturally emerging from the helping process?
- How well do I help clients identify viable goals when the future is both risky and uncertain?
- How effectively do I help clients choose the right mix of adaptive and stretch goals?
- How well do I help clients explore the consequences of the goals they are setting?
- How do I help clients make a bias toward goal-accomplishing action part of their style?

## Task 3: Help Clients Find the Incentives to Commit Themselves to a Better Future: "What Am I Willing to Pay for What I Want?"

Task 3, developing commitment, is a dimension of the entire goal-setting process. Task 3 should be on your mind as you help clients explore possibilities and set goals. Doing a good job with Tasks 1 and 2 goes a long way in helping clients develop commitment. Still, even though clients set goals, that does not mean that they are willing to pay for them. Once clients state what they want and set goals, the battle is joined, as it were. It is as if the client's "old self" or old lifestyle begins vying for resources with the client's potential "new self" or new lifestyle. On a more positive note, history is full of examples of people whose strength of will to accomplish some goal has enabled them to do seemingly impossible things.

> A woman with two sons in their 20s was dying of cancer. The doctors thought she could go at any time. However, one day she told the doctor that she wanted to live to see her older son get married in six months' time. The doctor talked vaguely about "trusting in God" and "playing the cards she had been dealt." Against all odds, the woman lived to see her son get married. Her doctor was at the wedding. During the reception, he went up to her and said, "Well, you got what you wanted. Despite the way things are going, you must be deeply satisfied." She looked at him wryly and said, "But, doctor, my second son will get married someday."

Although the job of counselors is not to encourage clients to heroic efforts, counselors should not undersell clients, either.

In this task, which is usually intermingled with the other two tasks of Stage II, counselors help their clients pose and answer such questions as:

- Why should I pursue this goal?
- Is it worth it? What incentives do I have for pursuing the goals I have set?
- Is this where I want to invest my limited resources of time, money, and energy?
- What other agendas are competing for my time and attention?

Again, there is no formula. Some clients, once they establish goals, race to accomplish them. At the other end of the spectrum are clients who, once they decide on goals, stop dead in the water. Furthermore, the same client might speed toward the accomplishment of one goal and drag her feet on another. Or start out fast and then slow to a crawl. The job of the counselor is to help clients explore, develop, and face up to their commitments.

There is a difference between initial commitment to a goal and an ongoing commitment to a strategy, or plan, to accomplish the goal. The proof of initial commitment lies in goal-accomplishing action. For instance, one client, who chose as a goal a less abrasive interpersonal style, began to engage in an "examination of conscience" each evening to review what his interactions with people had been like that day. In doing so, he discovered—somewhat painfully—that in some of his interactions he actually moved beyond abrasiveness to contempt. That forced him back to a deeper analysis of the problem situation and the blind spots associated with it. Being dismissive of people he did not like or who he deemed "not important" had become ingrained in his interpersonal lifestyle.

## Guidelines for Helping Clients Commit to Goals

There is a range of things you can do to help clients in their initial commitment to goals and the kind of action that is a sign of that commitment. Counselors can help clients by helping them make goals appealing, by helping them enhance their sense of ownership, and by helping them deal with competing agendas.

**Help clients set goals that are worth more than they cost.** Here we revisit the so-called economics of helping. Some goals that can be accomplished carry too high a cost in relation to their payoff. It may sound accountant-like to ask whether any given goal is "cost-effective," but the principle is important. Skilled counselors help clients budget, rather than squander, their resources—work, time, emotional energy.

Eunice discovered that she had a terminal illness. In talking with several doctors, she found out that she would be able to prolong her life a bit through a combination of surgery, radiation treatment, and chemotherapy. However, no one suggested that these would lead to a cure. She also learned what each form of treatment and each combination would cost, not so much in monetary terms, but in added anxiety and pain. Ultimately, she decided against all three, since no combination of them promised much for the quality of the life that was being prolonged. Instead, with the help of a doctor who was an expert in hospice care, she developed a scenario that would ease both her anxiety and her physical pain as much as possible.

It goes without saying that another patient might have made a different decision. Costs and payoffs are relative. Some clients might value an extra month of life no matter what the cost.

Since it is often impossible to determine the cost/benefit ratio of any particular goal, counselors can add value by helping clients understand the consequences of choosing a particular goal. For instance, if a client sets her sights on a routine job with minimally adequate pay, this outcome might well take care of some of her immediate needs but prove to be a poor choice in the long run. Helping clients foresee the consequences of their choices may not be easy. A woman with cancer felt she was no longer able to cope with the sickness and depression that came with her chemotherapy treatments. She decided abruptly one day to end the treatment, saying that she didn't care what happened. No one helped her explore the consequences of her decision. Eventually, when her health deteriorated, she had second thoughts about the treatments, saying, "There are still a number of things I must do before I die." But it was too late. Some reasonable challenge on the part of a helper might have helped her make a better decision.

**Help clients set appealing goals.** Just because goals will help in managing a problem situation, develop an opportunity, and are cost-effective does not mean that they will automatically appeal to the client. Setting appealing goals is common sense, but it is not always easy to do. For instance, for many—if not most—addicts, a drug-free life is not immediately appealing, to say the least.

> A counselor tries to help Chester work through his resistance to giving up prescription drugs. He listens and is empathic. He also challenges the way Chester has come to think about drugs and his dependency on them. One day the counselor says something about "giving up the crutch and walking straight." In a flash Chester sees himself not as a drug addict but as a "cripple." A friend of his had lost a leg in a landmine explosion in Kosovo. He remembered how his friend had longed for the day when he could be fitted with a prosthesis and throw his crutches away. The image of "throwing away the crutch" and "walking straight" proved to be very appealing to Chester.

An incentive is a promise of a reward. As such, incentives can contribute to developing a climate of hope around problem management and opportunity development. A goal is appealing if there are incentives for pursuing it. Counselors need to help clients in their search for incentives throughout the helping process. Ordinarily, negative goals (giving up something that is harmful) need to be translated into positive goals (getting something that is helpful). It was much easier for Chester to commit himself to return to school than to give up prescription drugs, because school represented something he was getting. Images of himself with a degree and of holding some kind of professional job were solid incentives. The picture of him "throwing away the crutch" proved to be an important incentive for ending his drug use.

**Help clients embrace and own the goals they set.** Earlier, we discussed how important it is for clients to *own* the problems and unused opportunities they talk about. It is also important for them to own the goals they set. It is essential that the goals chosen are the client's, rather than the helper's goals or someone else's. Helpers can use various kinds of probes to help clients discover what they want to do to manage some dimension of a problem situation more effectively. For instance, Carl Rogers, in a film of a counseling session (Rogers, Perls, & Ellis, 1965), is asked by a woman what she should do about her relationship with her daughter. He says to her, "I think you've been telling me all along what you want to do." She knew what she wanted the relationship to look like, but she was asking for his approval. If he had given it, the goal would, to some degree, have become his goal instead of hers. At another time he asks, "What is it that you want me to tell you to do?" This question puts the responsibility for goal setting where it belongs—on the shoulders of the client.

> Cynthia was dealing with a lawyer to handle her impending divorce. Discussions about what would happen to the children had taken place, but no decision had been reached. One day she came in and said that she had decided on mutual custody. She wanted to work out such details as which residence, hers or her husband's, would be the children's principal one and so forth. The lawyer asked her how she had reached her decision. She said that she had been talking to her husband's parents—she was still on good terms with them—and that they had suggested this arrangement. The lawyer challenged Cynthia to take a closer look at her decision. "Let's start from zero," he said, "and you tell me what kind of living arrangements *you* want and why." He did not think that it was wise to help her carry out a decision that was not her own.

Choosing goals suggested by others enables clients to blame others if they fail to reach the goals. Also, if they simply follow other people's advice, they often fail to explore the down-the-road consequences.

Commitment to goals can take different forms: mere compliance, substantial buy-in, and full ownership. The least useful is mere compliance. "Well, I guess I'll have to change some of my habits if I want to keep my marriage afloat" does not augur well for sustaining changes in behavior. But it may be better than nothing. Buy-in is a level up from compliance. "Yes, these changes are essential if we are to have a marriage that makes sense for both of us. We say we want to preserve our marriage, but now we have to prove it to ourselves." This client has moved beyond mere compliance. But sometimes, like mere compliance, buy-in alone does not provide enough staying power because it depends too much on reason. "This is logical" is far different from "This is what I really want!" Ownership is a higher form of commitment. It means that the client can say, "This goal is not someone else's, it's not just a good idea; it is mine, it is what I want to do." Consider the following case.

A counselor worked with a manager whose superiors had intimated that he would not be moving much further in his career unless he changed his style in dealing with the members of his team and other key people with whom he worked in the organization. At first the manager resisted setting any goals. "What they want me to do is a lot of hogwash. It won't do anything to make the business better," was his initial response. One day, when asked whether accomplishing what "they" wanted him to do would cost him that much, he pondered a few moments and then said, "No, not really." That got him started. He moved beyond resistance.

With a bit of help from the counselor, he identified a few areas of his managerial style that could well be "polished up." Within a few months he got much more into the swing of things. The people who reported to him responded favorably to his changed behavior, and he was then able to say, "Well, I now see that this makes sense. But I'm doing it because it has a positive effect on the people in the department. It's the right thing to do." Buy-in had arrived. A year later he moved up another notch. He became much more proactive in finding ways to improve his style. He delegated more, gave people feedback, asked for feedback, held a couple of managerial retreats, joined a human resource task force, and routinely rewarded his direct reports for their successes. Now he began to say such things as "This is actually fun." Ownership had arrived. The people in his department began to see him as one of the best executives in the company. This process took over two years.

The manager did not have a personality transformation. He did not change his opinion of some of his superiors, and was right in pointing out that they didn't follow their own rules. But he did change his behavior, because he gradually discovered meaningful incentives to do so.

**Help clients develop self-contracts.** Self-contracts—that is, contracts that clients make with themselves—can help clients commit themselves to new courses of action. Although contracts are promises clients make to themselves to behave in certain ways and to attain certain goals, they are also ways of making goals more focused. It is not only the expressed or implied promise that helps, but also the explicitness of the commitment. Consider the following example in which one of Dora's sons disappears without a trace.

About a month after one of Dora's two young sons disappeared, she began to grow listless and depressed. She was separated from her husband at the time the boy disappeared. By the time she saw a counselor a few months later, a pattern of depression was quite pronounced. Although her conversations with the counselor helped ease her feelings of guilt—for instance, she stopped engaging in self-blaming rituals—she remained listless. She shunned relatives and friends, kept to herself at work, and even distanced herself emotionally from her other son. She resisted developing images of a better future, because the only better future she would allow herself to imagine was one in which her son had returned.

Some strong challenging from Dora's sister-in-law, who visited her from time to time, helped jar her loose from her preoccupation with her own misery. Her sister-in-law screamed at Dora one night, "You're trying to solve one hurt, the loss of Bobby, by hurting Timmy and hurting yourself. I can't imagine in a thousand years that this is what Bobby would want!" Afterwards, Dora and the counselor discussed a recommitment to Timmy, to herself, to the extended family, and to their home. She agreed to getting a physical and began taking an anti-depressant, which helped with both the anxiety and insomnia she had been experiencing. Through a series of contracts, she began to reintroduce patterns of behavior that had been characteristic of her before the tragedy. For instance, she contracted to opening her life up to relatives and friends once more, to creating a much more positive atmosphere at home, to encouraging Timmy to have his friends over, and so forth. Contracts worked for Dora because, as she said to the counselor, "I'm a person of my word." The anti-depressant was withdrawn in due course.

When Dora first began implementing these goals, she felt she was just going through the motions. However, what she was really doing was acting herself into a new mode of thinking. Contracts helped Dora in both her initial commitment to a goal and her movement to action. In counseling, contracts are not legal documents but human instruments to be used if they can be helpful. They often provide both the structure and the incentives some clients need.

**Help clients deal with competing agendas.** Clients often set goals and formulate programs for constructive change without taking into account competing agendas—other things in their lives that soak up time and energy, such as job, family, and leisure pursuits. The world is filled with distractions. For instance, one manager wanted to begin developing computer and Internet-related skills, but the daily push of business and a divorce set up competing agendas and sapped his resources. He was unable to accomplish any of the goals on his self-development agenda. Programs for constructive change often involve a rearrangement of priorities. If a client is to be a full partner in the reinvention of his marriage, then he cannot spend as much time with the boys. Or, the under-employed blue-collar worker might have to put aside some parts of her social life if she wants a more fulfilling job. She eventually discovers a compromise. A friend introduces her to the job search possibilities on the Internet. She discovers that she can work full-time to support herself, do a better job looking for new employment on the Internet than by using traditional methods, and still have some time for a reasonable social life.

This is not to suggest that competing agendas are frivolous. Sometimes clients have to choose between right and right. The woman who wants to expand her horizons by getting involved in social settings outside the home still has to figure out how to handle the tasks at home. This is a question of balance, not frivolity. The single parent who wants a promotion at work needs to balance her new responsibilities with her involvement with her children. A counselor who had worked with a two-career couple as they made a decision to have a child helped them think of competing agendas once the pregnancy started. A year after the baby was born, they saw the counselor again for a couple of sessions to work on some issues that had come up. However, they started the session by saying, "Are we glad that you talked about competing agendas when we were struggling with the decision to become parents! After the baby was born, we went back time and time again and reviewed what we said about managing competing and conflicting priorities. It helped stabilize us for the last two years."

**EVALUATION QUESTIONS FOR TASK 3**

**How effectively do I help clients ask themselves the following kinds of questions as they struggle with committing themselves to a program of constructive change?**

- What is my state of readiness for change in this area at this time?
- How badly do I want what I say I want?
- How hard am I willing to work?
- To what degree am I choosing this goal freely?
- How highly do I rate the personal appeal of this goal?
- How do I know I have the courage to work on this?
- What's pushing me to choose this goal?
- What incentives do I have for pursuing this agenda for change?
- What rewards can I expect if I work on this agenda?
- If other people are in any way imposing this goal, what am I doing to make it my own?
- What difficulties am I experiencing in committing myself to this goal?
- In what way might my commitment not be a true commitment?
- What can I do to get rid of the disincentives and overcome the obstacles?
- What can I do to increase my commitment?
- In what ways can the goal be reformulated to make it more appealing?
- How reasonable is the timeframe for pursuing this goal?
- What do I have to do to stay committed?
- What resources can I use to help me?

## Stage II and Action

The work of Stage II—developing possibilities for a better future, crafting some goals, and focusing on commitment—is just what some clients need. It frees them from thinking solely about problem situations and unused resources and enables them to begin fashioning a better future. Once they identify some of their wants and needs and consider a few possible goals, they move into action.

> Francine is depressed because her aging and debilitated father has been picking on her, even though she has put off marriage in order to take care of him. Some of the things he says to her are quite hurtful. The situation has begun to affect her productivity at work. A counselor suggests to her that the hurtful things her father says to her are not her father, but his illness speaking. This gives her a whole new perspective and frees her to think about other possibilities. Once she spends a bit of time brainstorming answers to the counselor's question, "What do you want for both yourself and your

father," she says things like, "I'd like both of us to go through this with our dignity intact," and "I'd like to be living the kind of life he would want me to have if his mind wasn't so clouded." Once she brainstorms some possibilities for a better arrangement with her father, she needs little further help. Her usual resourcefulness returns. She gets on with life.

Some clients, once they set a clear goal or a set of goals, are off to the races. "Now that I know what I want, I'm going to set things is motion."

For some clients, the search for incentives for commitment is the trigger for action. Once they see "what's in it for me"—a kind of upbeat and productive selfishness, if you will—they move into action.

Callahan is seeing a business consultant because he is very distressed. He owns and runs a small business. A few of his employees have gotten together and filed a workplace discrimination suit against him. The "troublemakers" he calls them, meaning a few women, a couple of Hispanics, and three African Americans. The consultant finds out that Callahan believes that they are "decent workers." The fact is that they are more than decent. Callahan tells the consultant that he is paying them "scale." In fact he is underpaying them. Callahan also says that he doesn't expect his supervisors to "bend over backwards to become friends with the people reporting to them." The reality is that some of his supervisors are not only not friendly, but even abusive.

The counselor convinces Callahan to attend an excellent program on diversity "before some court orders you to." A couple of weeks after returning from the program, he has a session with the consultant. He says that he had never even once considered the advantages of diversity in the workplace. All the term had meant to him was "a bunch of politicians looking for votes." Now that he saw the business reasons for diversity, he knew there were a few things he could do, but he still needed the consultant's help and guidance. "I don't want to look like a soft jerk." Callahan's newly-acquired "human touch" is far from being soft. He remains a rather rough-and-tough business guy.

Callahan didn't change his stripes overnight, but finding a package of incentives certainly helped him move toward much needed action. Who knows, the whole situation might have even made a dent in his deeply-ingrained prejudices.

## The Shadow Side of Goal Setting

Despite the advantages of setting goals, some helpers and clients seem to conspire to avoid goal setting as an explicit process. It is puzzling to see counselors helping clients explore problem situations and unused opportunities, and then stopping short of asking them what they want and helping them set goals. As Bandura (1990, p. xii) put it, "Despite this unprecedented level of empirical support [for the advantages of goal setting], goal theory has not been accorded the prominence it deserves in mainstream psychology." Years ago, the same concern was expressed differently. A U.S. developmental psychologist was talking to a Russian developmental psychologist. The Russian said, "It seems to me that American researchers are constantly seeking to explain how a child came to be what he is. We in the USSR are striving to discover how he can become what he not yet is" (see Bronfenbrenner, 1977, p. 258). One of the main reasons that counselors do not help clients develop realistic life-enhancing goals is that they are not trained to do so.

There are other reasons. First of all, some clients see goal setting as very rational, perhaps too rational. Their lives are so messy, and goal setting seems so sterile. Both helpers and clients object to this overly rational approach. There is a dilemma. On the one hand, many clients need, or would benefit by, a rigorous application of the problem-management process, including goal setting. On the other, they resist its rationality and discipline. They find it alien. Second, goal setting means that clients have to move out of the relatively safe harbor of discussing problem situations and of exploring the possible roots of those problems in the past, and move into the uncharted waters of the future. This may be uncomfortable for client and helper alike. Third, clients who set goals and commit themselves to them move beyond the victim-of-my-problems game. Victimhood and self-responsibility make poor bedfellows. Some clients choose victimhood. Fourth, goal setting involves clients' placing demands on themselves, making decisions, committing themselves, and moving to action. If I say, "This is what I want," then, at least logically, I must also say, "Here is what I am going to do to get it. I know the price and I'm willing to pay it." Since this demands work and pain, clients will not always be grateful for this kind of "help." Fifth, goals, though liberating in many respects, also hem clients in. If a woman chooses a career, then it might not be possible for her to have the kind of marriage she would like. If a man commits himself to one woman, then he can no longer play the field.

There is some truth in the ironic statement "There is only one thing worse than not getting what you want, and that's getting what you want." The responsibilities accompanying getting what you want—a drug-free life, a renewed marriage, custody of the children, a promotion, the peace and quiet of retirement, freedom from an abusing husband—often open up a new set of problems. Even good solutions create new problems. It is one thing for parents to decide to give their children more freedom; it is another thing for them to watch them use that freedom. Finally, there is a phenomenon called post-decisional depression. Once choices are made, clients begin to have second thoughts that often keep them from acting on their decisions.

As to action, some clients move into action too quickly. The focus on the future liberates them from the past, and the first few possibilities are very attractive. They fail to get the kind of focus and direction provided by Stage II. So they go off half-cocked. Failing to weigh alternatives and shape goals often means that they have to do the process all over again.

Effective helpers know what lurks in the shadows of goal setting, both for themselves and for their clients, and are prepared to manage their own part of the process and help clients manage theirs. The answer to all of this lies in helpers' being trained in the entire problem-management process and in their sharing a picture of the entire process with the client. Then goal setting, described in the client's language, will be a natural part of the process. Artful helpers weave goal setting, under whatever name, into the flow of helping. They do so by moving easily back and forth among the stages of the helping process, even if they have only a few sessions with a client.

## EXERCISES ON PERSONALIZING

1a. Take a problem situation you currently face, for instance, the one you described in Chapter 1. Write out ten possibilities for a better future. Or draw ten pictures. Not actions, but accomplishments or outcomes.

1b. Take a relationship you currently have that is troubled or boring or lifeless. What would this relationship look like if it were better? Name five or six possibilities. What would you be doing? What would the other person be doing?

2a. Chose two of the most promising possibilities you came up with in step 1a. Develop and shape these possibilities into goals by applying the criteria for effective goals or outcomes outlined in Task 2.

2b. Do the same for the possibilities for a better relationship developed in step 1b.

3a. What incentives would you need to pursue the goal, or goals you came up with in step 2a vigorously? What obstacles would you have to overcome?

3b. What incentives would you need to pursue the goal, or goals you came up with in step 2b vigorously? What obstacles would you have to overcome?

# CHAPTER 10

# Stage III:
# Helping Clients Develop Strategies and Plans for Accomplishing Goals

Introduction to Stage III

Task 1: Help Clients Review Possible Strategies to Achieve Goals: "What Strategies Will Help Me Get What I Want?"

    Guidelines for Helping Clients Search for Multiple Paths to Goals

Task 2: Help Clients Choose Strategies That Best Fit Their Resources: "What Courses of Action Make Most Sense for Me?"

    The Case of Bud

    Guidelines for Helping Clients Choose Best-Fit Strategies

    Strategy Sampling

    The Balance-Sheet Method

    Linking Task 2 to Action

    The Shadow Side of Developing and Selecting Strategies

Task 3: Help Clients Pull Strategies Together into a Manageable Plan: "How Do I Put Everything Together and Move Forward?"

    No Plan of Action: The Case of Frank

    How Plans Add Value to Clients' Search for a Better Life

    Shaping the Plan: Two Cases

    Humanizing the Planning Process

    The Shadow Side of Planning

Exercises on Personalizing

# Introduction to Stage III

Planning, in its broadest sense, includes all the tasks of Stages II and III—that is, it deals with solutions with a big *S* and a small *s*. In a narrower sense, planning deals with identifying, choosing, and organizing the strategies needed to accomplish goals. Whereas Stage II is about outcomes, Stage III is about the activities or the work needed to produce these outcomes.

Clients, when helped to explore what is going wrong in their lives, often ask, "Well, what should I do about it?" That is, they focus on actions they need to take in order to *solve* things. But action, though essential, is valuable only to the degree that it leads to problem-managing and opportunity-developing outcomes. Accomplishments or outcomes, also essential, are valuable only to the degree that they have a constructive impact on the life of the client. The Skilled Helper model is a planning-action-outcome-impact model. The following example demonstrates the distinction between action, outcomes, and impact.

> Lacy, a 40-year-old single woman, is making a great deal of progress in staying sober. She controls her drinking through her involvement with an AA program. She engages in certain activities—for instance, she attends AA meetings, follows the twelve steps, stays away from situations that would tempt her to drink, and calls fellow AA members when she feels depressed or when the temptation to drink is pushing her hard. The outcome is that she has stayed sober for over seven months. This, she feels, is quite an accomplishment. The impact of all this is substantial. She feels better about herself and she has both the energy and the enthusiasm to do things that she has not done in years—developing a circle of friends, getting involved in church activities, and doing a bit of traveling.
>
> But Lacy is also struggling with a troubled relationship with a man. In fact, her drinking was, in part, an ineffective way of avoiding the problems in the relationship. She knows that she no longer wants to tolerate the psychological abuse she has been getting from her male friend, but she's afraid of the vacuum she will create by cutting off the relationship. She is, therefore, trying to determine what she wants, almost fearing that ending the relationship might turn out to be the best option.
>
> She has changed a number of behaviors in attempting to manage the relationship. For instance, she has become much more assertive with her friend. She now cuts off contact whenever her companion becomes abusive. And she no longer lets him make all the decisions about what they are going to do together. But the relationship remains troubled. Even though she is doing many things, there is no satisfactory outcome. She has not yet determined what the outcome should be; that is, she has not determined what kind of relationship she would like and if it is possible to have such a relationship with this man. Nor has she determined to end the relationship.
>
> Finally, after one seriously abusive episode, she tells him that she is ending the relationship. She does what she has to do to sever all ties with him. The outcome is that the relationship ends and stays ended. The impact is that she feels liberated but lonely. The helping process needs to be recycled to help her with this new problem.

Stage III has three interrelated tasks all aimed at determining what kind of action on the part of the client will lead to the accomplishment of problem-managing goals.

**Task 1: Help clients review possible strategies to achieve goals.** Often there are multiple paths to any given goal. "What strategies will help me get what I want?"

**Task 2: Help clients chose strategies that best fit their resources.** Help clients choose strategies tailored to their preferences and resources. "What courses of action make most sense for me?"

**Task 3: Help clients pull strategies together into a manageable plan.** Help clients formulate actionable plans. "How do I put everything together and move forward?" "What should my campaign for constructive change look like? What do I need to do first? Second? Third?" Stage III deals with the *game plan*.

Figure 10.1 illustrates these three tasks. As in other stages, these three tasks are usually intermingled. Strategy is the art of identifying and choosing realistic courses of action for achieving goals, and doing so under adverse conditions, such as war. The problem situations in which clients are immersed constitute adverse conditions; clients often are at war with themselves and the world around them. Helping clients develop strategies to achieve goals can be a most thoughtful, humane, and fruitful way of being with them. This step in the counseling process is another that helpers sometimes avoid because it is too "linear" or "technical." They do their clients a disservice. Clients with realistic goals but no clear idea of how to accomplish them still need help.

## Task 1: Help Clients Review Possible Strategies to Achieve Goals: "What Strategies Will Help Me Get What I Want?"

Counselors, understanding what could be called the technology of planning, can add value by helping clients find ways of accomplishing goals (getting what they need and want) in a systematic, yet flexible and personalized, way. Once more it is a question of helping clients stimulate their imaginations and engage in divergent thinking. Most clients do not instinctively seek different routes to goals and then choose the ones that make most sense. Clients often make better choices if they have a number of options.

### Guidelines for Helping Clients Search for Multiple Paths to Goals

There are many different ways of helping clients search for ways of accomplishing problem-managing goals. Here are some of them:

**Help clients brainstorm strategies for accomplishing goals.** Brainstorming, discussed in Chapter 9, plays an important part in strategy development. Consider the case of Karen, who came to realize that heavy drinking was ruining her life. Her goal was to stop drinking. She felt that it simply would not

**The Skilled-Helper Model**

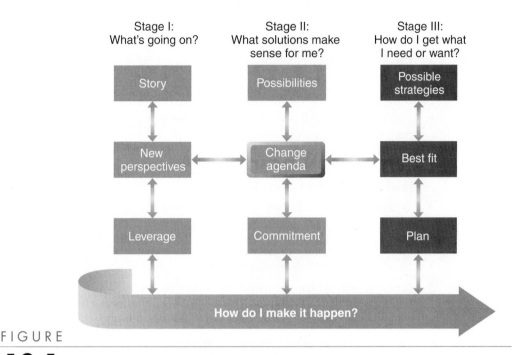

FIGURE

**10.1**                              The Helping Model—Stage III

be enough to cut down; she had to stop. To her, the way forward seemed simple enough: Whereas before she drank, now she wouldn't. Because of the novelty of not drinking, she was successful for a few days; then she fell off the wagon. This happened a number of times until she finally realized that she could use some help. Stopping drinking, at least for her, was not as simple as it seemed.

A counselor at a city alcohol and drug treatment center helps her explore a number of techniques that could be used in an alcohol-management program. Together they come up with the following possibilities:

- Just stop cold turkey and get on with life.
- Join Alcoholics Anonymous.
- Move someplace declared "dry" by local government.
- Take a drug that causes nausea if followed by alcohol.
- Replace drinking with other rewarding behaviors.
- Join some self-help group other than Alcoholics Anonymous.
- Get rid of all liquor in the house.

- Take the "pledge" not to drink; to make it more binding, take it in front of a minister.
- Join a residential hospital detoxification program.
- Avoid friends who drink heavily.
- Change other social patterns, for instance, find places other than bars and cocktail lounges to socialize.
- Try hypnosis to reduce the urge to drink.
- Use behavior modification techniques to develop an aversion for alcohol, for instance, pair painful, but safe, electric shocks with drinking or even thoughts about drinking.
- Change self-defeating patterns of self-talk, such as "I have to have a drink" or "One drink won't hurt me."
- Become a volunteer to help others stop drinking.
- Read books and view films on the dangers of alcohol.
- Stay in counseling as a way of getting support and challenge for stopping.
- Share intentions to stop drinking with family and close friends.
- Spend a week with an acquaintance who does a great deal of work in the city with alcoholics, and go with him on his rounds.
- Try to discover the reasons for drinking in the first place.
- Have a discussion with members of the family about the impact drinking has on them.
- Reduce the craving for alcohol by eating lo-cal sweets.
- Get a hobby or an avocation that demands time and energy.
- Substitute a range of self-enhancing activities, such as exercise or surfing the Web for drinking.

This list contained many more items than Karen would have thought of, had she not been stimulated by the counselor to take a census of possible strategies. One of the reasons that clients are clients is that they are not very creative in looking for ways of getting what they want. Once goals are established, getting them accomplished is not just a matter of hard work. It is also a matter of imagination.

If a client is having a difficult time coming up with strategies, the helper can "prime the pump" by offering a few suggestions. Driscoll (1984, p. 167) put it well.

> Alternatives are best sought cooperatively, by inviting our clients to puzzle through with us what is or is not a more practical way to do things. But we must be willing to introduce the more practical alternatives ourselves, for clients are often unable to do so on their own. Clients who could see for themselves the more effective alternatives would be well on their way to using them.

Although the helper may need to suggest alternatives, he or she can do so in such a way that the principal responsibility for evaluating and choosing possible strategies stays with the client. For instance, there is the *prompt and fade* technique. The counselor can say, "Here are some possibilities. . . . Let's review them and see whether any of them make sense to you, or suggest further possibilities." Or, "Here are some of the things that people with this kind of problem situation have tried. . . . How do they sound to you?" The *fade* part of this technique keeps it from being mere advice giving. It remains clear that the client must think these strategies over, choose the right ones, and commit to them.

Let's look at a case. Elton, a graduate student in counseling psychology, is plagued with perfectionism. Although he is an excellent student, he worries about getting things right. After he writes a paper or practices counseling, he agonizes over what he could have done better. This kind of behavior puts him on edge when he practices counseling with his fellow trainees. They tell him that his "edge" makes them uncomfortable and interferes with the flow of the helping process. One student says to him, "You make me feel as if I'm not doing the right things when I play the role of the client."

Elton realizes that "less is more," that is, becoming less preoccupied with the details of helping will make him a more effective helper. His goal is to become more relaxed in the helping sessions, free his mind of his imperatives to be perfect, and learn from mistakes rather than expending an excessive amount of effort trying to avoid them. He and his supervisor are talking about ways he can free himself of these inhibiting imperatives.

SUPERVISOR: What kinds of things can you do to become more relaxed?

ELTON: I need to focus my attention on the client and the client's goals instead of being preoccupied with myself.

SUPERVISOR: So a basic shift in your orientation right from the beginning will help.

ELTON: Right . . . And this means getting rid of a few inhibiting beliefs.

SUPERVISOR: Such as . . .

ELTON: That technical perfection in the helping model is more important than the relationship with the client. I get lost in the details of the model and have forgotten that I'm a human being with another human being.

SUPERVISOR: So "re-humanizing" the helping process in your own mind will help. . . . Any other internal behaviors need changing?

ELTON: Another belief is that I have to be the best in the class. That's my history, at least in academic subjects. Being as effective as I can be in helping a client has nothing to do with competing with my fellow students. Competing is a distraction. I know it's in my bones. It might have been all right in high school, but . . .

SUPERVISOR: Okay, so the academic game mentality doesn't work here . . .

ELTON (interrupting): That's precisely it. Even the practicing we do with one another is real life, not a game. You know that a lot of us talk about real issues when we practice.

SUPERVISOR: You've been talking about getting your attitudes right, and the impact that can have on helping sessions. Are there any external behaviors that might also help?

ELTON (pauses): I'm hesitating because it strikes me how I'm in my head too much, always figuring me out. . . . On a much more practical basis, I like what Jerry and Philomena do. Before each meeting with their "clients" in their practice sessions, they spend five or ten minutes reviewing just where the client is in the overall helping process and determining what they might do in the next session to add value and move things forward. That puts the focus where it belongs, on the client.

SUPERVISOR: So a mini-prep for each session can help you get out of your world and into the client's.

ELTON: Also in debriefing the training videos we make each week . . . I now see that I always start by looking at my behavior instead of what's happening with the client. . . . Oh, there's another thing I can do. I can share just what we've been discussing here with my training partner.

SUPERVISOR: I'm not sure whether you bring up the perfectionism issues when you're the "client" in the practice sessions or in the weekly lifestyle group meetings.

ELTON (hesitating): Well, not really. I'm just coming to realize how pervasive it is in my life. . . . To tell you the truth I think I haven't brought it up because I'd rather have my fellow trainees see me as competent, not a perfectionist. . . . Well, the cat is out of the bag with you, so I guess it makes sense to put it on my lifestyle group agenda.

This dialogue, which includes empathy, probes, and challenge on the part of the supervisor, produces a number of strategies that Elton can use to develop a more client-focused attitude. He ends by saying that all these can be reinforced through his interactions with his training partner.

**Develop a framework for stimulating clients' thinking about strategies.** How can helpers find the right probes to help clients develop a range of strategies? Simple frameworks can help. Consider the following case.

Jackson has terminal cancer. He has been in and out of the hospital several times over the past few months, and he knows that he probably will not live more than a year. He would like the year to be as full as possible, and yet he wants to be realistic. He hates being in the hospital, especially a large hospital, where it is so easy to be anonymous. One of his goals is to die outside the hospital. He would like to die as benignly as possible and retain possession of his faculties as long as possible. How is he to achieve these goals?

You can use probes and prompts to help clients discover possible strategies by using a framework to help them investigate resources in their lives. One such framework focuses on people, models, communities, places, things, organizations, programs, and personal resources.

*Individuals.* What individuals might help clients achieve their goals? Jackson gets the name of a local doctor who specializes in the treatment of chronic cancer-related pain. The doctor teaches people how to use a variety of techniques to manage pain. A friend of his has mentioned that his father got excellent hospice care and died at home. Also, he thinks that talking every once in a while with a friend whose wife died of cancer, a man he respects and trusts, will help him find the courage he needs.

*Models and exemplars.* Who is presently doing what the client wants to do? One of Jackson's fellow workers died of cancer at home. Jackson visited him there a couple of times. That's what gave him the idea of dying at home, or at least outside the hospital. He noticed that his friend never allowed himself to engage in poor-me talk. He refused to see dying as anything but part of living. This touched Jackson deeply at the time, and now, reflecting on that experience may help him develop the same kind of upbeat attitudes.

*Communities.* What communities of people are there through which clients might identify strategies for implementing their goals? Even though Jackson has not been a regular churchgoer, he does know that the parish within which he resides has some resources for helping those in need. A brief investigation reveals that the parish has developed a relatively sophisticated approach to providing various services for the sick. He also does an Internet search and discovers that there are of number of self-help groups for people like himself.

*Places.* Are there particular places that might help? Jackson immediately thinks of Lourdes, the shrine to which Catholic believers flock with all sorts of human problems. He doesn't expect miracles, but he feels that he might experience life more deeply there. It's a bit wild, but why not a pilgrimage? He still has the time and also enough money to do it. He also finds a high-tech place—an Internet chat room for cancer patients *and* their caregivers. This helps him get out of himself and, at times, become a helper instead of a client.

*Things.* What things exist that can help clients achieve their goals? Jackson has read about the use of combinations of drugs to help stave off pain and the side effects of chemotherapy. He has heard that certain kinds of electric stimulation can ward off chronic pain. Perhaps he could have a pain-pump installed. He explores all these possibilities with his doctor and even arranges for second opinions.

*Organizations.* Jackson runs across an organization that helps young cancer patients get their wishes. He volunteers. In his role as helper, he finds he receives as much help and motivation and solace as he gives.

*Programs.* Are there any ready-made programs for people in the client's position? He learns that a new hospice in his part of town has three programs. One helps people who are terminally ill stay in the community as long as they can. A second makes provision for part-time residents. The third provides a residential program for those who can spend little or no time in the community. The goals of these programs are practically the same as Jackson's.

## EVALUATION QUESTIONS FOR TASK 1

### How effectively do I do the following?

- Use probes, prompts, and challenges to help clients identify possible strategies
- Help clients engage in divergent thinking with respect to strategies
- Help clients brainstorm as many ways as possible to accomplish their goals
- Use some kind of framework in helping clients be more creative in identifying strategies
- Help clients identify and begin to acquire the resources they need to accomplish their goals
- Help clients identify and develop the skills they need to accomplish their goals
- Help clients see the action implications of the strategies they identify

## Task 2: Help Clients Choose Strategies That Best Fit Their Resources: "What Courses of Action Make Most Sense for Me?"

Task 2 is about ways of helping clients choose the strategies that are best for them. Clients need to choose one or more strategies (a set or "package") that best fit their situation and resources. Of course, clients often move back and forth between Task 1 and Task 2. There is no one right way of coming up with the best package.

### The Case of Bud

Some clients, once they are helped to develop a range of strategies to implement goals, move forward on their own; that is, they choose the best strategies, put together an action plan, and implement it. Others need some help. It is useless to have clients brainstorm if they don't know what to do with the action strategies they generate. Consider the case of Bud. This is an amazing example of a client who did four life-changing things. He:

- focused on one broad aim, emotional stability,
- translated this aim into a number of immediate, practical goals,
- discovered two broad strategies for accomplishing these goals—finding ongoing emotional support and making a career out of helping others, and
- translated these strategies into outcomes that had an amazingly constructive impact on his life.

Here is Bud's story.

One morning, Bud, then 18 years old, woke up unable to speak or move. He was taken to a hospital, where he was diagnosed as a catatonic schizophrenia. After repeated admissions to hospitals, where he underwent both drug and electroconvulsive therapy (ECT), his diagnosis was changed to paranoid schizophrenia. He was considered incurable.

A quick overview of Bud's earlier years suggests that much of his emotional distress was caused by poorly-managed life problems and the lack of human support. He was separated from his mother for four years when he was young. They were reunited in a city new to both of them, and there he suffered a great deal of harassment at school because of his "ethnic" looks and accent. There was simply too much stress and change in his life. He protected himself by withdrawing. He was flooded with feelings of loss, fear, rage, and abandonment. Even small changes became intolerable. His catatonic attack occurred in the autumn on the day of the change from daylight saving to standard time. It was the last straw.

In the hospital Bud became convinced that he and many of his fellow patients could do something about their illnesses. They did not have to be victims of themselves or of the institutions designed to help them. Later, reflecting on his hospital stays and the drug and ECT treatments, he said he found his "help" so debilitating that it was no wonder that he got crazier. Somehow Bud, using his own inner resources, managed to get out of the hospital. Eventually, he got a job, found a partner, and got married.

One day, Bud felt himself becoming agitated and thought he was choking to death. He had been trying to cope with a series of problems within his family and at work. His doctor sent him to the hospital "for more treatment." There Bud had the good fortune to meet Sandra, a psychiatric social worker who was convinced that many of the hospital's patients were there because of lack of support before, during, and after their bouts of illness. She helped him see his need for social support, especially at times of stress. She also discovered in the course of the inpatient counseling groups that she ran that Bud had a knack for helping others. Bud's broad aim was still emotional stability and he wanted to do whatever was necessary to achieve it. Finding human support and helping others cope with their problems, broad strategies and instrumental goals in themselves, were his best strategies for achieving the stability he wanted.

After leaving the hospital, Bud started a self-help group for ex-patients like himself. In the group, he was a full-fledged participant. Sandra, the social worker, also coached Bud's wife on how to provide support for him at times of stress. As to helping others, Bud not only founded a self-help group but eventually turned it into a network of self-help groups for ex-patients.

Bud did not spend a lot of time identifying and exploring the origins of his illness. Rather, it seems, he was solution-focused by nature. Establishing goals and finding ways to accomplish them were second-nature to him. Indeed, he did not use the language of goals and strategies that is being used here. He just did it. Although expecting clients to be self-starters like Bud is presumptuous, selling clients short is just as bad.

## Guidelines for Helping Clients Choose Best-Fit Strategies

Strategies to achieve goals should be, like goals themselves, specific, robust, prudent, realistic, sustainable, flexible, cost-effective, and in keeping with the client's values. As in Stage II, these criteria form a checklist for choosing the best paths to a goal. Let's take a look at a few of these criteria as applied to choosing strategies.

**Specific strategies.**  Strategies for achieving goals should be specific enough to drive behavior. In the preceding example, Bud's two broad strategies for achieving emotional stability, tapping into human support and helping others, were translated into quite specific strategies—keeping in touch with Sandra, getting help from his wife, participating in a self-help group, starting a self-help group, and founding and running a self-help organization. Bud's strategies proved to be powerful. They not only helped him gain stability but also gave him a new perspective on life.

**Substantive strategies.**  Strategies are robust to the degree that they challenge the client's resources and, when implemented, actually achieve the goal. All of Bud's strategies were robust, especially the strategy of starting and running a self-help organization. The work involved all his energies. In a sense, he didn't have time to be sick.

**Realistic strategies.** If clients choose strategies that are beyond their resources, they are doing themselves in. Strategies are realistic when they can be carried out with the resources the client has, are under the client's control, and are un-encumbered by obstacles. Bud's strategies would have appeared unrealistic to most helpers. But this highlights an important point. Just as we should help clients set stretch goals whenever possible, so we should not underestimate what actions clients are capable of. In the following case, Desmond moves from unrealistic to realistic strategies for getting what he wants.

Desmond was in a halfway house after leaving a state mental hospital. From time to time he still had bouts of depression that would incapacitate him for a few days. He wanted to get a job because he thought that a job would help him feel better about himself, become more independent, and manage his depression better. He answered job advertisements in a rather random way and was constantly turned down after being interviewed. He simply did not yet have the kinds of resources needed to put himself in a favorable light in job interviews. Moreover, he was not yet ready for a regular, full-time job.

On his own, Desmond does not do well in choosing strategies to achieve even modest goals. But here's what happened next.

> A local university received funds to provide outreach services to halfway houses in the metropolitan area. The university program included finding companies that were willing, on a win-win basis, to work with halfway-house residents. A counselor from the program helped Desmond to get in contact with companies that had specific programs to help people with psychiatric problems. He found two that he thought would fit his needs. Some of their best workers had a variety of disabilities, including psychiatric problems. After a few interviews, Desmond got a job in a hotel that fitted his situation and capabilities. The entire work culture was designed to provide the kind of support he needed.

Helping clients get the balance right is important. Robust strategies that make clients stretch for a valued goal can be most rewarding. Bud's case is an exceptional example of that.

**Strategies in keeping with the client's values.**  Make sure that the strategies chosen are consistent with the client's values. Let's return to the case of the priest who had been unjustly accused of child molestation.

> In preparing for the court case, the priest and his lawyer had a number of discussions. The lawyer wanted to do everything possible to destroy the credibility of the accusers. He had dug into their past and dredged up some dirt. The priest objected to these tactics. "If I let you do this," he said, "I descend to their level. I can't do that." The priest discussed this with his counselor, his superiors, and another lawyer. He stuck to his guns. They prepared a strong case but without the sleaze.
>
> After the trial was over and he was acquitted, the priest said that his discussion about the lawyer's preferred tactics was one of the most difficult issues he had to face. Something in him said that since he was innocent, any means to prove his innocence was allowed. Something else told him that this was not right. The counselor helped him clarify and challenge his values but made no attempt to impose either his own or the lawyer's values on his client.

Of course, when clients want to pursue a course of action that conflicts with your own deeply-held values, you have some choices to make. For instance, you can help them explore the consequences of strategies that you see as immoral or you can withdraw.

## Strategy Sampling

Some clients find it easier to choose strategies if they first sample some of the possibilities. Consider this case.

> Two business partners were in conflict over ownership of the firm's assets. Their goals were to see justice done, to preserve the business, and, if possible, to preserve their relationship. A colleague helped them sample some possibilities. Under her guidance they engaged in a kind of sampling. They met with a lawyer and discussed the process and consequences of bringing their dispute to the courts; they had a meeting with a consultant-counselor who specialized in these kinds of disputes; and they visited an arbitration firm.

In this case, the sampling procedure had the added effect of giving them time to let their emotions simmer down. They agreed to go the consultant-counselor route.

Karen, the woman who, with the help of her counselor, brainstormed a wide range of strategies for disengaging from alcohol, decided to sample some of the possibilities.

> Surprised by the number of program possibilities there were to achieve the goal of getting liquor out of her life, Karen decided to sample some of them. She went to an open meeting of Alcoholics Anonymous, attended a meeting of a women's lifestyle-issues group, visited the hospital that had the residential treatment program, and

joined up for a two-week trial physical-fitness program at a YMCA. She tried them out and then discussed them with her counselor. Her search for the programs that were best for her did occupy her energies and strengthened her resolve to do something about her alcoholism.

Of course, some clients could use strategy sampling as a way of putting off action. That was certainly not the case with Bud. His attending the meeting of a self-help group after leaving the hospital was a form of strategy sampling. Although he was impressed by the group, he thought that he could start a group limited to ex-patients that would focus more directly on the kinds of issues he and other ex-patients were facing.

## The Balance-Sheet Method

Balance sheets deal with the acceptability and unacceptability of both benefits and costs—that is, the consequences, of choosing both goals and strategies. A balance-sheet approach, applied to choosing strategies for achieving goals, poses questions such as the following:

- What are the benefits of choosing this strategy? for myself? for significant others?
- To what degree are these benefits acceptable? to me? to significant others?
- In what ways are these benefits unacceptable? to me? to significant others?
- What are the costs of choosing this strategy? for myself? for significant others?
- To what degree are these costs acceptable? to me? to significant others?
- In what ways are these costs unacceptable? to me? to significant others?

Once more, we have a checklist. But realism in using the balance sheet is essential. It is not meant to be used with every client to work out the pros and cons of every course of action. Tailor the balance sheet to the needs of the client. Choose the parts of the balance sheet that will add most value with *this* client pursuing *this* goal or set of goals. In fact, one of the best uses of the balance sheet is not to use it directly at all. Keep it in the back of your mind whenever clients are making decisions. Use it as a filter to listen to clients. Then turn relevant parts of it into probes to help clients focus on issues they may be overlooking. "How will this decision affect the significant people in your life?" is a probe that originates in the balance sheet. "Is there any downside to that strategy?" might help a client who is being a bit too optimistic. There is no one right formula. Box 10.1 outlines the kinds of questions you can help clients answer as they choose best-fit strategies.

## Linking Task 2 to Action

Many clients, once they begin to see what they can do to get what they want, begin acting immediately. They don't need a formal plan. Here are a couple of examples.

## Questions on Best-Fit Strategies

BOX
**10.1**

- Which strategies will be most useful in helping me get what I need and want?
- Which strategies are best for this situation?
- Which strategies best fit my resources?
- Which strategies will be most economic in the use of resources?
- Which strategies are most powerful?
- Which strategies best fit my preferred way of acting?
- Which strategies best fit my values?
- Which strategies will have the fewest unwanted consequences?

Jeff had been in the army for about ten months. He found himself both overworked and, perhaps not paradoxically, bored. He had a couple of sessions with one of the educational counselors on the base. During these sessions Jeff began to see quite clearly that not having a high school diploma was working against him. The counselor mentioned that he could finish high school while in the army. Jeff realized that this possibility had been pointed out to him during the orientation talks, but he hadn't paid any attention to it. He had joined the army because he wasn't interested in school and, being unskilled, couldn't find a job. Now he decided that he would get a high school diploma as soon as possible.

Jeff obtained the authorization needed from his company commander to go to school. He found out what courses he needed and enrolled in time for the next school session. It didn't take him long to finish. Once he received his high school degree, he felt better about himself and found that opportunities for more interesting jobs opened up for him in the army. Achieving his goal of getting a high school degree helped him manage the problem situation.

Jeff was one of those fortunate ones who, with a little help, quickly set a goal (the "what") and identify and implement the strategies (the "how") to accomplish it. Note, too, that his goal of getting a diploma was also a means to other goals—feeling good about himself and getting better jobs in the army.

Grace's road to problem management was quite different from Jeff's. She needed much more help.

As long as she could remember, Grace had been a fearful person. She was especially afraid of being rejected and of being a failure. As a result, she had an impoverished social life. She had held a series of jobs that were safe but boring. She became so depressed that she made a half-hearted attempt at suicide, probably more an expression of anguish and a cry for help than a serious attempt to get rid of her problems by getting rid of herself.

During her stay in the hospital, Grace had a few therapy sessions with one of the staff psychiatrists. The psychiatrist was supportive and helped her handle both the guilt she felt because of the suicide attempt and the depression that had led to the attempt. Just talking to someone about things she usually kept to herself seemed to help. She began to see her depression as a case of "learned helplessness." She saw quite clearly how she had let her choices be dictated by her fears. She also began to realize that she had a number of underused resources. For instance, she was intelligent and, though not good-looking, attractive in other ways. She had a fairly good sense of humor, though she seldom gave herself the opportunity to use it. She was also sensitive to others and basically caring.

After Grace was discharged from the hospital, she returned for a few outpatient sessions. She got to the point where she wanted to do something about her general fearfulness and her passivity, especially the passivity in her social life. A psychiatric social worker taught her relaxation and thought-control techniques that helped her reduce her anxiety. As she became less anxious, she was in a better position to do something about establishing some social relationships. With the social worker's help, she set goals of acquiring a couple of friends and becoming a member of some social group. However, she was at a loss as to how to proceed, since she thought that friendship and a fuller social life were things that should happen "naturally." She soon came to realize that many people had to work at acquiring a more satisfying social life. There was nothing automatic about it at all.

The social worker helped Grace identify various kinds of social groups that she might join. She was then helped to see which of these would best meet her needs without placing too much stress on her. She finally chose to join an arts and crafts group at a local YMCA. The group gave her an opportunity to begin developing some of her talents and to meet people without having to face demands for intimate social contact. It also gave her an opportunity to take a look at other, more socially oriented programs sponsored by the Y. In the arts and crafts program she met a couple of people she liked and who seemed to like her. She began having coffee with them once in a while and then an occasional dinner.

Grace still needed support and encouragement from her helper, but she was gradually becoming less anxious and feeling less isolated. Once in a while she would let her anxiety get the better of her. She would skip a meeting at the Y and then lie about having attended. However, as she began to let herself trust her helper more, she revealed this self-defeating game. The social worker helped her develop coping strategies for those times when anxiety seemed to be higher.

Grace's problems were more severe than Jeff's and she did not have as many immediate resources. Therefore, she needed both more time and more attention to develop goals and strategies.

Some clients are filled with great ideas for getting things done but never seem to do anything. They lack the discipline to evaluate their ideas, choose the best, and turn them into action. Often this kind of work seems too tedious to them, even though it is precisely what they need. Consider the following case.

Clint came away from the doctor feeling depressed. He was told that he was in the high-risk category for heart disease and that he needed to change his lifestyle. He was cynical, very quick to anger, and not prone to trusting others. Venting his suspicions and hostility did not make them go away; it only intensified them. Therefore, one critical lifestyle change was to change this pattern and develop the ability to trust others. He developed three broad goals: reducing mistrust of others' motives, reducing the frequency and intensity of such emotions as rage, anger, and irritation, and learning how to treat others with consideration. Clint read through the strategies suggested to help people pursue these broad goals. They included:

- Keeping a hostility log to discover the patterns of irritation in his life
- Finding someone to talk to about the problem, someone to trust
- "Thought stopping," catching oneself in the act of indulging in hostile thoughts or in thoughts that lead to hostile feelings
- Talking sense to oneself when tempted to put others down
- Developing empathic thought patterns—that is, walking in the other person's shoes
- Learning to laugh at his own silliness
- Using a variety of relaxation techniques to counter negative thoughts
- Finding ways of practicing trust
- Developing active listening skills
- Substituting assertive for aggressive behavior
- Getting perspective, seeing each day as his last
- Practicing forgiving others without being patronizing or condescending

Clint prided himself on his rationality (though his brand of "rationality" was one of the things that got him into trouble). So, as he read down the list, he chose strategies that could form an "experiment," as he put it. He decided to talk to a counselor (for the sake of objectivity), keep a hostility log (data gathering), and use the tactics of thought stopping and talking sense to himself whenever he felt that he was letting others get under his skin. The counselor noted to himself that none of these necessarily involved changing Clint's attitudes toward others. However, he did not challenge Clint at this point. His best bet was that through "strategy sampling" Clint would learn more about his problem, that he would find that it went deeper than he thought. Clint set himself to his experiment with vigor.

Clint chose strategies that fit his values. The problem was that the values themselves needed reviewing. But Clint did act, and action gave him the opportunity to learn.

## The Shadow Side of Developing and Selecting Strategies

In choosing courses of action, clients often fail to evaluate the risks involved and determine whether the risk is balanced by the probability of success. Gelatt, Varenhorst, and Carey (1972) suggested four ways in which clients may try to deal with the factors of risk and probability: wishful thinking, playing it safe, avoiding the worst outcome, and achieving some kind of balance. The first three are often pursued without reflection and therefore lie in the "shadows."

**Wishful thinking.** In this case, clients choose a course of action that might (they hope) lead to the accomplishment of a goal regardless of risk, cost, or probability. For instance, Serena wants her ex-husband to increase the amount of support he is paying for the children. She tries to accomplish this by constantly nagging him and trying to make him feel guilty. She doesn't consider the risk (he might get angry and stop giving her anything), the cost (she spends a great deal of time and emotional energy arguing with him), or the probability of success (he does not react favorably to nagging). The wishful-thinking client operates blindly, engaging in some course of action without taking into account its usefulness. At its worst, this is a reckless approach. Effective helpers find ways of challenging wishful thinking. "Serena, let's review what you've been doing to get Tom to pay up and how successful you've been."

**Playing it safe.** In this case, the client chooses only safe courses of action, ones that have little risk and a high degree of probability of producing at least limited success. For instance, Liam, a manager in his early 40s, is very dissatisfied with the way his boss treats him at work. His ideas are ignored, the delegation he is supposed to have is preempted, and his boss does not respond to his attempts to discuss career development. His goals center around his career. He wants to let his boss know about his dissatisfaction and he wants to learn what his boss thinks about him and his career possibilities. However, he fails to bring these issues up when his boss is "out of sorts." On the other hand, when things are going well, Liam doesn't want to "upset the applecart." He drops hints about his dissatisfaction, even joking about them at times. He tells others in hopes that word will filter back to his boss. During formal appraisal sessions he allows himself to be intimidated by his boss. However, in his own mind, he is doing whatever could be expected of a "reasonable" man. He does not know how safe he is playing it. The helper says, "Liam, you're playing pretty safe with your boss. And, while it's true that you haven't upset him, you're still in the dark about your career prospects."

**Avoiding the worst outcome.** In this case, clients choose means that are likely to help them avoid the worst possible result. They try to minimize the maximum danger, often without identifying what that danger is. For instance, Crissy, dissatisfied with her marriage, sets a goal to be "more assertive." However, even though she has never said this either to herself or to her counselor, the maximum danger for her is to lose her partner. Therefore, her "assertiveness" is her usual pattern of compliance, with some frills. For instance, every once in a while she tells her husband that she is going out with friends and will not be around for supper. He, without her knowing it, actually enjoys these breaks. At some level of her being, she realizes that her absences are not putting him under any pressure. She continues to be assertive in this way. But she never sits down with her husband to review where they stand with each other. That might be the beginning of the end. At the beginning of one session, the counselor says, "What if some good friend were to say to you, 'Bill has you just where he wants you.' How would you react?" Crissy is startled, but after discussing how she might respond to a friend's challenge, she comes away from the session much more realistic.

**Striking a balance.** In the ideal case, clients choose strategies for achieving goals that balance risks against the probability of success. This "combination" approach is the most difficult to apply, for it involves the right kind of analysis of problem situations and opportunities, choosing goals with the right edge, being clear about one's values, ranking a variety of strategies according to these values, and estimating how effective any given course of action might be. Even more to the point, it demands challenging the blind spots that might distort these activities. Since some clients have neither the skill nor the will for this combination approach, it is essential that their counselors help them engage in the kind of dialogue that will help them face up to this impasse.

 **EVALUATION QUESTIONS FOR TASK 2**

**How well am I doing the following as I try to help clients choose goal-accomplishing strategies that are best for them? How effective am I in:**

- Helping clients choose strategies that are clear and specific, that best fit their capabilities, that are linked to goals, that have power, and that are suited to clients' styles and values?
- Helping clients engage in and benefit from strategy sampling?
- Helping clients in selected cases use the balance sheet as a way of choosing strategies by outlining the principal benefits and costs for self, others, and relevant social settings?
- Helping clients manage the shadow side of selecting courses of action— that is, wishful thinking, playing it too safe, focusing on avoiding the worst possible outcome rather than on getting what they want, and wasting time by trying to spell out a perfectly balanced set of strategies?
- Helping clients use the act of choosing strategies to stimulate problem-managing action?

## Task 3: Help Clients Pull Strategies Together into a Manageable Plan: "How Do I Put Everything Together and Move Forward?"

"What should my campaign for constructive change look like? What do I need to do first? Second? Third?" Stage III deals with the *game plan.*

After identifying and choosing strategies to accomplish goals, clients need to organize these strategies into a plan. A plan is a clear step-by-step action program of accomplishing a goal or a set of goals. Of course, the formulation of a plan need not be a separate step. A plan can emerge through the dialogue about possible and best-fit strategies. Since many clients are not good at formulating plans, helpers can add great value through the planning dialogue.

## No Plan of Action: The Case of Frank

The lack of a plan keeps some clients mired in their problem situations. Consider the case of Frank, a vice president of a large West Coast corporation.

Frank was a go-getter. He was very astute about business and had risen quickly through the ranks. Vince, the president of the company, was in the process of working out his own retirement plans. From a business point of view, Frank was the heir apparent. But there was a glitch. Vince was far more than a good manager; he was a leader. He had a vision of what the company should look like five to ten years down the line. Early on he saw the power of the Internet and used it wisely to give the business a competitive edge.

Though tough, Vince related well to people. People constituted the human capital of the company. He knew that products *and* people kept customers happy. He also took to heart the results of a millennium survey of some two million employees in the United States. One of the sentences in the summary of the survey results haunted him—"People join companies but leave supervisors." In competing for the best workers, he couldn't afford supervisors who alienated their team members.

Frank was quite different. He was a "hands-on" manager, meaning, in his case, that he was slow to delegate tasks to others, however competent they might be. He kept second-guessing others when he did delegate, reversed their decisions in a way that made them feel put down, listened poorly, and took a fairly short-term view of the business—"What were last week's figures like?" He was not a leader but an operations man. His direct reports called him a micromanager.

One day, Vince sat down with Frank and told him that he was considering him as his successor down the line, but that he had some concerns. "Frank, if it were just a question of business acumen, you could take over today. But my job, at least in my mind, demands a leader." Vince went on to explain what he meant by a leader and to point out the things in Frank's style that had to change.

So Frank did something that he never thought he would do. He began seeing a coach. Roseanne had been an executive with another company in the same industry but had opted to be a consultant for family reasons. Frank chose her because he trusted her business acumen. That's what meant most to him. They worked together for over a year, often over lunch and in hurried meetings early in the morning or late in the evening. And, indeed, he valued their dialogues about the business.

Frank's ultimate aim was to become president. If getting the job meant that he had to try to become the kind of leader his boss had outlined, so be it. Since he was very bright, he came up with some inventive strategies for moving in that direction. But he could never be pinned down to an overall program with specific milestones by which he could evaluate his progress. Roseanne pushed him, but Frank was always "too busy" or would say that a formal program was "too stifling." That was odd, since formal planning was one of his strengths in the business world.

Frank remained as astute as ever in his business dealings. But he merely dabbled in the strategies meant to help him become the kind of leader Vince wanted him to be. Frank had the opportunity to both correct some mistakes and, even better, develop and expand his managerial style. But he blew it. At the end of two years, Vince appointed someone else as president of the company.

Frank never got his act together. He never put together the kind of change program needed to become the kind of leader Vince wanted as president. Why? Frank had two significant blind spots that the consultant did not help him over-come. First, he never really took Vince's notion of leadership seriously. So he wasn't really ready for a change program. He thought the president's job was his, that business acumen alone would win out in the end. Second, he thought he could change his management style at the margins, whereas more substan-tial changes were called for.

Roseanne never challenged Frank as he kept "trying things" that never led anywhere. If she had said something like this, "Come on, Frank, you know you don't really buy Vince's notion of leadership. But you can't just give lip service to it. Vince will see right through it. We're just messing around. You don't want a program because you don't believe in the goal. Let's do something or call these meetings off," maybe things would have been different. In a way she was a co-conspirator because she, too, relished their business discussions. When Frank didn't get the job, he left the company, leaving Roseanne to ponder her success as a consultant but her failure as a coach.

## How Plans Add Value to Clients' Search for a Better Life

Some clients—once they know what they want and some of the things they have to do to get what they want—get their act together, formally or informally develop a plan and move forward. Other clients need help. Since some clients (and some helpers) fail to appreciate the power of a plan, it is useful to review the advantages of planning.

Not all plans are formal. "Little plans," whether called such or not, are for-mulated and executed throughout the helping process. Tess, an alcoholic who wants to stop drinking, feels the need for some support. She contacts Lou, a friend who has shaken a drug habit, tells him of her plight, and enlists his help. He readily agrees. Objective accomplished. This *little plan* is part of her overall change program. Change programs are filled with setting *little objectives* and developing and executing *little plans* to achieve them.

Formal planning usually focuses on the sequence of the "big steps" clients must take in order to get what they need or want. Clients are helped answer the question, "What do I need to do first, second, and third?" A formal plan in its most formal version takes strategies for accomplishing goals, divides them into workable steps, puts the steps in order, and assigns a timetable for the accom-plishment of each step.

Formal planning, provided that it is adapted to the needs of individual clients, has a number of advantages.

**Plans help clients develop needed discipline.** Many clients get into trouble in the first place because they lack discipline. Planning places reasonable demands on clients to develop discipline. Desmond, the halfway-house resident discussed earlier, needed discipline and benefitted greatly from a formal job-seeking program. Indeed, ready-made programs such as the Twelve Step Program of Alcoholics Anonymous are in themselves plans that demand or at least encourage self-discipline.

**Plans keep clients from being overwhelmed.** Plans help clients see goals as doable. They keep the steps toward the accomplishment of a goal *bite-size*. Amazing things can be accomplished by taking bite-size steps toward substantial goals. Bud, the ex-psychiatric patient who ended up creating a network of self-help groups for ex-patients, started with the bite-size step of participating in one of those groups himself. He did not become a self-help organizer overnight. It was a step-by-step process.

**Plans provide an opportunity to evaluate the realism and adequacy of goals.** This is an example of the interaction that should take place among the stages of the helping process. When Walter, a middle manager who had many problems in the workplace, began tracing out a plan to cope with the loss of his job and with a lawsuit filed against him by his former employer, he realized that his initial goals—getting his job back and filing and winning a countersuit—were unrealistic. His revised goals included getting his former employer to withdraw the suit and getting into better shape to search for a job by participating in a self-help group of managers who had lost their jobs.

**Plans make clients aware of the resources they will need to implement their strategies.** When Dora was helped by a counselor to formulate a plan to pull her life together after the disappearance of her younger son, she realized that she lacked the social support needed to carry out the plan. She had retreated from friends and even relatives, but now she knew she had to get back into community. Normalizing life demanded ongoing social involvement and support. She added a goal of finding the support needed to get back into community to her constructive-change program.

**Formulating plans helps clients uncover unanticipated obstacles to the accomplishment of goals.** Lenny, a U.S. soldier who had accidentally killed an innocent bystander during his stint in Iraq, was seeing a counselor because of the difficulty he was having returning to civilian life. Only when he began pulling together and trying out plans for normalizing his social life did he realize how ashamed he was of what he had done in the military. He felt so flawed because of what had happened that it was almost impossible to involve himself intimately with others. Helping him deal with his shame became one of the most important parts of his healing process.

Formulating plans will not solve all our clients' problems, but it is one way of making time an ally instead of an enemy. Sometimes clients engage in aimless activity in their efforts to cope with problem situations. Plans help clients make the best use of their time.

## Shaping the Plan: Two Cases

Plans need *shape* to drive action. A formal plan identifies the activities or actions needed to accomplish a goal or a subgoal, puts those activities into a logical but flexible order, and sets a time frame for the accomplishment of each key step. Therefore, there are three simple questions that are key to shaping a plan:

- What specific actions are needed to accomplish the goal or any part of it?
- In what sequence should these be done?
- What is the time frame? What should be done today, what tomorrow, what next month?

If clients choose goals that are complex or difficult, then it is useful to help them establish subgoals as a way of moving step-by-step toward the ultimate goal. For instance, once Bud decided to start an organization of self-help groups composed of ex-patients from mental hospitals, there were a number of subgoals he needed to accomplish before the organization would become a reality. His first step was to set up a test group. This subgoal provided the experience needed for further planning. A later step was to establish some kind of charter for the organization. "Charter in place" was one of the subgoals leading to his main goal. Bud reviewed the charters of other self-help organizations as one step in the process of developing his own charter.

In general, the simpler the plan the better. However, simplicity is not an end in itself. The question is not whether a plan or program is complicated, but whether it is well shaped and designed to produce results. If complicated plans are broken down into subgoals and the strategies or activities needed to accomplish them, they are as capable of being achieved as simpler ones. In schematic form, shaping looks like this:

Subprogram 1 (a set of activities) leads to subgoal 1.
Subprogram 2 leads to subgoal 2.
Subprogram n (the last in the sequence) leads to the accomplishment of the ultimate goal.

**The case of Wanda.** Take the case of Wanda, a client who set a number of goals in order to manage a complex problem situation. One of her goals was finding a job. The plan leading to this goal had a number of steps, each of which led to the accomplishment of a subgoal. The following subgoals were part of Wanda's job-finding program. They are stated as accomplishments (the outcome, or results, approach).

Subgoal 1: Resume written.
Subgoal 2: Kind of job wanted determined.
Subgoal 3: Job possibilities canvassed.
Subgoal 4: Best job prospects identified.
Subgoal 5: Job interviews arranged.
Subgoal 6: Job interviews completed.
Subgoal 7: Offers evaluated.

The accomplishment of these subgoals led to the accomplishment of the overall goal of Wanda's plan—that is, getting the kind of job she wanted.

Wanda also had to set up a step-by-step process, or program, to accomplish each of these subgoals. For instance, the process for accomplishing the subgoal "job possibilities canvassed" included such things as doing an Internet search on one or more of the many of the job-search sites, reading the *Help Wanted* sections of the local papers, contacting friends or acquaintances who could provide leads, visiting employment agencies, reading the bulletin boards at school, and talking with someone in the job-placement office. Sometimes the sequencing of activities is important, sometimes not. In Wanda's case, it was important for her to have her resume completed - before she began to canvass job possibilities, but when it came to using different methods for identifying job possibilities, the sequence did not make any difference.

**The case of Harriet: The economics of planning.**  Harriet, an undergraduate student at a small state college, wanted to become a counselor. Although the college offered no formal program in counseling psychology, with the help of an advisor she identified several undergraduate courses that would provide some of the foundation for a degree in counseling. One was called "Social Problem-Solving Skills"; a second, "Effective Interpersonal Communication Skills"; a third, "Developmental Psychology: The Developmental Tasks of Late Adolescence and Early Adulthood." Harriet took the courses as they came up. Unfortunately, the Social Problem-Solving Skills was the first course. The good news is that it included a great deal of practice in the skills in question. The bad news was that it assumed competence in interpersonal communication skills. Too late, she realized that she was taking the courses out of optimal sequence. She would have gotten much more from the courses had she taken the communication skills course first.

Harriet also volunteered for the dormitory peer-helper program run by the Center for Student Services. The Center's counselors were very careful in choosing people for the program, but they didn't offer much training. It was a learn-as-you-go approach. Harriet realized that the developmental psychology course would have helped her enormously in this program. It would have helped her understand both herself and her peers better. She finally realized that she needed a better plan. She dropped out of the peer counselor program, sat down with one of the Center's psychologists, reviewed the school's offerings with him, discovered which courses would help her most, and determined the proper sequencing of these courses. He also suggested a couple of courses she could take in a local community college. Harriet's opportunity-development program would have been much more efficient had it been better shaped in the first place. Box 10.2 is a list of questions you can use to help clients think systematically about crafting a plan to get what they need and want.

### Questions on Planning

**Here are some questions you can help clients ask themselves in order to come up with a viable plan for constructive change:**

- Which sequence of actions will get me to my goal?
- Which actions are most critical?
- How important is the order in which these actions take place?
- What is the best timeframe for each action?
- Which step of the program needs substeps?
- How can I build informality and flexibility into my plan?
- How do I gather the resources, including social support, needed to implement the plan?

## Humanizing the Planning Process

If helpers skip the goal-setting and planning steps clients need, they short-change them. On the other hand, if they are pedantic, mechanistic, or awkward in their attempts to help clients engage in these steps—failing to give these processes a human face—they run the risk of alienating the people they are trying to serve. A client might well say, "I'm getting a lot of boring garbage from my counselor." Here, then, are some principles to guide the constructive-change process outlined in Stages II and III.

**Build a planning mentality into the helping process right from the start.** A constructive-change mind-set should permeate the helping process right from the beginning. Helpers need to see clients as self-healing agents capable of changing their lives, not just as individuals mired in problem situations. Even while listening to a client's story, the helper needs to begin thinking of how the situation can be remedied and, through probes, find out what approaches to change clients are thinking about—no matter how tentative these ideas might be. As mentioned earlier, helping clients act in their real world right from the beginning of the helping process helps them develop some kind of initial planning mentality. If helping is to be solution-focused, thinking about strategies and plans must be introduced early. When a client tells of some problem, the helper can ask early on, "What kinds of things have you done so far to try to cope with the problem?"

Cora, a battered spouse, did not want to leave her husband because of the kids. Right from the beginning, the helper saw Cora's problem situation from the point of view of the whole helping process. While she listened to Cora's story, she began to see possible goals and strategies. Within the helping sessions, the counselor helped Cora learn a great deal about how battered women typically respond to their plight and how dysfunctional some of those responses are. Cora also learned how to stop blaming herself for the violence and to overcome her fears of developing more active coping strategies. At home she confronted her husband and stopped submitting to the violence in a vain attempt to avoid further abuse. She also joined a local self-help group for battered women. There she found social support and learned how to both invoke police protection and have recourse to the courts. Further sessions with the counselor helped her gradually change her identity from battered woman to survivor and, eventually, to doer. She moved from simply facing problems to developing opportunities.

Constructive-change scenarios like this must be in the helper's mind from the start, not as preset programs to be imposed on clients, but as part of a constructive-change mentality.

**Adapt the constructive-change process to the style of the client.** Setting goals, devising strategies, and making and implementing plans can be done formally or informally. There is a continuum. Some clients actually like the detailed work of devising plans; it fits their style.

Gitta sought counseling as she entered the "empty nest" period of her life. Although there were no specific problems, she saw too much emptiness as she looked into the future. The counselor helped her see this period of life as a normal experience rather than a psychological problem. It was a developmental opportunity and challenge (see Raup & Myers, 1989). It was an opportunity to reset her life. After spending a bit of time discussing some of the maladaptive responses to this transitional phase of life, they embarked on a review of possible scenarios. Gitta loved brainstorming, getting into the details of the scenarios, weighing choices, setting strategies, and making formal plans. She had been running her household this way for years. So the process was familiar even though the content was new.

Another case.

Connor, in rebuilding his life after a serious automobile accident, very deliberately planned both a rehabilitation program and a career change. Keeping to a schedule of carefully planned actions not only helped him keep his spirits up, but also helped him accomplish a succession of goals. These small triumphs buoyed his spirits and moved him, however slowly, along the rehabilitation path.

Both Gitta and Connor readily embraced the positive psychology approach embedded in constructive change programs. They thrived on both the work and the discipline to develop plans and execute them. Many, if not most, people, however, are not like Gitta and Connor. The distribution is skewed toward the I-hate-all-this-detail-and-won't-do-it end of the continuum.

Kirschenbaum (1985) challenged the notion that planning should always provide an exact blueprint for specific actions, their sequencing, and the time frame. He proposed three questions:

- How specific do the activities have to be?
- How rigid does the order have to be?
- How soon does each activity have to be carried out?

He suggested that, at least in some cases, being less specific and rigid about actions, sequencing, and deadlines can "encourage people to pursue their goals by continually and flexibly choosing their activities." That is, flexibility in planning can help clients become more self-reliant and proactive. Rigid planning strategies can lead to frequent failure to achieve short-term goals.

Consider the case of Yousef, a single parent with a mentally retarded son. He was challenged one day by a colleague at work. "You've let your son become a ball and chain and that's not good for you or him!" his friend said. Yousef smarted from the remark, but eventually—and reluctantly—sought counseling. He never discussed any kind of extensive change program with his helper but, with a little stimulation from her, he began doing little things differently at home. When he came home from work especially tired and frustrated, he had a friend in the apartment building stop by. This helped him to refrain from taking his frustrations out on his son. Then, instead of staying cooped up over the weekend, he found simple things to do that eased tensions, such as going to the zoo and to the art museum with a woman friend and his son. He discovered that his son enjoyed these pastimes immensely, despite his limitations. In short, he discovered little ways of blending caring for his son with a better social life. His counselor had a constructive-change mentality right from the beginning, but did not try to engage Yousef in overly formal planning activities.

**Devise a plan for the client and then work with the client on tailoring it to his or her needs.** The more experienced helpers become, the more they learn about the elements of program development and the more they come to know what kinds of programs work for different clients. They build up a stockpile of useful programs and know how to stitch pieces of different programs together to create new programs. In this way, they can use their knowledge and experience to fashion a plan for clients who lack the skills or the temperament to pull together a plan for themselves. Of course, their objective is not to foster dependence but to help clients grow in self-determination. For instance, they can first offer a plan as a sketch, or in outline form, rather than as a detailed program. Helpers then work with clients to fill out the sketch and adapt it to their clients' needs and style. Consider the following case:

> Katrina, a woman who dropped out of high school but managed to get a high school equivalency diploma, was overweight and reclusive. Over the years she had restricted her activities because of her weight. Sporadic attempts at dieting had left her even heavier. Because she was chronically depressed and had little imagination, she was not able to come up with any kind of coherent plan. Once her counselor understood the dimensions of Katrina's problem situation, she pulled together an outline of a change program that included such things as blame reduction, the redefinition of beauty, decreasing self-imposed social restrictions, and cognitive restructuring activities aimed at lessening depression. She also provided information about obesity and suggestions for dealing with it, drawn from health care sources. She presented these in a simple format, adding detail only for the sake of clarity. She added further detail as Katrina got involved in the planning process and in making choices.

Although this counselor pulled together elements of a range of already existing programs, counselors are, of course, free to make up their own programs based on their expertise and experience. The point is to give clients something to work with, something to get involved in. The elaboration of the plan emerges through dialogue with the client, keeping in mind the kind of detail the client can handle. In the end, there is no such thing as a good plan in and of itself. Results, not planning or hard work, are the final arbiter.

**Help clients develop the working knowledge and skills needed to implement plans.** It often happens that people get into trouble or fail to get out of it because they lack the needed life skills or coping skills to deal with problem situations. If this is the case, then helping clients find ways of learning the life skills they need to cope more effectively is an important broad strategy. Indeed, the use of skills training as part of therapy—what Carkhuff, years ago (1971), called "training as treatment"—might be essential for some clients. Challenging clients to engage in activities for which they don't have the skills is compounding, rather than solving, the problem. What kinds of working knowledge and skills does this client need to get where he or she wants to go? Consider the following case:

> Jerzy and Zelda fell in love. They married and enjoyed a relatively trouble-free honeymoon period of about two years. However, the problems that inevitably arise from living together in such intimacy asserted themselves. They found, for instance, that they counted too heavily on strong positive feelings for each other and now, in their absence, could not communicate about finances, sex, and values. They lacked certain critical interpersonal communication skills. Furthermore, they lacked understanding of each other's developmental needs. Jerzy had little working knowledge of the developmental demands of a 20-year-old woman; Zelda had little working knowledge of the kinds of cultural blueprints that were operative in the lifestyle of her 29-year-old husband. The relationship began to deteriorate. Since they had few problem-solving skills, they didn't know how to handle their situation.

Jerzy and Zelda needed skills. This is hardly surprising. Lack of requisite interpersonal communication and other life skills is often at the heart of relationship breakdowns.

**Help clients develop the support systems needed to implement plans.** Many clients do not know how to mobilize the resources needed to implement plans. One of the most important resources is social support. If clients are to pursue goals *out there,* as they say, in their real lives, they also benefit greatly from support. However, Robert Putnam (2000) provides a great deal of evidence indicating that such support may not always be easy to find. His central thesis is that in North American society the supply of "social capital"—both informal social connectedness and formal civic engagement—has fallen dangerously low. Putnam shows that we belong to fewer organizations that meet, know our neighbors less, meet with friends less frequently, and even socialize with our families less often. This is the environment in which clients must do the work of constructive change.

However, social support is a key element in change (see Basic Behavioral Science Task Force of the National Advisory Mental Health Council, 1996, p. 628).

Social support has . . . been examined as a predictor of the course of mental illness. In about 75% of studies with clinically depressed patients, social-support factors increased the initial success of treatment and helped patients maintain their treatment gains. Similarly, studies of people with schizophrenia or alcoholism revealed that higher levels of social support are correlated with fewer relapses, less frequent hospitalizations, and success and maintenance of treatment gains.

In a study on weight loss and maintaining the loss (Wing & Jeffery, 1999) clients who enlisted the help of friends were much more successful than clients who took the solo path. This is called "social facilitation" and is quite different from dependence. Social facilitation, a positive psychology approach, is energizing, while dependence is often depressing. Therefore, a culture of social isolation does not bode well for clients. Of course, all of this reinforces what we already know through common sense. Which of us has not been helped through difficult times by family and friends?

The National Advisory Mental Health Council study just mentioned showed that people who are highly distressed and, therefore, most in need of social support may be the least likely to receive it because their expressions of distress drive away potential supporters. Which of us has not avoided, at one time or another, a distressed friend or colleague? Therefore, distressed clients can be helped to learn how to modulate their expressions of distress. Who wants to help whiners? On the other hand, potential supporters can learn how to deal with distressed friends and colleagues, even when they let themselves become whiners.

The Task Force study suggested two general strategies for fostering social support: helping clients mobilize or increase support from existing social networks, and "grafting" new ties onto impoverished social networks. Both of these come into play in the following case.

> Casey, a bachelor whose job involved frequent travel literally around the world, fell ill. He had many friends, but they were spread around the world. Because he was neither married nor in a marriage-like relationship, he had no primary caregiver in his life. He received excellent medical care, but his psyche fared poorly.
>
> Once out of the hospital, he recuperated slowly, mainly because he was not getting the social support he needed. In desperation, he had a few sessions with a counselor, sessions that proved to be quite helpful. The counselor challenged him to ask for help from his local friends. He had underplayed his illness with them because he didn't want to "be a burden." He discovered that his friends were more than ready to help. But since their time was limited, he, with some hesitancy, "grafted" onto his rather sparse hometown social network some very caring people from the local church. He was fearful that he would be deluged with piety, but instead he found people like himself. Moreover, they were, in the main, socially intelligent. They knew how much or how little care to give. In fact, most of the time their care was simple friendship. Finally, he hired a couple of students from a local university to do word processing and run errands for him from time to time. They also provided some social support.

As the Task Force authors note, it's important not only that people be available to provide support but also that those needing support perceive that it is available. This may mean, as in Casey's case, working with the client's attitudes and openness to receive support.

## The Shadow Side of Planning

Some years ago I lent a friend of mine an excellent, though somewhat detailed, book on self-development. About two weeks later he came back, threw the book on my desk, and said, "Who would go through all of that!" I retorted, "I suppose anyone really interested in self-development." That was the righteous, not the realistic, response. Planning in the real world seldom looks like planning in textbooks. Textbooks do provide useful frameworks, principles, and processes, but they are simply not used. Most people are too impatient to do the kind of planning just outlined. One of the reasons for the dismal track record of discretionary change mentioned earlier is that, even when clients do set realistic goals, they lack the discipline to develop reasonable plans. The detailed work of planning is too burdensome.

It is your job as a helper to get planning out of the textbook and into the fresh air. Find ways of insinuating a planning mentality into the helping process right from the beginning. Take the principles of effective planning out of the textbook, adapt them to the needs of each client, and make them engaging.

## EVALUATION QUESTIONS FOR TASK 3

**Helpers can ask themselves the following questions as they help clients formulate the kinds of plans that actually drive action.**

1.  To what degree do I prize and practice planning in my own life?

2.  How effectively have I adopted the whole-model mind-set in helping, seeing each session and each intervention in the light of the entire helping process?

3.  How quickly do I move to planning when I see that it is what the client needs to manage problems and develop opportunities better?

4.  What do I do to help clients overcome resistance to planning? How effectively do I help them identify the incentives for, and the payoff of, planning?

5.  How effectively do I help clients formulate subgoals that lead to the accomplishment of overall preferred-scenario goals?

6.  How practical am I in helping clients identify the action needed to accomplish subgoals, sequence those actions, and establish realistic timeframes for them?

7.  How well do I adapt the specificity and detail of planning to the needs of each client?

8.  How easily do I move back and forth among the different stages and tasks of the helping model as the need arises?

9.  How readily do clients actually move to action because of my work with them in planning?

10. How human is the technology of constructive change in my hands?

11. How well do I adapt the constructive-change process to the style of the client?

 EXERCISES ON PERSONALIZING

1. Name something you want because you think it would help manage some problem situation, develop some opportunity, or because you think it would be life-enhancing. "We can't have children ourselves, but we would like to adopt a child." "I'd like to start my own business. There's an entrepreneur in me dying to get out." Brainstorm ways of achieving your goal. List as many things as you can think of. What are the things you can do to get what you want? How many different ways are there to your goal?

2. After you have finished your list, choose the most promising and practical action in light of the resources available to you. If your list is short and the strategies you come up with do not seem adequate, think about modifying your goal.

3. Finally, combine the strategies you have chosen into a program or plan. Sequence the steps. Set timelines.

# CHAPTER 11

# The Action Arrow: Helping Clients Overcome Obstacles, Execute Plans, and Get Results

Obstacles to Change

Helping Clients Become Change Agents in Their Own Lives

Getting Along Without a Helper

An Amazing Case of Resilience and Perseverance

The Shadow Side of Self Change

Exercises on Personalizing

**The Skilled-Helper Model**

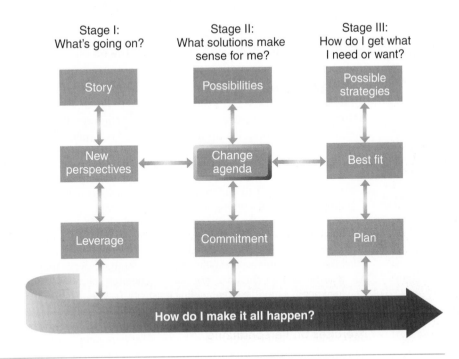

FIGURE

**11.1**    The Helping Model—Complementing Planning with Action

The Action Arrow of the helping model (Figure 11.1) points out the difference between planning and action. Stages I, II, and III all revolve around planning for change, not change itself. Planning without action is a lot of blah, blah, blah. The need to incorporate action into planning, and planning into action, has been emphasized throughout the book. That is, the little actions needed to get the change process moving right from the start have been noted and illustrated. We now take a more formal look at results-producing action, including obstacles to action and ways to overcome them. Let's start with the obstacles to action and then move on, more positively, to ways of helping clients develop a bias toward action as they implement their change programs.

## Obstacles to Change

Discipline and self-control play an important part in implementing change programs. Kirschenbaum (1987) found that many things can contribute to not getting started or giving up: low initial commitment to change, weak self-efficacy, poor outcome expectations, the use of self-punishment rather than self-reward, depressive thinking, failure to cope with emotional stress, lack of consistent self-monitoring, giving in to social pressure, failure to cope with initial relapse, and paying attention to the wrong things—for instance, focusing on the difficulty of the problem situation rather than on the attractiveness of the opportunity. Let's take a closer look at four obstacles to change: helper passivity, client inertia, client entropy, and the decision not to change.

**Passive helpers.**  Sometimes helpers are part of the problem rather than a catalyst for solutions. Driscoll (1984, pp. 91–97) discussed the temptation of helpers to respond to the passivity of their clients with a kind of passivity of their own. This "sorry, it's up to you" attitude is a mistake.

> A client who refuses to accept responsibility thereby invites the therapist to take over. In remaining passive, the therapist foils the invitation, thus forcing the client to take some initiative or to endure the silence. A passive stance is therefore a means to avoid accepting the wrong sorts of responsibility. It is generally ineffective, however, as a long-run approach. Passivity by a therapist leaves the client feeling unsupported and thus further impairs the already fragile therapeutic alliance. Troubled clients, furthermore, are not merely unwilling but generally and in important ways unable to take appropriate responsibility. A passive countermove is therefore counterproductive, for neither therapist nor client generates solutions, and both are stranded together in a muddle of entangling inactivity. (p. 91)

To help others act, helpers must be agents and doers in the helping process, not mere listeners and responders. The best helpers keep looking for ways to enter the worlds of their clients, to get them to become more active in the sessions, to get them to own more of the helping process, to help them see the need for action—action in their heads and action outside their heads in their everyday lives. Although they don't push reluctant clients too hard, thus turning reluctance into resistance, neither do they sit around waiting for reluctant clients to act.

**Client inertia: Reluctance to get started.**  Inertia is the human tendency to put off problem-managing action. With respect to inertia, I sometimes say to clients who I suspect are reluctant to act, something like this, "The action program you've come up with seems to be a sound one. The main reason that sound action programs don't work, however, is that they are never tried. Don't be surprised if you feel reluctant to act or are tempted to put off the first steps. This is quite natural. Ask yourself what you can do to get by that initial barrier." The sources of inertia are many, ranging from pure sloth to paralyzing fear. Understanding what inertia is like is easy. We need only look at our own behavior. The list of ways in which people avoid taking responsibility is endless. We'll examine several of them here: passivity, learned helplessness, disabling self-talk, and getting trapped in vicious circles.

    *Client passivity.*  One of the most important ingredients in the generation and perpetuation of mediocrity is passivity, the failure of people to take responsibility for themselves in one or more developmental areas of life or in various life situations that call for action. Passivity takes many forms: doing nothing—that is, not responding to problems and options; uncritically accepting the goals and solutions suggested by others; acting aimlessly; becoming paralyzed—that is, shutting down; and becoming violent—that is, blowing up.

> When Zelda and Jerzy first noticed small signs that things were not going right in their relationship, they did nothing. They noticed certain irritations with each other, mused on them for a while, and then forgot about them. They lacked the communication skills to engage each other immediately and to explore what was happening. Zelda and Jerzy had both learned to remain passive before the little crises of life, not realizing how much their passivity would ultimately contribute to their downfall. A long series of unmanaged problems led to an increasing number of major blowups. Then, without much discussion, they decided to end their marriage.

Passivity in dealing with little things can prove very costly because little things have a way of turning into big things.

**Learned helplessness.** Seligman's concept of "learned helplessness" (1975, 1991; Peterson, Maier, & Seligman, 1995; see also Mikulincer, 1994) and its relationship to depression has received a great deal of attention since he first introduced it. Some clients learn to believe from an early age that there is nothing they can do about certain life situations. There are degrees in feelings of helplessness—from mild forms of "I'm not up to this" to feelings of total helplessness coupled with deep depression. Learned helplessness, then, is a step beyond mere passivity.

The trick is helping clients learn what is and what is not in their control. A man with a physical disability may not be able to do anything about the disability itself, but he does have some control over how he views his disability and the power to pursue certain life goals despite it. The opposite of helplessness is "learned optimism" (Seligman, 1998), resourcefulness, and resilience. If helplessness can be learned, so can resourcefulness. Indeed, increased resourcefulness is one of the principal goals of successful helping.

**Disabling self-talk.** Clients often talk themselves out of things, thus talking themselves into passivity. They say to themselves such things as "I can't do it," "I can't cope," "I don't have what it takes to engage in that program; it's too hard," and "It won't work." Such self-defeating conversations with themselves get people into trouble in the first place, and then prevent them from getting out. Helpers can add great value by helping clients challenge the kind of self-talk that interferes with action.

**Vicious circles.** Pyszczynski and Greenberg (1987) developed a theory about self-defeating behavior and depression. They said that people whose actions fail to get them what they want can easily lose a sense of self-worth and become mired in a vicious circle of guilt and depression.

> Consequently, the individual falls into a pattern of virtually constant self-focus, resulting in intensified negative affect, self-derogation, further negative outcomes, and a depressive self-focusing style. Eventually, these factors lead to a negative self-image, which may take on value by providing an explanation for the individual's plight and by helping the individual avoid further disappointments. The depressive self-focusing style then maintains and exacerbates the depressive disorder. (p. 122)

It does sound depressing. One client, Amanda, fits this theory perfectly. She had aspirations of moving up the career ladder where she worked. She was very enthusiastic and dedicated, but she was unaware of the "gentleman's club" politics of the company in which she worked and didn't know how to work the system. She kept doing the things that she thought should get her ahead. They didn't. Finally, she got down on herself, began making mistakes in the things that she usually did well, and made things worse by constantly talking about how she "was stuck," thus alienating her friends. By the time she saw a counselor, she felt defeated and depressed. She was about to give up. The counselor focused on the entire circle—low self-esteem producing passivity which produced even lower self-esteem—and not just the self-esteem part. Instead of just trying to help her change her inner world of disabling self-talk, he also helped her intervene in her life to become a better problem solver. Small successes in problem solving led to the start of a so-called benign circle—success producing greater self-esteem leading to greater efforts to succeed.

**Disorganization.** Tico lived out of his car. No one knew exactly where he spent the night. The car was chaos, and so was his life. He was always going to get his career, family relations, and love life in order, but he never did. Living in disorganization was his way of putting off life decisions. Ferguson (1987, p. 46) painted a picture that may well remind us of ourselves, at least at times.

> When we saddle ourselves with innumerable little hassles and problems, they distract us from considering the possibility that we may have chosen the wrong job, the wrong profession, or the wrong mate. If we are drowning in unfinished housework, it becomes much easier to ignore the fact that we have become estranged from family life. Putting off an important project—painting a picture, writing a book, drawing up a business plan—is a way of protecting ourselves from the possibility that the result may not be quite as successful as we had hoped. Setting up our lives to insure a significant level of disorganization allows us to continue to think of ourselves as inadequate or partially-adequate people who don't have to take on the real challenges of adult behavior.

Many things can be behind this unwillingness to get our lives in order, like defending ourselves against a fear of succeeding.

Driscoll (1984, pp. 112–117) has provided us with a great deal of insight into this problem. He described inertia as a form of control. He says that if we tell some clients to jump into the driver's seat, they will compliantly do so—at least, until the journey gets too rough. The most effective strategy, he claimed, is to show clients that they have been in the driver's seat right along: "Our task as therapists is not to talk our clients into taking control of their lives, but to confirm the fact that they already are and always will be." That is, inertia, in the form of staying disorganized, is itself a form of control. The client is actually successful, sometimes against great odds, at remaining disorganized and thus maintaining the status quo.

**Entropy: The tendency of things to fall apart.** Entropy is the tendency to give up action that has been initiated. Change programs, even those that start strong, often dwindle and disappear. All of us have experienced the problems involved in trying to implement programs. We make plans, and they seem realistic to us. We start the steps of a program with a good deal of enthusiasm.

However, we soon run into tedium, obstacles, and complications. What seemed so easy in the planning stage now seems quite difficult. We become discouraged, flounder, recover, flounder again, and finally give up, rationalizing to ourselves that we did not want to accomplish those goals anyway.

Phillips (1987, p. 650) called this the "ubiquitous decay curve." Attrition, noncompliance, and relapse are the name of the game. A married couple trying to reinvent their marriage might eventually say to themselves, "We had no idea that it would be so hard to change ingrained ways of interacting with each other. Is it worth the effort?" Their motivation is on the wane. Wise helpers know that the decay curve is part of life and help clients deal with it. With respect to entropy, a helper might say, "Even sound action programs begun with the best of intentions tend to fall apart over time, so don't be surprised when your initial enthusiasm seems to wane a bit. That's only natural. Rather, ask yourself what you need to do to keep yourself at the task."

**Choosing not to change.** Some clients who seem to do well in analyzing problems, developing goals, and even identifying reasonable strategies and plans end up by saying—in effect, if not directly—something like this: "Even though I've explored my problems and understand why things are going wrong—that is, I understand myself and my behavior better, and I realize what I need to do to change—right now I don't want to pay the price called for by change. Right now the price is too high."

> Harris, a bachelor, witty, jovial, beloved by his friends, moved into his sixties with an unsustainable lifestyle. He smoked in private ("but not that much"). His diet was poor ("I can't stand fish"). He was overweight, obese in fact ("I'm a bit chunkier now"). He never exercised ("you know, I never have"). He was given to outbursts of anger ("so many people these days have no manners at all"). He almost always had one drink too many ("this wine is so delicious"). He paid little attention to his health ("what do doctors really know?"). He was shocked when a doctor told him he needed bypass surgery "immediately." After surgery ("I never want to go through that again") he cleaned up his act. Strict diet. Exercise regime. One glass of wine per day. And so forth. A year later he had returned to his old lifestyle, much to the consternation of his friends ("yeah, I know"). Six months after that he had a debilitating stroke. He died of a heart attack, age 64. At the funeral, one of his friends described his death as a kind of "benign suicide." Harris chose not to change.

The question of human motivation seems almost as enigmatic now as it must have been at the dawn of history. So often we seem to choose our own misery. Worse, we choose to stew in it rather than endure the relatively short-lived pain of change. Or, as in Harris's case, learn to love a healthier lifestyle. Helpers can and should challenge clients to search for incentives and rewards for managing their lives more effectively. They should also help clients understand the consequences of not changing. But in the end, it is the client's choice.

But enough of the negative. Let's move on to the positive side of things—helping clients, sometimes against all odds, become effective *change agents* in their own lives, people who find ways of breaking through the obstacles to constructive change.

## Helping Clients Become Change Agents in Their Own Lives

Some clients, once they have a clear idea of what to do to handle a problem situation or develop some opportunity, go ahead and do it, whether they have a formal plan or not. They need little or no further support and challenge from their helpers. They either find the resources they need within themselves or get support and challenge from the significant others in the social settings of their lives. However, other clients, while choosing goals and coming up with strategies for implementing them, are, for whatever reason, stymied when it comes to action. Most clients fall between these two extremes.

In the implementation phase, strategies for accomplishing goals need to be complemented by tactics and logistics. A *strategy* is a practical plan to accomplish some objective. *Tactics* is the art of adapting a plan to the immediate situation. This includes being able to change the plan on the spot to handle unforeseen complications. *Logistics* is the art of being able to provide the resources needed for the implementation of a plan in a timely way.

> During the summer, Rebecca wanted to take an evening course in statistics so that the first semester of the following school year would be lighter. Having more time would enable her to act in one of the school plays, a high priority for her. But she didn't have the money to pay for the course and, at the university she planned to attend, prepayment for summer courses was the rule. Rebecca had counted on paying for the course from her summer earnings, but she would not have the money until later. Consequently, she did some quick shopping around and found that the same course was being offered by a community college not too far from where she lived. Her tuition there was minimal, since she was a resident of the area the college served.

In this example, Rebecca keeps to her overall plan (strategy). However, because of an unforeseen circumstance—the demand for prepayment—she adapts the plan (tactics) by locating another resource (logistics).

Since many well-meaning and motivated clients are simply not good tacticians, counselors can add value by using the following principles to help them engage in focused and sustained goal-accomplishing action.

**Help clients identify possible obstacles and resources.** Years ago, Kurt Lewin (1969) codified common sense by developing what he called "force-field analysis." In ordinary language, this is simply a review by the client of the major restraining forces or obstacles to, and the major facilitating forces for, implementing action plans. The slogan is "forewarned is forearmed."

*Restraining forces.* The identification of possible obstacles to the implementation of a program helps make clients forewarned. A number of these obstacles have been outlined earlier in this chapter.

> Raul and Maria were a childless couple living in a large Midwestern city. They had been married for about five years and had not been able to have children. They finally decided that they would like to adopt a child, so they consulted a counselor familiar with adoptions. The counselor helped them work out a plan of action that

included helping them examine their motivation, reviewing their suitability to be adoptive parents, contacting an agency, and preparing themselves for an interview. After the plan of action had been worked out, Raul and Maria, with the help of the counselor, identified two possible obstacles or pitfalls: The negative feelings that often arise on the part of prospective parents when they are being scrutinized by an adoption agency, and the feelings of helplessness and frustration caused by the length of time and uncertainty involved in the process.

The assumption here is that, if clients are aware of some of the wrinkles that can accompany any given course of action, they will be less disoriented when they encounter them. Identifying possible obstacles is, at its best, a straightforward census of likely pitfalls, rather than a self-defeating search for every possible thing that could go wrong.

Obstacles can come from within the clients themselves, from others, from the social settings of their lives, and from larger environmental forces. Once an obstacle is spotted, ways of coping with it need to be identified. Sometimes simply being aware of a pitfall is enough to help clients mobilize their resources to handle it. At other times a more explicit coping strategy is needed. For instance, the counselor arranged a couple of role-playing sessions with Raul and Maria in which she assumed the role of the examiner at the adoption agency and took a "hard line" in her questioning. These rehearsals helped them stay calm during the actual interviews. The counselor also helped them locate a mutual-help group of parents working their way through the adoption process. The members of the group shared their hopes and frustrations and provided support for one another. In short, Raul and Maria were trained to cope with the restraining forces they might encounter on the road toward their goal.

*Facilitating forces.* In a more positive vein, counselors can help their clients identify unused resources that facilitate action.

> Nora found it extremely depressing to go to her weekly dialysis sessions. She knew that without them she would die, but she wondered whether it was worth living if she had to depend on a machine. The counselor helped her see that she was making life more difficult for herself by letting herself think such discouraging thoughts. He helped her learn how to think thoughts that would broaden her vision of the world instead of narrowing it down to herself, her discomfort, and the machine. Nora was a religious person and found in the Bible a rich source of positive thinking. She initiated a new routine: The day before she visited the clinic, she began to prepare herself psychologically by reading from the Bible. Then, as she traveled to the clinic and underwent treatment, she meditated slowly on what she had read.

In this case, the client substituted positive thinking, an underused resource, for poor-me or it's-not-worth-it thinking. Helping clients brainstorm facilitating forces raises the probability that they will act in their own interests. They can be simple things. George was avoiding an invasive diagnostic procedure. After a brainstorming session he decided to get a friend to go with him. This meant two things. Once he asked for his friend's help, he "had to go through with it." Second, his friend's very presence distracted him from his fears. Or, consider Lucy. She had a history of letting her temper get the better of her. This was especially the case when she returned home after experiencing crises at work. Her mother-in-law and children became the target of her wrath. After a counseling session, she took two photographs with her to work. One was a wedding-day picture which included her mother-in-law. The second was a recent picture of her three children. When she parked the car at work, she placed the pictures on the driver's seat. Then, when she got in the car in the evening, the first thing she saw was two photographs of her life at its best. This made her reflect on the way home on how she wanted to enter the house.

**Help clients develop "implementation intentions."**  Commitment to goals must be followed by commitment to courses of action. Gollwitzer (1999) has researched a simple way to help clients cope with the common problems associated with translating goals into action—failing to get started, becoming distracted, reverting to bad habits, and so forth. Strong commitment to goals is not enough. Equally strong commitment to specific actions to accomplish goals is required. Good intentions, Gollwitzer points out, don't deserve their poor reputation. Strong intentions—for instance, "I'm strongly intending to study for an hour every weekday before dinner"—are "reliably observed to be realized more often than weak intentions" (p. 493). Implementation intentions specify the when, where, and how of actions leading to goal attainment.

> Gwendolyn, an aide in a nursing home, has a very difficult patient, Enid. Often enough, she loses patience and yells at Enid. This does not help at all. So Gwendolyn, in talking to her supervisor about the situation, develops an implementation intention: "When Enid becomes abusive, I will not respond immediately. I'll tell myself that it's her illness that's talking. Then I'll respond with patience and kindness." Her ongoing goal is to control her anger and other negative responses to all patients, not just Enid. Since Enid has been a particularly difficult patient, Gwendolyn needs to refresh her intentions frequently. However, her initial strong intention to substitute anger and impatience with kindness and equanimity means that, in most cases, her responses become more or less automatic. The environmental cue—patient anger, abuse, lack of consideration, and whatever—triggers the appropriate response in Gwendolyn. In a way, poor patient behavior becomes a stimulus, or opportunity, for her responses.

While mere "good intentions" might not be sufficient, really robust intentions, carefully considered, can stimulate constructive action.

**Help clients avoid imprudent action.**  For some clients, the problem is not that they refuse to act but that they act imprudently. Rushing off to try the first strategy that comes to mind is often imprudent.

> Elmer injured his back and underwent a couple of operations. After the second operation he felt a little better, but then his back began troubling him again. When the doctor told him that further operations would not help, Elmer was faced with the problem of handling chronic pain. It soon became clear that his psychological state affected the level of pain. When he was anxious or depressed, the pain always seemed much worse. Elmer talked this problem through with his counselor.

One day he read about a pain clinic located in a Western state. Without consulting the counselor or anyone else, he signed up for a six-week program. Within ten days he was back, feeling more depressed than ever. He had gone to the program with extremely high expectations because his needs were so great. The program was a holistic one that helped the participants develop a more realistic lifestyle. It included programs dealing with nutrition, stress management, problem solving, and quality of interpersonal life. Group counseling was part of the program, and training was part of the group experience. For instance, the participants were trained in behavioral approaches to the management of pain.

The trouble was that Elmer had arrived at the clinic, which was located on a converted farm, with unrealistic expectations. He had not really studied the materials that the clinic had sent to him. He had bought a "packaged" program without studying the package carefully. Since he had expected to find marvels of modern medicine that would magically help him, he was extremely disappointed when he found that the program focused mainly on reducing and managing, rather than eliminating, pain.

Elmer's goal was to be completely free of pain, but he failed to explore the realism of his goal. A more realistic goal would have centered on the reduction and management of pain. Elmer's counselor failed to help him avoid two mistakes—setting an unrealistic goal and, in desperation, acting on the first strategy that came along. Obviously, action cannot be prudent if it is based on flawed assumptions—in this case, Elmer's assumption that he could be pain free.

**Help clients develop contingency plans.** If counselors help clients brainstorm both possibilities for a better future (goals) and strategies for achieving those goals (courses of action), then clients will have the raw materials, as it were, for developing contingency plans. Contingency plans answer the question, "What will I do if the plan of action I choose is not working? What is plan B?" Contingency plans help make clients more effective tacticians. We live in an imperfect world. Often enough, goals have to be fine-tuned or even changed. The same is true for strategies for accomplishing goals.

Jackson, the man dying of cancer, decided to become a resident in the hospice he had visited. The hospice had an entire program in place for helping patients like Jackson die with dignity. Once there, however, he had second thoughts. He felt incarcerated. Fortunately, he had worked out alternative scenarios with his helper. One was living at the home of an aunt he loved and who loved him dearly, with some outreach services from the hospice. He moved out of the hospice into his aunt's home. This worked well for him. When his condition began to worsen rapidly, he returned to the hospice and died there.

Contingency plans are especially needed when clients choose a high-risk program to achieve a critical goal. Having backup plans also helps clients develop more responsibility. If they see that a plan is not working, then they have to decide whether to try the contingency plan. Backup plans need not be complicated. A counselor might merely ask, "If that doesn't work, then what will you do?"

**Help clients overcome procrastination.**  Some clients make reasonable plans but then find all sorts of reasons for not starting to implement them. There are many reasons for procrastination. Take the case of Eula.

> Eula, disappointed with her relationship with her father in the family business, decided that she wanted to start her own. She thought that she could capitalize on the business skills she had picked up in school and in the family business. Her goal, then, was to establish a small software firm that created products for the family-business market.
>
> But a year went by and she still did not have any products ready for market. A counselor helped her see two things. First, her activities—researching the field, learning more about family dynamics, going to information-technology seminars, getting involved for short periods with professionals such as accountants and lawyers who did a great deal of business with family-owned firms, drawing up and redrafting business plans, and creating a brochure—were helpful, but they did not produce products. The counselor helped Eula see that at some level of her being she was afraid of starting a new business. She had a lot of half-finished products. Overpreparation and half-finished products were signs of that fear. So she plowed ahead, finished a product, and brought it to market on the Internet. To her surprise, it was successful. Not a roaring success, but it meant that the cork was out of the bottle. Once she got one product to market, she had little problem developing and marketing others.

Put "managing procrastination" into an Internet search engine and you will find dozens of excellent ways of helping clients move beyond procrastination.

**Help clients find incentives and rewards for sustained action.**  Clients avoid engaging in action programs when the incentives and the rewards for not engaging in the program outweigh the incentives and the rewards for doing so.

> Michael, a policeman on trial for use of excessive force with a young offender, had a number of sessions with a counselor from an HMO that handled police health insurance. In the sessions, the counselor learned that although this was the first time Michael had run afoul of the law, it was in no way the first expression of a brutal streak within him. He was a bully on the beat and a despot at home, and had run-ins with strangers when he visited bars with his friends. Some of this came out during the trial.
>
> Up to the time of his arrest, he had gotten away with all of this, even though his friends had often warned him to be more cautious. His badge had become a license to do whatever he wanted. His arrest, and now the trial, shocked him. Before, he had seen himself as invulnerable; now, he felt very vulnerable. The thought of being a cop in prison understandably horrified him. He was found guilty, was suspended from the force for several months, and received probation on the condition that he continue to see the counselor.
>
> Beginning with his arrest, Michael had modified his aggressive behavior a great deal, even at home. Of course, fear of the consequences of his aggressive behavior was a strong incentive to change that behavior. The next time, the courts would show no sympathy. The counselor took a tough approach to this tough cop. He confronted Michael for "remaining an adolescent" and for "hiding behind his badge." He called the power Michael exercised over others "cheap power." He challenged the "decent person" to "come out from behind the screen." He told Michael point blank that the fear he was experiencing was probably not enough to keep him out of trouble in the

future. After probation, the fear would fade and Michael could easily fall back into his old ways. Even worse, fear was a "weak man's" crutch.

In a more positive vein, the counselor saw in Michael's expressions of vulnerability the possibility of a much more decent human being, one "hiding" under the tough exterior. The real incentives, he suggested, came from the "decent guy" buried inside. He had Michael paint a picture of a "tough but decent" cop, family man, and friend. He had Michael come up with "experiments in decency"—at home, on the beat, with his buddies—to get first-hand experience of the rewards associated with decency.

The counselor was not trying to change Michael's personality. Indeed, he didn't believe in personality transformations. But he pushed him hard to find and bring to the surface a different, more constructive, set of incentives to guide his dealings with people. The new incentives had to drive out the old.

The incentives and the rewards that help a client get going on a program of constructive change in the first place may not be the ones that keep the client going.

Dwight, a man in his early 30s who was recovering from an accident at work that had left him partially paralyzed, had begun an arduous physical rehabilitation program with great commitment. Now, months later, he was ready to give up. The counselor asked him to visit the children's ward. Dwight was both shaken by the experience and amazed at the courage of many of the kids. He was especially struck by one teenager who was undergoing chemotherapy. "He seems so positive about everything," Dwight said. The counselor told him that the boy was tempted to give up, too. Dwight and the boy saw each other frequently. Dwight put up with the pain. The boy hung in there. Three months later the boy died. Dwight's response, besides grief, was, "I can't give up now; that would really be letting him down."

Dwight's partnership with the teenager proved to be an excellent incentive. It helped him renew his resolve. While the counselor joined with Dwight in celebrating his new-found commitment, he also worked with Dwight to find back-up incentives for those times when current incentives seem to lose their power. One was the possibility of marrying and starting a family, despite residual limitations due to the accident.

Constructive-change activities that are not rewarded tend, over time, to lose their vigor, decrease, and even disappear. This process is called *extinction*. It was happening with Luigi.

Luigi, a middle-aged man, had been in and out of mental hospitals a number of times. He discovered that one of the best ways of staying out was to use some of his excess energy helping others. He had not returned to the hospital once during the three years he worked at a soup kitchen. However, finding himself becoming more and more manic over the past six months, and fearing that he would be rehospitalized, he sought the help of a counselor.

Luigi's discussions with the counselor led to some interesting findings. He discovered that, whereas in the beginning he had worked at the soup kitchen because he wanted to, he was now working there because he thought he should. He felt guilty about leaving and also thought that doing so would lead to a relapse. In sum, he had not lost his interest in helping others, but his current work was no longer interesting or challenging. As a result of his sessions with the counselor, Luigi began to work for a group that provided housing for the homeless and the elderly. He poured his energy into his new work and no longer felt manic.

The lesson here is that incentives cannot be put in place and then be taken for granted. They need tending.

**Help clients develop action-focused contracts and agreements.** Earlier we discussed self-contracts as a way of helping clients commit themselves to what they want—that is, their goals. Self-contracts are also useful in helping them both initiate and sustain problem-managing action and the work involved in developing opportunities. Here is an example. In this case, several parties had to commit themselves to the provisions of the contract.

> A boy in the seventh grade was causing a great deal of disturbance by his outbreaks in class. These outbreaks included verbal jousting and profanity with his friends. The apparent purpose of the outbreaks was to position himself among his friends. He seemed to want to cultivate a reputation for being unafraid of the teachers and principal. The punishments handed down by them were dwarfed by what he saw as the admiration of his friends. After the teacher discussed the situation with the school counselor, the counselor called a meeting of all the stakeholders—the boy, his parents, the teacher, and the principal. The counselor offered a simple contract. When the boy disrupted the class, he would spend the next day working by himself under the direction of a teacher's aide. This would take him away from his friends. The day after, he would return to the classroom. There would be no further punishment. Concurrently, the counselor would work with him on what real leadership behavior in the classroom might look like.
>
> The first month, the boy spent a fair number of days with the teacher's aide. The second month, however, he missed only two days, and the third month, only one. The truth is that he really wanted to be in school with his classmates. That's where the action was. And so he paid the price of self-control to get what he wanted. He also began to discover more constructive forms of leadership behavior. For instance, on occasion he challenged what the teacher was saying about a particular topic. Sometimes this led to a very lively classroom debate. Even the teacher liked this.

The counselor had suspected that the boy found socializing with his classmates rewarding. But now he had to pay for the privilege of socializing. Reasonable behavior in the classroom was not too high a price.

**Help clients to be resilient after mistakes and failures.** Clients, like the rest of us, stumble and fall as they try to implement their constructive-change programs. However, everyone has some degree of *resilience* within (Hartling, 2005; Lee, 2005; Luthar, Ciccetti, & Becker, 2000; Masten, 2001) that enables them to get up, pull themselves together, and move on once more. The ability to bounce back is an essential life capability. Holaday and McPhearson (1997) have compiled a list of common factors that influence resilience. Although they focused on severe-burn victims, what they have to say about resilience applies more generally. They distinguish between *outcome* resilience and *process* resilience. While resilience in general is the ability to overcome or adapt to significant stress or adversity, outcome resilience implies a return to a previous state. This is *bounce back* resilience. Erin goes through the trauma of divorce, but within a few months she bounces back. Her friends say to her, "You seem to be your old self now." She replies, "Both older and wiser." Process resilience, on the other hand, represents the continuous effort to cope with difficulties that are embedded in the client's life. The sufferers that Holaday and McPhearson studied would say such things as, "Resilience? It's my spirit and it's the reason I'm here," and resilience "is deep inside of you, it's already there, but you have to use it," and "To do well takes a lot of determination, courage, and struggling, but it's your choice" (p. 348).

You can encourage both kinds of resilience in clients. Take outcome or bounce-back resilience. Kerry finds himself in a financial mess because of a tendency to be a spendthrift and because of a few poor financial decisions. Although he makes a couple of more mistakes, he works his way out of the mess. Once he has his financial house in order, including putting himself on a strict but somewhat flexible budget, things stabilize. It's not difficult for him to walk the financial straight and narrow because the mess has been too painful to repeat. He bounces back and feels little inclination to get into financial trouble again.

Here are a couple of examples of process resilience. Oscar finds that controlling his anger is a constant struggle. He can't say, at least at this stage, "Well, that's it. I don't have to worry about anger any more." He has to remain vigilant in order to spot early warning signals that he's headed for a blow-up. He also counts on caring challenge from his wife, a colleague at work, and a good friend to keep plugging away. And then there is Nadia. Because she suffers from chronic fatigue syndrome, she has to dig deep within herself every day to find the will to go on. Like many people suffering from this condition, she wants to do her best and even make a good impression. On the days she's successful in pulling herself together, the people she meets cannot believe that she is ill. Running into this kind of disbelief on the part of otherwise intelligent people is a two-edged sword. It means that even people close to her do not understand the nature of her malady, but it also means that she is succeeding in her effort to normalize her life whenever she can.

There is a range of "resilience levers" in every client. Your job is to help them discover the levers, pull them, bounce back, or keep going. Resilience is "deep inside you" and inside your clients. It's part of their self-healing nature. In a sense, you help clients keep their hopes alive. But hope must itself be realistic. As Groopman (2004, p. xiv) notes, "Hope is the elevating feeling we experience when we see—in the mind's eye—a path to a better future. Hope acknowledges the significant obstacles and deep pitfalls along that path. True hope has no room for delusion." Box 11.1 outlines the kinds of questions you can help clients ask themselves as they struggle to implement their plans.

Colan (2003) developed an "art of adherence" formula—focus, competence, and passion—to help people consistently drive their plans. In the end, however, there is no formula for everyone. Your job is to help your clients find their own formula for "sticking to it."

## Questions on Implementing Plans

**Here are some questions you can help clients ask themselves as they struggle to implement their plans.**

- Now that I have a plan, how do I move into action?
- What kind of self-starter am I? How can I improve?
- What obstacles lie in my way? Which are critical?
- How can I manage these obstacles?
- How do I keep my efforts from flagging?
- What do I do when I feel like giving up?
- What kind of support will help me keep going?

## Getting Along Without a Helper

In most cases, helping is a relatively short-term process. But even in longer-term therapy, clients must eventually get on with life without their helpers. Ideally, the counseling process not only helps clients deal with specific problem situations and unused opportunities, but also, as outlined in Chapter 1, equips them with the working knowledge and skills needed to manage those situations more effectively on their own.

Since adherence to constructive-change programs is often difficult, social support and challenge in their everyday lives can help clients move to action, persevere in action programs, and both consolidate and maintain gains. When it comes to social support and challenge, there are a number of possible scenarios at the implementation stage and beyond:

- Counselors help clients with their plans for constructive change and then clients, using their own initiative and resources, take responsibility for the plans and pursue them on their own.
- Clients continue to see a helper regularly in the implementation phase.
- Clients see a helper occasionally either "on demand" or in scheduled stop-and-check sessions.
- Clients participate in some kind of self-help group, together with one-to-one counseling sessions, which are eventually eliminated.
- Clients develop social relationships that provide both ongoing support and challenge for the changes they are making in their lives.

Support was discussed earlier and, since the literature tends to focus on caring support, a few words about caring challenge in everyday life are in order.

**Challenging relationships.** It was suggested earlier that support without challenge can be hollow and that challenge without support can be abrasive. Ideally, the people in the lives of clients provide a judicious mixture of support and challenge.

> Harry, a man in his early 50s, was suddenly stricken with a disease that called for immediate and drastic surgery. He came through the operation quite well, even getting out of the hospital in record time. For the first few weeks he seemed, within reason, to be his old self. However, he had problems with the drugs he had to take following the operation. He became quite sick and took on many of the mannerisms of a chronic invalid. Even after his medical team found the right mix of drugs, he persisted in invalid-like behavior. Whereas right after the operation he had walked tall, he now began to shuffle. He also talked constantly about his symptoms and generally used his "state" to excuse himself from normal activities.
>
> At first Harry's friends were in a quandary. They realized the seriousness of the operation and tried to put themselves in his place. They provided all sorts of support. But gradually they realized that he was adopting a style that would alienate others and keep him out of the mainstream of life. Support was essential, but it was not enough. They used a variety of ways to challenge his behavior, mocking his "invalid" movements, engaging in serious one-to-one talks, turning a deaf ear to his discussion of symptoms, and routinely including him in their plans.

Harry did not always react graciously to his friends' challenges, but in his better moments he admitted that he was fortunate to have such friends. Counselors can help clients, as they attempt to change their behavior, find people willing to provide a judicious mixture of support and challenge.

**Feedback from significant others.** Gilbert (1978, p. 175), in his book on human competence, claimed that "improved information has more potential than anything else I can think of for creating more competence in the day-to-day management of performance." Feedback is certainly one way of providing both support and challenge. If clients are to be successful in implementing their action plans, they need adequate information about how well they are performing. Sometimes they know themselves; other times they need a more objective view. The purpose of feedback is not to pass judgment on the performance of clients, but rather to provide guidance, support, and challenge. There are two kinds of feedback.

*Confirmatory feedback.* Through confirmatory feedback, significant others such as helpers, relatives, friends, and colleagues let clients know that they are on course—that is, moving successfully through the steps of an action program toward a goal.

*Corrective feedback.* Through corrective feedback, significant others let clients know that they have wandered off course and what they need to do to get back on.

Corrective feedback, whether from helpers or people in the client's everyday life, should incorporate the following principles:

- Give feedback in the spirit of caring.
- Remember that mistakes are opportunities for growth.

- Use a mix of both confirmatory and corrective feedback.
- Be concrete, specific, brief, and to the point.
- Focus on the client's behaviors rather than on more elusive personality characteristics.
- Tie behavior to goals.
- Explore the impact and implications of the behavior.
- Avoid name-calling.
- Provide feedback in moderate doses. Overwhelming the client defeats the purpose of the entire exercise.
- Engage the client in dialogue. Invite the client not only to comment on the feedback, but also to expand on it. Lectures don't usually help.
- Help the client discover alternative ways of doing things. If necessary, prime the pump.
- Explore the implications of changing versus not changing.

The spirit of these "rules" should also govern confirmatory feedback. Very often people give very detailed corrective feedback and then just say "nice job" when a person does something well. All feedback provides an opportunity for learning. Consider the following statement from a father talking to his high-school son who stood up for the rights of a friend who was being bullied by some of his classmates:

> "Jeb, I'm proud of you. You stood your ground even when they turned on you. They were mean. You weren't. You gave your opinion calmly, but forcefully. You didn't apologize for what you were saying. You were ready to take the consequences. It's easier now that a couple of them have apologized to you, but at the time you didn't know they would. You were honest to yourself. And now the best of them appreciate it. It made me think of myself. I'm not sure that I would have stood my ground the way you did . . . But I'm more likely to do so now."

While brief, this is much more than "I'm proud of you, son." Being specific about behavior and pointing of the impact of the behavior turn positive feedback into a learning experience. One of the main problems with feedback is finding people in the client's day-to-day life who see the client in action enough to make it meaningful, who care enough to give it, and who have the skills to provide it constructively.

## An Amazing Case of Resilience and Perseverance

We end with a very real case of a woman who certainly did not choose not to change. Quite the contrary. Her case is a good example of a no-formula approach to developing and implementing a program for constructive change.

> Vicky readily admits that she has never fully "conquered" her illness. Some 20 years ago, she was diagnosed as manic-depressive. The picture looked something like this: She would spend about six weeks on a high; then the crash would come, and for

about six weeks she'd be in the pits. After that she'd be normal for about eight weeks. This cycle meant many trips to the hospital. Some seven years into her illness, during a period in which she was in and out of the hospital, she made a decision. "I'm not going back into the hospital again. I will so manage my life that hospitalization will never be necessary." This nonnegotiable goal was her manifesto.

Starting with this declaration of intent, Vicky spelled out what she wanted: (1) she would channel the energy of her "highs"; (2) she would consistently manage or at least endure the depression and agony of her "lows"; (3) she would not disrupt the lives of others by her behavior; (4) she would not make important decisions when either high or low. Vicky, with some help from a rather nontraditional counselor, began to do things to turn those goals into reality. She used her broad goals to provide direction for everything she did.

Vicky learned as much as she could about her illness, including cues about crisis times and how to deal with both highs and lows. To manage her highs, she learned to channel her excess energy into useful—or at least nondestructive—activity. Some of her strategies for controlling her highs centered on the telephone. She knew instinctively that controlling her illness meant not just managing problems but also developing opportunities. She was so grateful for the "miracle" of cell phones. During her free time, she would spend long hours on the phone with a host of friends, being careful not to overburden any one person. Phone marathons became part of her lifestyle. She made the point that a big phone bill was infinitely better than a stay in the hospital. She called the telephone her "safety valve."

At the time of her highs, she would do whatever she had to do to tire herself out and get some sleep, for she had learned that sleep was essential if she was to stay out of the hospital. This included working longer shifts at the business. She developed a cadre of supportive people, including her husband. She took special care not to overburden him. She made occasional use of a drop-in crisis center but preferred avoiding any course of action that reminded her of the hospital.

It must be noted that the central driving force in this case was Vicky's decision to stay out of the hospital. Her determination drove everything else. This case also exemplifies the spirit of action that ideally characterizes the implementation stage of the helping process. Here is a woman who, with occasional help from a counselor, took charge of her life. She set some simple goals and devised a set of simple strategies for accomplishing them. She never looked back. And she was never hospitalized again. Some will say that she was not "cured" by this process. But her goal was not to be cured, but to lead as normal a life as possible in the real world. Some would say that her approach lacked elegance. But it certainly did not lack results.

## The Shadow Side of Self Change

Polivy and Herman (2002) have written an engaging article on the "false hopes" of self-change.

Despite repeated failure at attempts to change aspects of their behavior, people make frequent attempts at self-change. The generally negative outcome of many such self-change efforts [think of smoking, dieting, and exercise programs] makes it difficult to understand why so many individuals persist at these attempts. We have described this cycle of failure and renewed effort as a "false hope syndrome" characterized by unrealistic expectations about the likely speed, amount, ease, and consequences of self-change attempts. In this article, we . . . review the reasons why so many people fail in their self-change attempts and then examine how people interpret these failings in such a way that they are lead to keep trying repeatedly despite overwhelming odds (p. 677).

They outline four main sources of failure. First, people often believe that they can change more than they actually can. That is, their goals are unrealistic. False expectations lead people to reject more modest goals that they could actually achieve. A step in the right direction, as it were. Second, people often believe that they will be able to change quickly, when what is needed is a slower, but steady pace. Third, many people think that change will be easy. It is easy—on paper or in fantasy. Fourth, people are often unrealistic about how life-enhancing successful change efforts will be. People who go through heroic efforts to stop smoking or lose weight find out that they still have interpersonal problems or are still underemployed. "When I'm thin, men (women) will flock to my feet." To make matters worse, those who fail in their self-change efforts find all sorts of "reasons" for their failures—external factors or a lack of "will power." As if will power is something that can be summoned, well, at will. They then discount the failures and try again.

In the end, Polivy's and Herman's article—worth reading in its entirety— leaves us, paradoxically, with a sense of hope. Most of us would find comfort in recognizing ourselves, our humanity, in what they say. Then, if we move from smart to wise, we might actually pursue self-change in a more reasonable way. Sometimes, when I suspect a client is running afoul of the false-hope syndrome, I say (gently, of course) something like this: "All right, I hear what you are trying to accomplish. But, for the sake of argument, I say you're not going to do it. Now prove to me that I'm wrong." What I mean, of course, is, "Prove it to yourself."

## Evaluation Questions for the Action Arrow

### How well do I do the following as I try to help clients make the transition to action?

- Understand how widespread both inertia and entropy are, and how they are affecting any given client
- Help clients become effective tacticians
- Help clients form "implementation intentions" especially when obstacles to goal attainment are foreseen
- Help clients avoid both procrastination and imprudent action
- Help clients develop contingency plans
- Help clients discover and manage obstacles to action
- Help clients discover resources that will enable them to begin acting, to persist, and to accomplish their goals

- Help clients find the incentives and the rewards they need to persevere in action
- Help clients acquire the skills they need to act and to sustain goal-accomplishing action
- Help clients develop a social support and challenge system in their day-to-day lives
- Prepare clients to get along without a helper
- Come to grips with what kind of agent of change I am in my own life
- Face up to the fact that not every client wants to change
- Spot and help clients deal with the "false-hope" syndrome

 **EXERCISES ON PERSONALIZING**

1. You know what a self-starter is. What kind of self-starter are you? What keeps you from being the kind of self-starter you would like to be? Name someone whom you see as a self-starter. What is it about him or her that you could emulate (not just slavishly copy)?

2. Consider a change you wanted to make in your life but which never got off the ground, or which drifted around aimlessly, or which failed. In the light of the action-arrow principles outlined in this chapter, where did things go wrong? What would you now do differently?

3. Consider a change that you have successfully pulled off. How did it differ from the change you failed to pull off? What do you learn from comparing the two?

4. Consider a goal you are currently trying to achieve, or something you want, or something you are trying to do. What obstacles stand in the way? What can you do about them? What would help you move forward? If you find that you are stuck, go back to Stage II and see if you need to do more work on the goal.

ARGYRIS, C. (1999). *On Organizational Learning* (2nd ed.). Cambridge, MA: Blackwell.

ARKIN, R. M., Hermann, A. D. (2000, July). Constructing desirable identities: Self-presentation in psychotherapy and daily life. *Psychological Bulletin, 126,* 501–504.

ASHFORD, J. B., Lecroy, C. W. & Lortie, K. L. (2006). *Human behavior in the social environment: A multidimensional perspective* (3rd ed.). Pacific Grove, CA: Brooks/Cole.

ATKINSON, D. R., Worthington, R. L., Dana, D. M., & Good, G. E. (1991). Etiology beliefs, preferences for counseling orientations, and counseling effectiveness. *Journal of Counseling Psychology, 38,* 258–264.

AZAR, B. (1995). Breaking through barriers to creativity. *APA Monitor, 26.*

BANDURA, A. (1990). Foreword to E. A. Locke, & G. P. Latham (1990), *A theory of goal setting and task performance.* Englewood Cliffs, NJ: Prentice Hall.

Basic Behavioral Science Task Force of the National Advisory Mental Health Council (1996). Basic behavioral science research for mental health: Family processes and social networks. *American Psychologist, 51,* 622–630.

BERENSON, B. G., & Mitchell, K. M. (1974). *Confrontation: For better or worse.* Amherst, MA: Human Resource Development Press.

BERNE, E. (1964). *Games people play.* New York: Grove Press.

BOHART, A. C. & Greenberg, L. S. (Eds.). (1997). *Empathy reconsidered: New directions in psychotherapy.* Washington, D.C.: American Psychology Association.

BOHART, A. C. & Tallman, K. (1999). *How clients make therapy work.* Washington, D.C.: American Psychological Association.

BORDIN, E. S. (1979). The generalizability of the psychoanalytic concept of the working alliance. *Psychotherapy: Theory, Research and Practice, 16,* 252–260.

BORNSTEIN, R. F. (2005a, February). The dependent patient: Diagnosis, assessment, and treatment. *Professional Psychology: Research and Practice. 36,* 82–89.

BORNSTEIN, R. F. (2005b). *The dependent patient: A practitioner's guide.* Washington, D.C.: American Psychological Association.

BRODER, M. S. (2000). Making optimal use of homework to enhance your therapeutic effectiveness. *Journal of Rational-Emotive & Cognitive Behavior Therapy, 18,* 3–18.

BRAMMER, L. (1973). *The helping relationship: Process and skills.* Englewood Cliffs, NJ: Prentice Hall.

BRONFENBRENNER, U. (1977). Toward an experimental ecology of human development. *American Psychologist,* 513–531.

CARKHUFF, R. R. (1969a). *Helping and human relations: Vol. 1.* Selection and training. New York: Holt, Rinehart & Winston.

CARKHUFF, R. R. (1969b). *Helping and human relations: Vol. 2.* Practice and research. New York: Holt, Rinehart & Winston.

CARKHUFF, R. R. (1971). Training as a preferred mode of treatment. *Journal of Counseling Psychology, 18,* 123–131.

CARKHUFF, R. R., & Anthony, W. A. (1979). *The skills of helping: An introduction to counseling.* Amherst, MA: Human Resource Development Press.

CHAMBERLIN, J. (2005, February). Think again. *Monitor on Psychology.* Pp. 82–83.

CHANG, E. C. (Ed.). (2001). *Optimism and pessimism: Implications for theory, research, and practice.* Washington, D.C.: American Psychological Association.

CHANG, E. C., D'Zurilla, T. J., & Sanna, L. J. (Eds.). (2004). *Social problem solving: Theory, research, and training.* Washington, D.C.: American Psychological Association.

CLAIBORN, C. D. (1982). Interpretation and change in counseling. *Journal of Counseling Psychology, 29,* 439–453.

COHEN, E. D., & Cohen, G. S. (1999). *The virtuous therapist: Ethical practice of counseling and psychotherapy.* Pacific Grove, CA: Brooks/Cole.

COLAN, L. J. (2003). *Sticking to it: The art of adherence.* Dallas: Cornerstone Leadership Institute.

COLE, H. P., & Sarnoff, D. (1980). Creativity and counseling. *Personnel and Guidance Journal, 59,* 140–146.

CONYNE, R. K., & Cook, E. P. (Eds.). (2004). *Ecological counseling: An innovative approach in conceptualizing person–environment interaction.* Alexandria, VA.: American Counseling Association.

COREY, G. (1996). *Theory and practice of counseling and psychotherapy* (5th ed.). Pacific Grove, CA: Brooks/Cole.

COREY, G., Corey, M. S., & Callanan, P. (2003). *Issues and ethics in the helping professions* (6th ed.). Pacific Grove, CA: Brooks/Cole.

COVEY, S. R. (1989). *The seven habits of highly effective people.* New York: Simon & Schuster (Fireside edition, 1990).

CRANO, W. D. (2000, March). Milestones in the psychological analysis of social influence. *Group Dynamics, 4,* 68–80.

CUMMINGS, N. A. (1979). Turning bread into stones: Our modern antimiracle. *American Psychologist, 34,* 1119–1129.

CUMMINGS, N. A., (2000). *The first session with substance abusers.* San Francisco: Jossey-Bass.

DE JONG, P., & Berg, I. K. (2002). *Interviewing for solutions* (2nd ed.). Pacific Grove, CA: Brooks/Cole.

DEUTSCH, M. (1954). Field theory in social psychology. In G. Lindzey (Ed.), *The handbook of social psychology* (pp. 181–222). Cambridge, MA: Addison-Wesley.

DILLER, J. V. (2004). *Cultural diversity: A primer for the human services.* Brooks/Cole: Belmont, CA.

DINKMEYER, D. & McKay, G. D. (1994). *How you feel is up to you.* San Luis Obispo, CA: Impact Publishers.

DORNER, D. (1996). *The logic of failure: Why things go wrong and what we can do to make them right.* New York: Holt.

DRAYCOTT, S., & Dabbs, A. (1998). Cognitive dissonance 1: An overview of the literature and its integration into theory and practice of clinical psychology. *British Journal of Clinical Psychology, 37,* 341–353.

DRISCOLL, R. (1984). *Pragmatic psychotherapy.* New York: Van Nostrand Reinhold.

DRYDEN, W. (1995). *Brief rational emotive behaviour therapy.* New York: Wiley.

DUAN, C., & Hill, C. E. (1996). The current state of empathy research. *Journal of Counseling Psychology, 43,* 261–274.

D'ZURILLA, T. J., & Nezu, A. M. (1999). *Problem-solving therapy: A social competence approach to clinical intervention* (2nd ed.). NY: Springer.

D'ZURILLA, T., J., & Nezu, A. M. (2001). Problem-solving therapies. In K. S. Dobson (Ed.), *The handbook of cognitive-behavioral therapies* (2nd ed., pp. 211–245). NY: Guilford.

EDWARDS, C. E., & Murdock, N. L. (1994). Characteristics of therapist self-disclosure in the counseling process. *Journal of Counseling and Development, 72,* 384–389.

EGAN, G., & Bailey, A. (2001). *Talkworks at work.* London: BT.

EGAN, G. (2002). *The skilled helper: A systematic approach to effective helping (7th ed.)* Pacific Grove, CA: Brooks/Cole.

EGAN, G., & Cowan, M. A. (1979). *People in systems: A model for development in the human-service professions and education.* Pacific Grove, CA: Brooks/Cole.

ELIAS, M. J., & Butler, L. B., with Bruno, E. M., Papke, M. R., & Shapiro, T. F. (2005a). *Social decision making/social problem solving (Grades 2–3).* Champaign, IL: Research Press.

ELIAS, M. J., & Butler, L. B., with Bruno, E.M., Papke, M. R., & Shapiro, T. F. (2005b). *Social decision making/social problem solving (Grades 4–5).* Champaign, IL: Research Press.

ELIAS, M. J., & Butler, L. B. (2005c). Social decision making/problem solving for middle schools (Grades 4–5). Champaign, IL: Research Press.

ELIAS, M. J., & Tobias, S. E. (2002). *Social problem solving interventions in the schools.* New York: Guilford Publications.

ELLIS, A. (2004). *Rational emotive behavior therapy: It works for me—It can work for you.* Amherst, NY: Prometheus Books.

ELLIS, A., & MacLaren, C. (2004). *Rational emotive behavior therapy: A therapist's guide* (2nd ed.). Atascadero, CA: Impact.

EPSTEIN, M. H., Kutash, K., & Duchnowski, A. (2005). *Outcomes for children and youth with emotional disorders and their families.* (2nd ed.). Austin: TX: Pro-Ed.

FARBER, B. A., Berano, K. C., & Capobianco, J. A. (2004). Clients' perceptions of the process and consequences of self-disclosure in psychotherapy. *Journal of Counseling Psychology, 51,* 340–346.

FARRELLY, F., & Brandsma, J. (1974). *Provocative therapy.* Cupertino, CA: Meta Publications.

FERGUSON, T. (1987, January–February). Agreements with yourself. *Medical Self-Care,* 44–47.

FESTINGER, S. (1957). *A theory of cognitive dissonance.* New York: Harper & Row.

FISH, J. M. (1995, Spring). Does problem behavior just happen? Does it matter? *Behavior and Social Issues, 5,* 3–12.

FISHER, C. B. (2003). *Decoding the ethics code.* Thousand Oaks, CA: Sage Publications.

FRANCES, A., Clarkin, J., & Perry, S. (1984). *Differential therapeutics in psychiatry.* New York: Brunner/Mazel.

FREIRE, P. (1970). *Pedagogy of the oppressed.* New York: Seabury.

GELATT, H. B. (1989). Positive uncertainty: A new decision-making framework for counseling. *Journal of Counseling Psychology, 36,* 252–256.

GELATT, H. B., Varenhorst, B., & Carey, R. (1972). *Deciding: A leader's guide.* Princeton, NJ: College Entrance Examination Board.

GIANNETTI, E. (1997). *Lies we live by: The art of self-deception.* New York and London: Bloomsbury.

GILBERT, T. F. (1978). *Human competence: Engineering worthy performance.* New York: McGraw-Hill.

GLASSER, W. (2000). *Reality therapy in action.* New York: HarperCollins.

GOLEMAN, D. (1985). *Vital lies, simple truths: The psychology of self-deception.* New York: Simon & Schuster.

GOLEMAN, D. (1995). *Emotional intelligence.* New York: Bantam Books.

GOLEMAN, D. (1998). *Working with emotional intelligence.* New York: Bantam Books.

GOLLWITZER, P. M. (1999). Implementation intentions: Strong effects of simple plans. *American Psychologist, 54,* 493–503.

GREENBERG, L. S. (1986). Change process research. *Journal of Consulting and Clinical Psychology, 54,* 4–9.

GROOPMAN, J. (2004). The anatomy of hope: *How people prevail in the face of illness.* New York: Random House.

GUADALUPE, K. L., & Lum, D. (2005). *Multidimensional contextual practice: Diversity and transcendence.* Pacific Grove, CA: Brooks/Cole.

GUISINGER, S., & Blatt, S. J. (1994). Individuality and relatedness: Evolution of a fundamental dialectic. *American Psychologist, 49,* 104–111.

HALEY, J. (1976). *Problem solving therapy.* San Francisco: Jossey-Bass.

HALL, E. T. (1977). *Beyond culture.* Garden City, NJ: Anchor Press.

HALLECK, S. L. (1988). Which patients are responsible for their illnesses? *American Journal of Psychotherapy, 42,* 338–353.

HARTLING, L. M. (2005). Fostering resilience across the lifespan. In D. Comstock (Ed.), *Diversity and development: Critical contexts that shape our lives and relationships.* Belmont, CA: Thomson-Wadsworth.

HENDRICK, S. S. (1988). Counselor self-disclosure. *Journal of Counseling and Development, 66,* 419–424.

HIGHLEN, P. S., & Hill, C. E. (1984). Factors affecting client change in counseling. In S. D. Brown & R. W. Lent (Eds.), *Handbook of counseling psychology* (pp. 334–396). New York: Wiley.

HIGGINSON, J. G. (1999). Defining, excusing, and justifying deviance: Teen mothers' accounts for statutory rape. *Symbolic Interaction, 22,* 25–44.

HILL, C. E. (2004). *Helping skills: Facilitating exploration, insight, and action.* (2nd ed.). Washington, D.C: American Psychological Association.

HILL, C. E., Gelso, C. J., & Mohr, J. (2000). Client concealment and self-presentation in therapy: Comment on Kelly (2000). *Psychological Bulletin, 126,* 495–500.

HILL, C. E., Thompson, B. J., Cogar, M. C., & Denmann III, D. W. (1993). Beneath the surface of long-term therapy: Therapist and client report of their own and each other's covert processes. *Journal of Counseling Psychology, 40,* 278–287.

HOGAN-GARCIA, M. (2003). *Four skills of cultural diversity competence: A process for understanding and practice.* (2nd ed.). Pacific Grove, CA: Brooks/Cole.

HOLADAY, M., & McPhearson, R. W. (1997, May/June). Resilience and severe burns. *Journal of Counseling and Development, 75,* 346–356.

HOLMES, G. R., Offen, L, & Waller, G. (1997). See no evil, hear no evil, speak no evil: Why do relatively few male victims of childhood sexual abuse receive help for abuse-related issues in adulthood? *Clinical Psychology Review, 27,* 69–88.

HOWARD, G. S. (1991). Culture tales: A narrative approach to thinking, cross-cultural psychology, and psychotherapy. *American Psychologist, 46,* 187–197.

HOWELL, W. S. (1982). *The empathic communicator.* Belmont, CA: Thomson-Wadsworth.

HUTCHINSON, E.D. (2003), *Dimensions of human behavior: Person and environment* (2nd ed.). Pacific Grove, CA: Brooks/Cole.

ICKES, W. (1993). Empathic accuracy. *Journal of Personality, 61,* 587–610.

ICKES, W. (1997). Introduction. In W. Ickes (Ed.), *Empathic Accuracy* (pp. 1–16). New York: Guilford Press.

IVEY, A. E., & Ivey, M. B. (2003). *Intentional interviewing and counseling* (5th ed.). Pacific Grove, CA: Brooks/Cole.

KAGAN, J. (1996). Three pleasing ideas. *American Psychologist, 51,* 901–908.

KAHN, J. H., Kelly, A. E., Coulter, R. G. (1996). Client self-presentations at intake. *Journal of Counseling Psychology, 43,* 300–309.

KAZANTIS, N. (2000). Power to detect homework effects in psychotherapy outcome research. *Journal of Consulting & Clinical Psychology, 68,* 166–170.

KELLY, A. E. (2000a July). Helping construct desirable identities: A self-presentational view of psychotherapy. *Psychological Bulletin, 126,* 475–496.

KELLY, A. E. (2000b July). A self-presentational view of psychotherapy: Reply to Hill, Gelso and Mohr (2000) and to Arkin and Hermann (2000). *Psychological Bulletin, 126,* 505–511.

KELLY, A. E. (2002). *The psychology of secrets.* New York: Plenum.

KELLY, A. E, Kahn, J. H., & Coulter, R. G. (1996). Client self-presentations at intake. *Journal of Counseling Psychology, 43,* 300–309.

KIRSCHENBAUM, D. S. (1985). Proximity and specificity of planning: A position paper. *Cognitive Therapy and Research, 9,* 489–506.

KIRSCHENBAUM, D. S. (1987). Self-regulatory failure: A review with clinical implications. *Clinical Psychological Review, 7,* 77–104.

KIVLIGHAN, D. M., Jr., & Shaughnessy, P. (2000). Patterns of working alliance development: A typology of client's working alliance ratings. *Journal of Counseling Psychology, 47,* 362–371.

KNOX, S, Hess, S. A., Petersen, D. A., & Hill, C. E. (1997). A qualitative analysis of client perceptions of the effects of helpful therapist self-disclosure in long-term therapy. *Journal of Counseling Psychology, 44,* 274–283.

KOHUT, H. (1978). The psychoanalyst in the community of scholars. In P. H. Ornstein (Ed.), *The search for self: Selected writings of H. Kohut.* New York: International Universities Press.

KOTTLER, J. A. (2000). *Doing good: Passion and commitment for helping others.* Philadelphia: Brunner/Routledge.

KUSHNER, M. G., Sher, K. J. (1989). Fear of psychological treatment and its relation to mental health service avoidance. *Professional Psychology: Research and Practice, 20,* 251–257.

LANDRETH, G. L. (1984). Encountering Carl Rogers: His views on facilitating groups. *Personnel and Guidance Journal, 62,* 323–326.

LAZARUS, R. S. (2000). Toward better research on stressing and coping. *American Psychologist, 55,* 665–673.

LEE, R. M. (2005). Resilience against discrimination: Ethnic identity and other-group orientation as protective factors for Korean Americans. *Journal of Counseling Psychology, 52,* 36–44.

LEVITT H. M. & Rennie, D. L. (2004). The act of narrating: Narrating activities and the intentions that guide them. In L. Angus & J. McLeod (Eds.), *The handbook of narrative and psychotherapy: Practice, theory and research.* (pp 299–314). Thousand Oaks, CA: Sage Publications.

LEWIN, K. (1969). Quasi-stationary social equilibria and the problem of permanent change. In W. G. Bennis, K. D. Benne, & R. Chin (Eds.), *The planning of change.* New York: Holt, Rinehart & Winston.

LLEWELYN, S. P. (1988). Psychological therapy as viewed by clients and therapists. *British Journal of Clinical Psychology, 27,* 223–237.

LOCKE, E. A., & Latham, G. P. (1984). *Goal setting: A motivational technique that works.* Englewood Cliffs, NJ: Prentice Hall.

LOCKE, E. A., & Latham, G. P. (1990). *A theory of goal setting and task performance.* Englewood Cliffs, NJ: Prentice Hall.

LOWENSTEIN, L. (1993). Treatment through traumatic confrontation approaches: The story of S. *Education Today, 43,* 198–201.

LUTHAR, S. S., Cicchetti, D., & Becker, B. (2000). The construct of resilience: A critical evaluation and guidelines for future work. *Child Development, 71,* 543–562.

MAHRER, A. R., Gagnon, R., Fairweather, D. R., Boulet, D. B., & Herring, C. B. (1994). Client commitment and resolve to carry out postsession behaviors. *Journal of Counseling Psychology, 41,* 407–414.

MANTHEI, R. (1998). *Counseling: The skills of finding solutions to problems.* New York: Routledge.

MANTHEI, R., & Miller, J. (2000). *Good counseling: A guide for clients.* Auckland: Longman (Addison Wesley Longman New Zealand).

MARKUS, H., & Nurius, P. (1986). Possible selves. *American Psychologist, 41,* 954–969.

MARTIN, J. (1994). *The construction and understanding of psychotherapeutic change.* New York: Teachers College Press.

MASTEN, A. S. (2001). Ordinary magic: Resilience processes in development. *American Psychologist, 56,* 227–238.

MATHEWS, B. (1988). The role of therapist self-disclosure in psychotherapy: A survey of therapists. *American Journal of Psychotherapy, 42,* 521–531.

McAULIFFE, G. J., & Eriksen, K. P. (1999). Toward a constructivist and developmental identity for the counseling profession: The context-phase-stage-style model. *Journal of Counseling and Development, 77,* 267–279.

McKAY, G. D., & Dinkmeyer, D. (1994). *How you feel is up to you.* San Luis Obispo, CA: Impact.

McKAY, M., Davis, M., & Fanning, P. (1997). *Thoughts and feeling: Taking control of your moods and your life.* Oakland, CA: New Harbinger Publications.

McMILLEN, C., Zuravin, S., & Rideout, G. (1995). Perceived benefit from child sexual abuse. *Journal of Consulting and Clinical Psychology, 63,* 1037–1043.

MEHRABIAN, A., & Reed, H. (1969). Factors influencing judgments of psychopathology. *Psychological Reports, 24,* 323–330.

MEIER, S. T., & Letsch, E. A. (2000). What is necessary and sufficient information for outcome assessment? *Professional Psychology: Research and Practice, 31,* 409–411.

MIKULINCER, M. (1994). *Human learned helplessness: A coping perspective.* New York: Plenum Press.

MILLER, G. A., Galante, E., & Pribram, K. H. (1960). *Plans and the structure of behavior.* New York: Holt, Rinehart & Winston.

MILLER, L.M. (1984). American spirit: Visions of a new corporate culture. New York: Morrow.

MILLER, W. C. (1986). *The creative edge: Fostering innovation where you work.* Reading, MA: Addison-Wesley.

MILLER, W. R., & Rollnick, S. (1991). *Motivational interviewing: Preparing people to change addictive behavior.* New York: Guilford Press.

MURPHY, B. C., & Dillon, C. (2003). *Interviewing in action: Relationship, process, and change.* (2nd ed.). Pacific Grove, CA: Brooks/Cole.

MURPHY, J. J. (1997) *Solution-focused counseling in middle and high schools.* Alexandria, VA: American Counseling Association.

NEZU, A. M., Nezu, C. M., Friedman, S. H., Faddis, S., & Houts, P. S. (1998). *Helping cancer patients cope. A problem-solving approach.* Washington, DC: American Psychological Association.

NORCROSS, J. C. (Ed.). (2002). *Psychotherapy relationships that work.* New York: Oxford University Press.

NORCROSS, J. C., & Kobayashi, M. (1999). Treating anger in psychotherapy: Introduction and cases. *In Session, 55,* 275–282.

O'DONOHUE, W., & Ferguson, K. E. (Eds.). (2003). *Handbook of professional ethics for psychologists: Issues, questions, and controversies.* Thousands Oaks, CA: Sage Publications.

O'HANLON, W. H., Weiner-Davis, M., O'Hanlon, W.H. (2003). *In search of solutions: A new direction in psychotherapy.* New York: Norton.

PACK-BROWN, S. P., & Williams, C. B. (2003). *Ethics in a multicultural context.* Thousand Oaks, CA: Sage Publications.

PENNEBAKER, J. W. (1995). *Emotion, disclosure, and health: An overview.* In J. W. Pennebaker (Ed.), Emotion, Disclosure, and Health, 3–10.

PETERSON, C., Seligman, M. E. P., & Vaillant, G. E. (1988). Pessimistic explanatory style as a risk factor for physical illness: A thirty-five-year longitudinal study. *Journal of Personality and Social Psychology, 55,* 23–27.

PETERSON, C., Maier, S. F., & Seligman, M. E. P., (1995). Learned helplessness: A theory for the age of personal control. New York: Oxford University Press.

PHILLIPS, E. L. (1987). The ubiquitous decay curve: Service delivery similarities in psychotherapy, medicine, and addiction. *Professional Psychology: Research and Practice, 18,* 650–652.

PICHOT, T. (2003). *Solution-focused brief therapy: Its effective use in agency settings.* Binghamton, NY: Hawthorn Press.

PLUTCHIK, R. (2001). *Emotions in the practice of psychotherapy: Clinical implications affect theories.* Washington, D.C.: American Psychological Association.

POLIVY, J., & Herman, C. P. (2002, September). If at first you don't succeed: False hopes of self-change. *American Psychologist, 57,* 677–689.

PUTNAM, R. D. (2000). *Bowling alone.* New York: Simon & Schuster.

PYSZCZYNSKI, T., & Greenberg, J. (1987). Self-regulatory preservation and the depressive self-focusing style: A self-awareness theory of depression. *Psychological Bulletin, 102,* 122–138.

RAUP, J. L., & Myers, J. E. (1989). The empty nest syndrome: Myth or reality? *Journal of Counseling & Development, 68,* 180–183.

RICHMOND, V. P., & McCroskey, J. C. (2000). *Nonverbal behavior in interpersonal relations. (4th ed.).* Needham Heights, MA: Allyn & Bacon.

ROBERTSHAW, J. E., Mecca, S. J., & Rerick, M. N. (1978). *Problem-solving: A systems approach.* New York: Petrocelli Books.

ROGERS, C. R. (1951). *Client-centered therapy.* Boston: Houghton Mifflin.

ROGERS, C. R. (1957). The necessary and sufficient conditions of therapeutic personality change. *Journal of Consulting Psychology, 21,* 95–103.

ROGERS, C. R. (1965). *Client-centered therapy: Its current practice, implications, and theory.* Boston: Houghton Mifflin.

ROGERS, C. R. (1980). *A way of being.* Boston: Houghton Mifflin.

ROGERS, C. R., Perls, F., & Ellis, A. (1965). *Three approaches to psychotherapy* 1 [Film]. Orange, CA: Psychological Films, Inc.

ROLLNICK, S., & Miller, W. R. (1995). What is motivational interviewing? *Behavioural and Cognitive Psychotherapy, 23,* 325–334.

ROSEN, S., & Tesser, A. (1970). On the reluctance to communicate undesirable information: The MUM effect. *Sociometry, 33,* 253–263.

ROSEN, S., & Tesser, A. (1971). Fear of negative evaluation and the reluctance to transmit bad news. Proceedings of the 79th Annual Convention of the American Psychological Association, 6, 301–302.

ROWAN, T., & O'Hanlon, W. H., & O'Hanlon, B. (1999). *Solution-oriented therapy for chronic and severe mental illness.* New York: Wiley.

ROYSIRCAR, G., Sandhu, D. S., & Bibbens Sr., V. (Eds.). (2003). *Multicultural compentencies: A guidebook of practices.* Alexandria, VA: American Counseling Association.

SALOVEY, P., Rothman, A. J., Detweiler, J. B., & Steward, W. T. (2000, January). Emotional states and physical health. *American Psychologist, 55,* 110–121.

SATEL, S. (1996, May 8). Psychiatric Apartheid. *Wall Street Journal,* pp. A14.

SCHEEL, M. J., Hanson, W. E., & Razzhavaikina, T. I. (2004, Spring). The process of recommending homework in psychotherapy: A review of therapist delivery methods, client acceptability, and factors that affect compliance. *Psychotherapy: Theory, Research, Practice, Training, 41,* 38–55.

SCOTT, N. E., & Borodovsky, L. G. (1990). Effective use of cultural role taking. *Professional Psychology: Research and Practice, 21,* 167–170.

SELIGMAN, M. (1975). *Helplessness: On depression, development, and death.* San Francisco: W. H. Freeman.

SELIGMAN, M. (1991). *Learned optimism.* New York: Knopf.

SELIGMAN, M. (1998). *Learned optimism* (2nd ed.). New York: Pocket Books (Simon & Schuster).

SELIGMAN, M. & Csikszentmihalyi, M. (2000). Positive psychology: An introduction. *American Psychologist, 55,* 5–14.

SHARF, R. S. (2001). *Life's choices: Problems, solutions.* Pacific Grove, CA: Brooks/Cole.

SHARRY, J., Madden, B., & Darmody, M. (2003). *Becoming a solution detective: Identifying your clients' strengths in practical brief therapy.* Binghamton, NY: Hawthorn Press.

SIMON, J. C. (1988). Criteria for therapist self-disclosure. *American Journal of Psychotherapy, 42,* 404–415.

SLATTERY, J. M. (2004). *Counseling diverse clients: Bringing context into therapy.* Pacific Grove, CA: Brooks/Cole.

SLIFE, B. D., Reber, J. S., & Richardson, F. C. (Eds.). (2005). Critical thinking about psychology: Hidden assumptions and plausible alternatives. Washington, D.C.: American Psychological Association.

SNYDER, C. R. (1995). Conceptualizing, measuring, and nurturing hope. *Journal of Counseling and Development, 73,* 355–360.

SNYDER, C. R. (Ed.). (2000). *Handbook of hope: Theory, measures, and applications.* San Diego: Academic Press.

SNYDER, C. R., & Higgins, R. L. (1988). Excuses: Their effective role in the negotiation of reality. *Psychological Bulletin, 104,* 23–35.

SNYDER, C. R., Higgins, R. L., & Stucky, R. J. (1983). *Excuses: Masquerades in search of grace.* New York: Wiley.

STERNBERG, R. J., & Grigorenko, E. L. (Eds). (2004). *Culture and competence.* Washington, D.C.: American Psychological Association.

STRICKER, G., & Fisher, A. (Eds.). (1990). *Self-disclosure in the therapeutic relationship.* New York: Plenum.

STRONG, S. R., & Claiborn, C. D. (1982). *Change through interaction: Social psychological processes of counseling and psychotherapy.* New York: Wiley.

SUE, D. W. (1990). Culture-specific strategies in counseling: A conceptual framework. *Professional Psychology: Research and Practice, 21*(6), 424–433.

SWINDLE, R. W. Jr., Heller, K, Pescosolido B, Kikuzawa S. (2000). Responses to 'nervous breakdowns' in America over a 40-year period: Mental health policy implications. *American Psychologist; 55,* 740–749.

TAUSSIG, I. M. (1987). Comparative responses of Mexican Americans and Anglo-Americans to early goal setting in a public mental health clinic. *Journal of Counseling Psychology, 34,* 214–217.

TAYLOR, S. E., Pham, L. B., Rivkin, I. D., & Armor, D. A. (1998). Harnessing the imagination: Mental stimulation, self-regulation, and coping. *American Psychologist, 53,* 429–439.

TESSER, A., & Rosen, S. (1972). Similarity of objective fate as a determinant of the reluctance to transmit unpleasant information: The MUM effect. *Journal of Personality and Social Psychology, 23,* 46–53.

TESSER, A., Rosen, S., & Batchelor, T. (1972). On the reluctance to communicate bad news (the MUM effect): A role play extension. *Journal of Personality, 40,* 88–103.

TESSER, A., Rosen, S., & Tesser, M. (1971). On the reluctance to communicate undesirable messages (the MUM effect): A field study. *Psychological Reports, 29,* 651–654.

TEYBER, E. (2000). *Interpersonal process in psychotherapy: A relational approach* (4th ed.). Pacific Grove, CA: Brooks/Cole.

TREVINO, J. G. (1996, April). Worldview and change in cross-cultural counseling. *Counseling Psychologist, 24,* 198–215.

TYLER, F. B., Pargament, K. I., & Gatz, M. (1983). The resource collaborator role: A model for interactions involving psychologists. *American Psychologist, 38,* 388–398.

WACHTEL, P. L. (1989, August 6). Isn't insight everything? Review of the book *How does treatment help: One the modes of therapatic action of psychoanalytic psychotherapy. New York Times,* p. 18.

WAHLSTEN, D. (1991). Nonverbal behavior and self-presentation. *Psychological Bulletin, 110,* 587–595.

WALSH, M. E., Galassi, J. P., Murphy, J. A., & Park-Taylor, J. (2002, September). A conceptual framework for counseling psychologists in schools. *Counseling Psychologist, 30,* 682–704.

WATKINS, C. E., Jr. (1990). The effects of counselor self-disclosure: A research review. *The Counseling Psychologist, 18,* 477-500.

WHEELER, D. D., & Janis, I. L. (1980). *A practical guide for making decisions.* New York: Free Press.

WEI, M., & Heppner, P. P. (2005). Counselor and client predictors of the initiate working alliance: A replication and extension to Taiwanese client-counselor dyads. *Counseling Psychologist, 33,* 51–71.

WEICK, K. E. (1979). *The social psychology of organizing* (2nd ed.). Reading, MA: Addison-Wesley.

WEINER, M. F. (1983). *Therapist disclosure: The use of self in psychotherapy* (2nd ed.). Baltimore: University Park Press.

WING, R. R. & Jeffery, R. W. (1999). Benefits of recruiting participants with friends and increasing social support for weight loss and maintenance. *Journal of Consulting & Clinical Psychology, 67,* 132–138.

YALOM, I. (1980). *Existential psychotherapy.* New York: Basic Books.

YUN. K. A. (1998). Relational closeness and production of excuses in events of failure. *Psychological Reports, 83,* 1059–1066.

Anthony, 152
Argyris, 133
Arkin, 188
Ashford, 66
Atkinson, 165
Azar, 214

Bachelor, 163
Bandura, 241
Becker, 286
Berano, 170
Berg, 205
Berne, 128
Berenson, 158
Beutler, 156
Bibbins, 32
Blatt, 32–33
Bohart, 30, 38
Bordin, 26
Brammer, 111
Brandsma, 36
Broder, 186
Bronfenbrenner, 242
Butler, 9

Callanan, 40
Capobianio, 170
Carey, 259
Carkhuff, Robert, 152, 270
Chamberlin, 22
Chang, 6, 11
Chiccetti, 286
Claiborne, 149
Clarkin, 194
Cogar, 69
Cohen, 40
Colan, 287
Cole, 214
Connor, 268
Conyne, 66, 177
Cook, 66, 177
Corey, 40

Coulter, 188
Covey, 31
Cowan, 66
Crano, William, 35
Csikszentmihalyis, 6
Cummings, 155

de Jong, 205
Denman, 69
Diller, 32
Dillon, 26
Dobbs, 137
Draycott, 137
Driscoll, 7, 248, 276, 278
Duchnowski, 7
Duan, 30
D'Zurilla, 11

Edwards, 149
Egan, 9, 42, 47, 66
Elias, 9, 11, 121
Ellis, Albert, 186
Epstein, 7
Eriksen, 66

Faddis, 11
Farber, 170
Farrelly, 36
Ferguson, 40, 278
Festinger, 137
Fish, 179
Fisher, 40, 149
Frances, 194
Freire, 36
Friedman, 11

Galanter, 8
Galassi, 66
Gatz, 37
Gelatt, 210, 259
Gelso, 188
Giannetti, Eduardo, 131

Gilbert, 289
Gitta, 268
Glasser, 179
Goleman, 31
Gollwitzer, 282
Greenberg, 30, 231, 277
Grigorenko, 32
Groopman, 287
Guadalupe, 32
Guisinger, 32-33

Hall, 70
Halleck, 128
Hamlet, 184
Handpick, 149, 151
Hanson, 185
Hiller, 3
Heppner, 26
Herman, 291–92
Hermann, 188
Hartling, 286
Hess, 149
Higgins, 128
Higginson, 128
Highlen, 49
Hill, 11, 30, 49, 69, 149, 188
Hogan-Garcia, 32
Holaday, 286-87
Holmes, 188
Houts, 11
Howard, 178
Hutchinson, 66, 177
Howell, 37

Ickes, 75
Ivey, 45

Janis, 189
Jeffrey, 270

Kagan, 179–181
Kahn, 188

Kazantzis, 185
Kelly, 188
Kikuzawa, 3
Kirschbaum, 268, 275
Knox, 149
Kobayashi, 61
Kohut, 31
Kottler, 164–165
Kushner, 10
Kutash, 7

Landreth, Yalom, 153
Latham, 210, 227
Lazarus, R., 6
Lecroy, 66
Lee, 286
Letsch, 8
Levitt, 188
Lewin, Kurt, 280
Llewelyn, 205
Locke, 210, 227
Lorti, 66
Lowenstein, 156
Lum, 32
Luthar, 286

MacLaren, 121
Mahra, 185
Maier, 277
Manther, 7, 36
Martin, 119
Masten, 286
Mathews, 149
McAuliffe, 66
McCroskey, 64
McMillen, 160
McPhearson, 286–87
Mecca, 214
Mehrabian, 178
Meier, 8
Messer, 163
Mikulincer, 277
Miller, 8, 36, 156
Miller, William, 184
Mitchell, 158

Mohr, 188
Murdoch, 149
Murphy, 26, 66, 205

Nezu, 11
Norcross, 26, 61

O'Donohue, 40
Offen, 188
O'Hanlon, 7, 205

Pack-Brown, 40
Pargament, 37
Park-Taylor, 66
Patel, 34
Pennebaker, 170
Perick, 214
Perry, 194
Pescosolido, 3
Peterson, 72, 277
Petersen, 149
Phillips, 279
Pichot, 205
Polivy, 291–92
Pribam, 8
Putnam, Robert, 270
Pyszcyznski, 277

Razzhavalkina, 185
Reber, 22
Reed, 178
Rennie, 188
Richardson, 22
Richmond, 64
Rideout, 160
Robertshaw, 214
Rogers, 26, 31
Rollnick, 156
Rosen, 163
Rowan, 7
Roysircar, 32

Sandhu, 32
Sanna, 11
Sarnoff, 214

Scheel, 185
Seligman, 6, 72, 277
Shakespeare, 184
Sharf, 205
Sharry, 205
Sher, 10
Simon, 149
Slattery, 32
Slife, 22
Snyder, 128
Sternberg, 32
Sticker, 149
Strong, 149
Stucky, 128
Sue, 93
Swindle, 3

Tallman, 38
Teyber, 26
Thompson, 69
Tobias, 11
Toynbee, A.J., 181
Trevino, 119
Tum, 32
Tyler, 37

Vaillant, 72
Varenhorst, 259

Wachtel, 134, 162
Waller, 188
Walsh, 66
Watkins, 149
Wei, 26
Weiner, 149
Weiner-Davis, 7, 205
Wheeler, 189
Williams, 40
Wing, 271

Zuravin, 160

Action,
  arrow: helping clients implement their plans and get results, 18–19
  arrow: helping clients overcome obstacles, execute plans, and get results, 275–280
  arrow: implementation of sharing highlights in the helping process, 86–87
  a bias toward a solution-focused value, 39–40
  and discretionary change, 39–40
  real-life focus, 40
  as a solution-focused value, 39–40
Active listening, 53–56
Advice, 94
Affect, 57–63
Agreement, 95
Awareness, degrees of, 130–132

Balance-sheet method, 256–257
Behavior, 57–59
  dysfunctional external, 153
  self-limiting internal, 122–123
Blind spots, 130–134
Bohart and Tallman / How Clients Make Therapy Work, 38
Brainstorming, 216–218, 246–255

Challenging,
  the basic concept, 119
  clients' imperfect understanding of the world, 124
  discrepancies, 124
  dishonesties of everyday life, 126
  dysfunctional external behavior, 123
  earning the right to, 158–159
  goals of, 119–120
  guidelines for effective, 157
  helper self-disclosure and its judicious use, 149–151

inadequate participation in the helping process, 128–130
  linking to action, 162–163
  relationships, 289
  self-limiting internal behavior, 122–123
  the shadow side of, 137–140
    cooperating in the helping session, then doing nothing, 138–140
    devaluing the issue, 138
    discrediting, 137–138
    helpers, 163
    persuading them to change their views, 138
    seeking support elsewhere, 138
  skills, 148–166
    guidelines for effective, 157–166
    immediacy: discussing the helper-client relationship, 151–154
    sharing advanced highlights: capturing the message behind the message, 142–147
  suggestions and recommendations, 154–155
  targets of, 120–130
  unused strengths and resources, 125–126
  use of probes for, 108–109
  ways of managing and expressing feelings and emotions, 123
  ways to, 159–166

Change,
  choosing not to change, 279
  discretionary vs. nondiscretionary, 39
  obstacles to, 275–279
Cliches, 94

Client,
  action, 171
  appreciate their self-healing nature, 38
  as agents of change in their own lives, 9
  can change if they choose, 36
  concerns, 177–179
  definition of, 3
  do not see them as overly fragile, 38
  do not see them as victims, 38
  future, help to find incentives for a better, 16–17
  help them to evaluate their progress, 19
  imperfect understanding of the world, 124
  inertia, passivity, and change, 276–277
  knowing but not caring, 131–132
  problems,
    assessing the severity of, 178–179
    begin with, 196–199
    beginning with sub-problems, 197–199
    invite clients to own their own, 189–190
    invite clients to state as solvable, 190–191
  share the helping process with, 36
  views of helping, 35–38
  view of solution-focused helping, 205–206
  ways to view them, 38
  when s/he needs a new perspective, 112
  with missed opportunities and unused potential, 5

with problem situations and
unused opportunities, 4
Co-creating outcomes, 47–49
Communication skills,
shadow side of, 114,
developing proficiency in,
115–116
vs. helping technologies,
114–115
Competing agendas, 239
Confirmation, 155–156
Confirmatory feedback, 289
Confrontation, 155–156
Connecting, 47
Contingency plans, 283
Contracts,
and agreements, 286
flawed, 43
Corrective feedback, 289–290
Creativity, 222

Degrees of awareness, 130–132
Dialogue. *See also* Therapeutic
dialogue
importance of in helping, 47–49
Disabling self-talk, 277
Discipline and self-control, 275–276
Discrepancies, 124
Discretionary change, 39
Dishonesties of everyday life, 126
Disorganization, 278
Distortions, 126–127
Distracting questions, 94
Divergent thinking, 215–219
Diversity,
coming to terms with your blind
spots, 33
interventions in a sensitive way, 33
multicultural, 32–33
use of empathetic highlights as a
way of bridging gaps, 91–92
work with individuals, 34
Dreaming, 206–207
"Dying better", 220–222

Effective vs. inadequate listening,
54–56
*Emotional Intelligence* / Goleman,
131
Empathetic highlights. *See also*
Verbal and Nonverbal Prompts
shadow side of sharing, 93–95
tactics for communicating, 92–93
Empathetic,
listening, 56

presence, 49
relationships, 96–97
Empathy, 30–31, 74–92
Encouragement, 156–157
Entropy, 278–279
Establishing rapport and
relationship building, 171
Excuse making, 128
Exemplars, 219–220
Experiences, 57–58

Fads, 22
Failures, 286–287
Faking, 95
Feedback from significant others,
289–291
Feelings and emotions, 123

Games, tricks, and smoke screens,
128
Genuineness, 34–35
Goals,
guidelines for helping clients
commit to, 235–239
guidelines for helping clients design and shape theirs, 223–235
guidelines for helping clients for
multiple paths to, 246–255
balance sheet method, 256–257
help clients craft problem-
managing, 15–16
shadow side developing and
selecting strategies, 259–263
strategies for choosing, 254–255
strategy sampling, 255–256
strategies and plans for accomplishing, 17–18
Goal-setting, 208–210, 241–243
Goleman / *Emotional Intelligence*, 131
Guidelines for helping clients explore problem situations and
unexplained opportunities,
171–173

Hearing, opportunities and
resources, 61
Helper,
-client relationship, 151–155
getting along without one, 288–290
longevity of, 10
model, 245
role of in solution-focused
helping, 205
and the shadow side of
challenging, 163

self-disclosure and its judicious
use, 149–151
solution-focused, 205–208
the three sets of, 3
Helpers,
human tendencies in these and
clients, 40–41
nonverbal behavior, 50–53
passive, 276
vague and violated values, 41–42
values of, 28–29
Helping,
client-centered, 23
clients challenge themselves,
118–130
clients clarify, 176–178
collaborative nature of, 26–27
describing the helpers, 3, 10
efficacy of, 9–10
failure to share the process, 42
how clients become change
agents in their own lives,
280–288
importance of dialogue in, 47–49
importance of empathetic
presence, 49
is it for everyone? 10
and learning, 37
managing the shadow side of,
21–23
models, 11–12, 22. *See also* Skilled
helper model
nature and purpose of, 3–6
as a natural, two-way-influence
process, 37
nonverbal behavior, 49
primary goal of,
importance of results, 7
a results-focused case, 7–8
process,
inadequate participation in the,
128–130
moving to the next stage, 191–193
relationship, 26–27, 40–42
secondary goal of, 8–9
session, 138–140
the shadow side of, 9, 22–23
as a social-influence process, 35–36
stages and tasks of,
Stage I: helping clients tell their
stories: an introduction to
Stage I, 168–204
Stage II: helping clients identify,
choose, and shape problem
managing goals, 204–243

Stage III: helping clients develop strategies and plans for accomplishing goals, 244–273
technologies, 114–115
use of empathetic highlights at every stage of the process, 85–86
success of, 9–10
values of, 27–29
   competence and commitment, 30
   and the client, 30
   diversity, 31–34
   empathy, 30–31
   genuineness, 34–35
   respect, 29
   self-responsibility, 35
*How Clients Make Therapy Work* / Bohart and Tallman, 38

Incentives and rewards, 284–286
Innovation, 222
Insights, 206
Interpretations, 94
Issues,
   to begin with, 196–197
   devaluing the, 138
   help clients to clarify key, 176–177

Learned helplessness, 277
Leverage, 14–15
   and action, 199
   guidelines in the search for, 193–195
   helping clients achieve, 169
   shadow side of, 202–203
*Lies We Live By: The Art Of Self Deception* / Giannett, Eduardo, 131
Life skills, 8–9
Listening, 53–73
   active, 53–54
   to clients, 56–65
   to clients' feelings, emotions, and moods, 61–63
   to clients' nonverbal messages and modifiers, 63–65
   effective vs. inadequate, 54–56
   empathetic, 56
   evaluative, 71
   fact-centered vs. person-centered, 72
   to feelings, emotions, and moods, 61–63

filtered, 70–71
helper's internal conversation, 68–70
inadequate, 54
interrupting, 72–73
non-, 55
to oneself: internal conversation, 68–69
partial, 55
processing what you hear: the search for meaning, 65
   hearing the slant or spin, 67
   identifying key messages and feelings, 65–66
   through context, 66–67
rehearsing, 55–56
shadow side of, 70–73
stereotyped-based, 71–72
sympathetic, 72
tape-recorder, 55
what's missing, 68
to words, 56–61

Managing the crisis, 195
Meaning, the thoughtful search for, 65–68
Mistakes, 286–287
Models, 219–220
Mutual influencing, 47

Nonverbal
   behavior, 49–53. *See also* Physically tuning into clients
      as a channel of communication, 49–50
      of helpers, 50–51
   shadow side of, 64–65
   vs. verbal prompts, 99–100

Open-ended questions, 102
Opportunities, 5–6
Overcoming procrastination, 284

Parroting, 94
Past, The,
   dealing with, 206
   helping clients talk productively about, 179–181
Physically tuning in to clients, 51–53. *See also* Nonverbal behavior
Planning. *See* Plans
Plans,
   adding value to clients' search for a better life, 263

shadow side of, 272–273
Positive psychology, 6
Possibilities,
   helping clients discover these for a better future, 15
   sources of, 219–220
Potential, 5–6
Prejudices, 120
Proactive, 184–188
Probes, 98–111
   the different forms of, 100
   guidelines for using, 102–109
   the relationship with sharing highlights and using, 109–111
Problematic mind-sets, 120
Problem-management,
   model, 11
   and opportunity-developing action,
      approach to helping, 11–12
      linking to insight, 134–137
Problems, 4–5
   assessing the severity of clients', 178–179
   future-oriented probes, 218–219
   nature of in solution-focused helping, 206
Problem situations and opportunities, 4–5
   guidelines for helping clients explore, 171–176
   and the use of probes to get a balanced view of, 106

Questions,
   as forms of probes, 100–101
   using, effectively, 101–102

Rational-Emotional-Behavioral (REBT) approach, 121
Real-life focus, 40
REBT. *See* Rational-Emotional-Behavioral approach
Rehearsing, 55–56
Relationship with the client, 152–154
Requests, 100–101
Resources,
   mobilization, 200–201
   searching for, as clients tell their stories, 181–183
Respect, 29–31
Responding skills,
   action arrow: implementation issues, 86–92

assertiveness, 77
empathetic highlights as a way of bridging diversity gaps, 91
know-how, 76
perceptiveness, 75–76
principles for sharing highlights, 85
Stage I: problem clarification and opportunity identification, 85–86
Stage II: evaluating goal options, 86
Stage III: choosing actions to accomplish goals, 86
recovering from inaccurate understanding, 90
responding to clients' feelings, emotions, and moods, 79–82
responding to the context, not just the words, 88
responding to key experiences and behaviors in clients' stories, 82–83
responding to points of view, interactions, proposals, and decisions, 83–85
responding selectively to core client messages, 87–88
sharing empathetic highlights: communicating understanding to clients, 77–79
using highlights as a mild social-influence process, 88–89
using highlights to stimulate movement in the helping process, 89–90
Rogers, Carl, 26, 56

Self-,
    challenge, 157–158
    -change, 291–293
    contracts, 238–239
    deception, 130–131
    defeating emotions, 201–202
    limiting beliefs and assumptions, 121–122
    responsibility, 35–38
Skilled helper model, 11
    basic framework for the process, 11–12
    Stage I: helping clients tell their stories, 13–14
    Stage II: helping clients identify, choose, and shape problem-managing goals, 15
    Stage III: helping clients develop strategies and plans for accomplishing goals, 17–23
Skilled Helping Around the World, 34
SOLER, 53
Solution-focused therapy, 7
Solutions, 207
Statements, as forms of probes, 100
Stories, 56
Storytelling, 13–14
    helping clients tell their stories, 169–173
    learning to work with all styles of, 171–172
    starting where the client starts, 175–176
Strategies,
    helping clients chose those that best fit their resources, 17–18
    helping clients pull these into a manageable plan, 18

Strategy sampling, 255–256
Stress reduction, 171
Summarizing, 98, 111–114
    getting the client to, 113–114
    providing focus and direction, 111–114
Sympathy, 95

Tape-recorder listening, 55
Therapeutic dialogue, 45, 50, 102
    the art of probing and summarizing, 98–117
    communicating empathy and checking understanding, 74–97
    helping clients challenge themselves, 118–140
    tuning into clients and active listening, 45–73
Turn taking, 47

Unawareness, simple, 130
Understanding,
    recovering from inaccurate, 90–91
    and responding skills, 75–97
Unused opportunities, 174

Values, vague and violated, 41–42
Verbal prompts, 99–100
Vicious circle of guilt and depression, 277–278
Virtuosity, 23
Vital Lies / Goleman, 131
Vocal prompts, 99–100